Survey Sampling and Multivariate Analysis for Social Scientists and Engineers

Survey Sampling and Multivariate Analysis for Social Scientists and Engineers

Peter R. Stopher
Northwestern University

Arnim H. Meyburg
Cornell University

Lexington Books
D.C. Heath and Company
Lexington, Massachusetts
Toronto

Library of Congress Cataloging in Publication Data

Stopher, Peter R.
  Survey sampling and multivariate analysis for social
scientists and engineers.

  Includes bibliographical references and index.
  1. Sampling (Statistics) 2. Multivariate analysis. 3. Social sciences—Statistical methods.
4. Engineering—Statistical methods. I. Meyburg,
Arnim H., joint author. II. Title.
QA276.6.S76   001.4'22   74-25056
ISBN 0-669-96966-4

Published simultaneously in Canada.

Printed in the United States of America.

International Standard Book Number: 0-669-96966-4

Library of Congress Catalog Card Number: 74-25056

To Lee and Valerie
Helen and Claire

Contents

List of Figures

List of Tables

Preface

Over the past several years, we have been reminded continually of the special needs of readers in engineering and the social sciences for an appropriate treatment of statistical methods and the collection of sample data. We suspect that these special needs are similar for many people working in other areas, such as business management and public administration, and some branches of the medical sciences. Furthermore, many professionals involved in both practice and research find a need for a useful reference on statistical methods that covers a not-too-theoretical discussion of the methods and some clear indications of how, when, and where to use them. Too often, both the courses and books available fail to provide the type of treatment that these various people need. In many of the specialist areas covered in these disciplines, there is a need to be able to use various statistical methods, for which it is necessary, first, to recognize where a particular method is appropriate; second, to know what assumptions and restrictions exist in the method; and third, to know how to apply the methods correctly. These needs cannot generally be met by treatments that concentrate on proofs of theorems and detailed treatment of all theoretical aspects of the methods; nor can they be met by a treatment that simply shows how to apply each method, without dealing with underlying assumptions, restrictions, or concerns with applicability.

While nobody would dispute the existence of very strong interrelationships between the collection of data and the subsequent statistical analysis of data, there do not appear to be any texts that attempt to address both of these issues. Instead, the texts on survey methods seem to assume that the analyst wishes to do nothing more than determine population values and place error bounds on those estimates, while books on multivariate analysis seem to assume the existence of perfect data sets and ignore the issue of the effects of imperfect data collection on these procedures. Many students and professionals, working in the disciplines mentioned above, are faced with the necessity of data collection other than through laboratory experiments. Due to time and resource constraints, information can only be gathered on a representative subset of the population under study. The quality and accuracy of this representative sample has to be assured through properly designed and executed sampling procedures, the topic of discussion in the early chapters of the book. The reliability of the results obtained through subsequent analysis of the data by means of multivariate-analysis techniques, the topic of the later chapters, is a direct function of the quality of the data input. The perception of this close and necessary relationship between the two phases of survey sampling and statistical analysis has prompted us to generate this book.

To serve the needs that we feel exist for a treatment that is neither overly theoretical nor overly empirical, we have attempted to cover what we feel to be the minimum of proofs and theory consonant with developing a basic understanding of the methods. We have supplied a substantial level of narrative, particularly for the multivariate-analysis methods, to provide the reader with a strong foundation in the concepts of the methods. We hope that the balance of treatment given here will help the reader to develop some of the insights and judgmental capabilities that are essential for these methods and that, as Kendall states in his book on *Multivariate Analysis,* "characterises the statistician, and for which pure mathematics is no substitute." We hope that the treatment and coverage of the material will prove useful and interesting to readers in a variety of disciplines, such as urban planning, psychological measurement, demography, consumer behavior, transportation planning and engineering, marketing, decision theory, and various specializations within engineering, the medical sciences, and management and business administration.

Acknowledgments

Probably no book is written without the help and contribution of many people other than the authors; this book is certainly no exception. We would like to express our thanks and appreciation to all those who contributed to this book. Though it is not possible to name all contributors, we would like to name a few.

We particularly express our thanks and appreciation to Patricia Apgar, Hilary Silberman, and Valerie Stopher, who typed the manuscript. We are again most fortunate to have the use of the drafting skills of Wilfred R. Sawbridge of Cornell University, who prepared all the drawings for the book.

We thank Dr. Richard Tso of McMaster University and Mr. Larry Lavery of Northwestern University, who commented extensively on an earlier version of the manuscript. We also thank Dr. Richard B. Westin of the University of Toronto, who provided detailed comments on chapter 15. Over the years, many students at Cornell University, McMaster University, and Northwestern University have contributed to the development of the book, through their responses to earlier drafts, their questions during classes, and their evaluations of the courses that underlie the book.

The authors also acknowledge the cooperation of Charles Griffin and Company, Limited, of London and High Wycombe for permission to quote and use examples from *Sampling Methods for Censuses and Surveys* by Frank S. Yates (3d ed., 1971); John Wiley & Sons for permission to quote and use examples from *Econometrics* by Gerhard Tintner (1952); Lexington Books, D.C. Heath and Company, for permission to quote from *Urban Transportation Modeling and Planning* by Peter R. Stopher and Arnim H. Meyburg (1975); the U.S. Department of Transportation for the use of a figure from *Applications of New Travel-Demand Forecasting Techniques* by Bruce D. Spear (1977); and the Elsevier Publishing Company for permission to use tables from and to base chapter 16 extensively on "Goodness-of-Fit Measures for Probabilistic Travel Choice Models" by Peter R. Stopher, published in *Transportation,* vol. 4, no. 1, 1975.

While we have endeavored to remove all errors from the text, it seems almost impossible to remove every last one. The authors accept complete responsibility for any errors that may still exist in the text.

1

Introduction

Purpose of This Book

Much of the work in engineering and the social sciences is concerned with the development and testing of hypotheses about phenomena in the real world and with drawing inferences from observations of those phenomena. Such work is based on the application of the principles of scientific enquiry, which focus on the formulation and testing of theories and hypotheses. The objective of these efforts may be twofold. First, the aim may be to attempt to understand some phenomenon in order to increase understanding of the world around us or to provide opportunities for control of the phenomenon. Second, the aim may be to produce a simulation of the phenomenon that would permit the engineer or scientist to predict changes in the phenomenon that might occur as a result of various processes. In either case, the product sought may be a representation of the phenomenon in some mathematical form that can be manipulated for understanding or prediction. Such a representation may be termed a mathematical model, a concept explained in more detail in chapter 7.

The process of scientific enquiry requires first the development of hypotheses about the phenomenon. Based on these hypotheses, observations are made of the phenomenon so that the hypotheses may be tested and refined or reformulated. This book is concerned primarily with the last two steps of this process; namely, observations and hypothesis testing. Clearly, the proper observation of phenomena constitutes a most important step in scientific enquiry, since the relevance, significance, and accuracy of hypothesis testing depends heavily on the quality of the underlying observations (data).

While hypotheses and simulations may take many different forms, the specific concern of the book is with the development of statistical models from the observations. The need is for rules and procedures for testing the fit of hypotheses to a set of data where those hypotheses are couched in mathematical terms.

Accordingly, the book provides a treatment of the techniques of data collection, methods of sampling, computation of population values and sampling errors, and a number of statistical-analysis techniques of interest to a broad group of social scientists and engineers. While there are many books available that deal with sampling methods and associated statistical procedures, for example, Yates,[1] Kish,[2] and Cochran,[3] and a number that deal with specific procedures of multivariate analysis and statistical model

1

building, such as Draper and Smith,[4] Kendall,[5] Finney,[6] Goldberger,[7] and Johnston,[8] it is unusual to find a book that links sampling and the (frequently) subsequent multivariate analysis. The authors attempt to make that link and, by so doing, stress the strong interactions that exist between the collection of sample data and the subsequent statistical analyses executed on the data. That this link is important should become apparent as the reader works through the book.

Again, among the books available on each of survey sampling and multivariate analysis, two basic treatments are generally found. On the one hand, the books may provide extensive theorem proving, with little attention paid to the application of the methods to scientific enquiry. Such texts are frequently understandable only to theoretical statisticians and econometricians and leave the reader in engineering or social science without any feel for the usefulness of the procedures. On the other hand, the books are so practice-oriented as to give the reader no feeling for the fundamental assumptions and hypotheses underlying the techniques. Such books are frequently given the epithet "cook books," since they provide recipes for solving certain data collection or curve-fitting problems but give no idea of the reasons why the recipes work nor give the reader sufficient information to know under what circumstances the recipes may be inapplicable.

This book seeks an intermediate position between detailed theorem proving and a simplistic practice-oriented approach to the use of survey sampling and multivariate-statistical techniques. Bearing in mind the variety of mathematical backgrounds that may be expected in engineering and the social sciences, matrix algebra is used as little as possible and proofs are generally provided for only the simplest case, using simple algebra when possible. A basic knowledge is assumed, however, of elementary probability and statistics, calculus, and algebra. Emphasis is given to concepts, and assumptions and limitations on the various methods discussed.

Scope of the Book

In survey sampling, the book covers such techniques as simple random sampling (both with and without replacement), stratified sampling (with uniform and variable sampling fractions), cluster sampling, and two-stage and multistage sampling. Each sampling technique is described together with methods for computing population estimates and sampling errors. The book also addresses the design of survey instruments for surveys of human populations and provides anecdotal examples from the field of transportation planning.

It is not the purpose here to provide an exhaustive treatment of these topics. A number of methods of computing values, refining survey techniques, and calculating or reducing sampling errors are not dealt with. It is hoped, instead, that the treatment provided will be sufficient to cater to the basic needs of social scientists and engineers and that they will be encouraged to consult appropriate specialist texts on survey sampling as needed.

In multivariate analysis, the book covers linear-regression analysis, simultaneous equations methods, canonical correlation, principal-components analysis and elementary factor analysis, discriminant analysis, probit analysis, and logit analysis. In all cases, the basic assumptions of the techniques are discussed and proofs provided of the analytical results. Illustrations are provided, principally from the field of transportation planning.

Again, it is not the purpose of the book to provide exhaustive treatment of these topics, many of which comprise entire textbooks by themselves, such as factor analysis,[9] probit analysis,[10] and linear-regression analysis.[11] The intention is to provide an introduction to the techniques, sufficient to acquaint the reader with the fundamentals of each analytic procedure, and a strong enough background on each to facilitate tackling more complex and detailed versions of the methods. Furthermore, there does not appear to be a text available at present that deals with all these techniques. To do this, however, requires that the treatment of each technique be at an introductory level, since to do otherwise would require the development of a multivolumed book and would probably defeat the objectives of this book.

It is the purpose here to provide a sufficient level of treatment of both survey sampling and multivariate-analysis methods to meet the needs of most professionals, students, and readers in the social sciences and engineering. In many cases, even those engaged in research should find these treatments sufficient for their needs. Scientific enquiry should always seek to find the simplest and most straightforward explanation for a phenomenon, resorting to more complexity only when it becomes apparent that the simple explanation is overly simplistic and inadequate. To this end, the book provides a more extensive discussion of the simpler procedures of survey sampling and multivariate analysis, while providing substantially less detail on more complex issues and problems. On the other hand, it also provides the first coherent discussions in a textbook of multivariate-analysis techniques, such as logit analysis and the correlation ratio, setting these in the context of applications in the social sciences and engineering.

The book, therefore, can serve both as a teaching device and as a resource for research purposes. It should also provide the reader with a sufficiently strong background in the concepts and limitations of the various methods of survey sampling and multivariate analysis, that he or she should not become confused or lose sight of the relevance and applicability of the methods when consulting more detailed and sophisticated treatments in other texts.

Fundamentals of Scientific Enquiry

In preparation for the material in the remainder of the book, some points relating to scientific enquiry are important. Scientific enquiry demands a logical, systematic, and objective approach to the analysis and understanding of phenomena. This approach results in the development of a process of hypothesis formulation, observation, and testing and refinement of hypotheses. Since the remainder of the book is concerned with the second and third of these activities, it is necessary to stress the importance of the first activity in the process of scientific enquiry.

The importance of beginning the process with careful hypothesis formulation cannot be overstressed. The development of prior postulates of the behavior of a phenomenon provides needed guidance on the decisions of what observations to make and what analytic processes to use for analyzing the observations and simulating the phenomenon. Data collection is an expensive process, no matter how well it is designed and handled. When data collection concerns observations on people, as it often does in the social sciences and some branches of engineering, there are significant costs associated with repetitive surveys. These costs are manifested in the increasing reluctance of people to be subjected to repeated surveys, thereby resulting in an increasing rate of refusal to assist in the survey and concomitantly higher cost for the survey. Therefore, collection of wrong or insufficient data should be avoided. This can be done only by making careful postulates of the process first, and defining from them the observations needed to test the postulates.

The proper development of hypotheses can be seen to have more far-reaching effects than just defining the needed data and the extent of the data. In many cases, a failure to develop prior hypotheses may even lead to observing the wrong phenomenon or making observations on the wrong unit of analysis. For example, in attempting to understand some aspects of human behavior, such as consumer purchasing behavior, failure to specify prior hypotheses might lead the analyst to observe household behavior rather than individual behavior, or vice versa. Such errors may often be irretrievable in that they can be corrected only by discarding the initial set of observations and making a fresh set.

The formulation of prior hypotheses is necessary to permit making decisions about the forms of statistical models to be examined. The importance of this will be appreciated better after the reader has studied the chapters dealing with the multivariate-analysis methods. Stated simply, however, the choice of analysis method has considerable impact on the form and content of the data to be collected. Having collected data in a specific form, several methods of multivariate analysis and, more importantly, several structures of models within any multivariate-analysis technique will be precluded. To avoid the costly pitfall of finding that the data collected cannot be used to test an appropriate model structure, it is again necessary to develop careful and thorough prior hypotheses.

Finally, it is worthwhile to consider the form that prior hypotheses should take. First, the hypotheses should postulate the unit of observation and modeling. They should specify whether, for social or behavioral phenomena, the appropriate unit to analyze is the individual, the household, or some aggregation of households, such as a suburb, township, city, or region. Second, hypotheses should specify the causes and effects of the process to be analyzed. Suppose one is studying the process of household car purchases. The effect to be observed could be the purchase of a replacement car, the purchase of an additional car, or the purchase of a first car. Among the causal agencies that could be considered are the income of the household; the number of household members who are employed; the size of the family; the quality, quantity, and service area of public transport; the desired status of the household; the proximity of shops and other facilities; and the proportion of income used for necessities such as food, clothing, and shelter. (It should be noted that the above causal agents are already based on the idea that the unit of analysis is the household. If this were to be changed to an individual, a census tract, or some other unit, the set of causal agencies and the form of the "effect" variable would all be changed.)

Third, the hypotheses must specify the way in which it is expected that cause and effect are linked. In other words, the form of the relationship expected between cause and effect should be specified. In the example, one might postulate a probabilistic relationship, in which the probability of acquiring an additional car is assumed to increase monotonically with increases in the set of causal variables. Alternatively, one might postulate a threshold model in which there are boundaries on the size of the combined causal agents, the crossing of which precipitates a decision to purchase another car. The form in which the causal agents operate must also be specified. For example, these agencies may be assumed to be additive and compensatory. That is, an increase in the value of one causal variable may offset (compensate for) a decrease in another one. Alternatively, the causal variables may be assumed to have a multiplicative effect, to be additive in the logarithms of the values, or to have any other plausible construct. Specifying these aspects of the process allows the analyst to select candidate multivariate-analysis techniques, to determine the data that must be collected to observe the phenomenon of concern and develop the desired models or simulations, and to determine the unit that must be observed.

It is extremely tempting to use statistical methods, whether in the design and execution of surveys or the selection and use of multivariate-analysis methods, as a means to seek for models of processes without taking the trouble to develop prior hypotheses. The literature in many of the social sciences and branches of engineering abounds with examples of such unscientific enquiry. Unfortunately, the widespread availability of high-speed computers and programs for various forms of statistical analysis seems to encourage this form of enquiry. The reader, however, is urged to eschew this temptation. Statistical method should not be considered as a form of convenient "fishing rod" with which to go and "catch" a model. In particular, it must be emphasized that statistical methods themselves are incapable of differentiating between coincidental but spurious relationships and genuine cause-and-effect relationships. The judgment of the analyst is crucial in making such distinctions. Failure to form coherent and thorough prior hypotheses reduces the extent to which the analyst can make such judgments.

As an example, a demographer reported having found that he could predict the rate of illegitimate births in Sweden as a direct function of the divorce rate in New Zealand! Had prior hypotheses been constructed, the analyst should never have examined these two measures as candidate cause-and-effect variables. However, having developed such a relationship, the temptation is to justify it, rather than using judgment, that might have been possible from drawing up prior hypotheses, to discard the relationship as spurious and coincidental.

Notes

1. Frank S. Yates, *Sampling Methods for Censuses and Surveys*, 3d ed., London: Charles Griffin and Co., 1971.

2. Leslie Kish, *Survey Sampling,* New York: John Wiley & Sons, 1965.

3. William G. Cochran, *Sampling Techniques,* 2d ed., New York: John Wiley & Sons, 1966.

4. N.R. Draper and H. Smith, *Applied Regression Analysis*, New York: John Wiley & Sons, 1968.

5. Maurice G. Kendall, *A Course in Multivariate Analysis,* London: Charles Griffin and Co., 1965.

6. D.J. Finney, *Probit Analysis,* Cambridge, England: Cambridge University Press, 1971.

7. Arthur S. Goldberger, *Econometric Theory,* New York: John Wiley & Sons, 1964.

8. J. Johnston, *Econometric Methods,* New York: McGraw-Hill Book Co., 1963.

9. H.H. Harman, *Modern Factor Analysis,* Chicago: University of Chicago Press, 1960; R.J. Rummel, *Applied Factor Analysis*, Evanston, Ill.: Northwestern University Press, 1973.

10. Finney, *Probit Analysis*.

11. Draper and Smith, *Applied Regression Analysis*.

2 Review of Data Needs and Sources of Error

Data Needs and Scientific Enquiry

Scientific enquiry focuses on the formulation and testing of theories and hypotheses. Frequently, theory or hypothesis formulation is preceded by observation. In that instance, the proper observation of phenomena constitutes the first important step in scientific enquiry. The relevance, significance, and accuracy of a proposed theory depends heavily on the quality of the underlying observations (data).

Scientific research in the social sciences and engineering is based on the analysis of observed data related to the phenomenon under investigation. In many instances, these data can be developed experimentally under laboratory conditions. However, in other areas of enquiry, notably in urban planning, transportation planning, and traffic engineering, laboratory experimentation is of limited relevance, if not completely infeasible. The data base for analysis in such areas has to be developed from real-world observations and enquiries.

It is obviously infeasible to base these analyses on all elements of a population which exhibit or could exhibit a characteristic that is of relevance to the phenomenon under investigation. Time, money, and manpower constraints would prohibit such an exhaustive data-collection effort. In general, data collection tends to be by far the most expensive element of an investigation. Therefore, the required information has to be obtained from a representative subset of the population, called a sample. On the other hand, the sample must be sufficiently large to permit statistically reliable inferences to be drawn for the population as a whole.

The statements above illustrate two conflicting objectives that must be satisfied in developing sampling procedures and drawing statistically sound inferences. Sample size needs to be reduced on the grounds of cost but increased for reliable inference. This represents the underlying rationale of the survey-sampling methods discussed in this and the next four chapters; namely, how to choose a sampling method and draw a sample that produces an acceptable compromise between these conflicting objectives. Most importantly, it must be realized that the inferences drawn can be no better than the quality of the data on which they are based. Indeed, if the data are of very poor quality, it is not worthwhile applying sophisticated multivariate-analysis techniques to them.

It should be obvious that error is inherent in the process of sampling, since the use of a sample implies that certain information or measures, pertaining to the nonsampled parts of the population, will not be obtained.

9

The error that is incurred could take two forms. First, it may be a random occurrence in the population, such that increasing the size of the sample reduces this component of error. Second, the error may be a systematic one that may remain of the same size or even increase as the sample size is increased. The first form is usually termed the sampling error, is assumed to exist in all samples, and can be calculated for specific methods of sampling. The second is termed bias, cannot usually be calculated when it exists, but must be avoided in proper sampling.

Many multivariate-analysis techniques are based on the assumption of a random error being present in the data. They assume, however, that there is no bias present. These techniques are based on the principle of minimizing the total random-error component in each observation, where a portion of that random error will arise from the sampling process. To do that, some assumptions are required about the characteristics of these errors, such as the shape of their distributions and the value of the mean error. These assumptions on the error component form the basis for the differential process and they always lie behind the testing for statistical validity and reliability of the inferences drawn from the data. It is important to understand what errors can be introduced by the sampling procedure in order not to violate the assumptions made in the multivariate-analysis techniques. The presence of bias, for example, would constitute such a violation.

By definition, the sampling error increases as the sample size decreases. The implication of this with respect to the multivariate-analysis techniques is that the smaller the sample size, the poorer will be the goodness-of-fit to any postulated relationship and the larger will be the unexplained residual variance in the data. Hence both the selection of the sample size and the conduct of the sampling process have profound effects upon any subsequent multivariate analysis of the data collected.

A further argument to illustrate the importance of presenting the information-gathering phase, namely, survey sampling, in conjunction with the use of the information in multivariate-analysis techniques relates to the suitability, relevance, and completeness of the data for the desired analysis. If these three criteria are not met, the analyst may find it impossible to conduct the desired tests of hypotheses. For example, suppose a relevant causal variable has been postulated as cost divided by income. If the income information is collected by broad categories, the causal variable will be undefinable in those data, that is, the income variable is unsuitable for the analysis. Again, suppose the appropriate measure of income is the net disposable income of the household, but the data collected are of the gross household income. In this case, the data collected are irrelevant to the analysis to be performed. Another example is provided by the situation in which it is desired to test a relationship across the entire population, but data are collected only from wage earners. In this case, the data are incomplete and will again result in an inability to carry out the desired hypothesis tests.

In summary, the design and execution of data-collection procedures are inextricably tied to the subsequent analysis, in this case multivariate analysis, so the two activities must be viewed together.

Survey Sampling

To make data collection economically feasible and operationally manageable, methods have been devised to draw reliable inferences about the whole, when information has only been collected for a subset, or sample, of the whole. For example, it would take about 3–4 months to collect data on a 3.5% sample of Chicagoans and another 9–10 months to put these data into usable form, such that they can be analyzed by electronic data-processing equipment. Were one to collect these data for the whole population, assuming the same budget allocation, the data collection and processing would take 10 and 25 years, respectively. On these grounds alone, the necessity for sampling becomes obvious.

A sample is defined as a specimen or a small portion representative of the quality of the whole. It is important to remember the key elements of the definition; namely, the fact that a sample constitutes only a small part of the entire mass and this small part, or sample, is representative of the entire mass.

If the concern is with a population of similar units, such as a handful of sand from a clean sandy beach or a spoonful of instant coffee from a coffee jar, little attention has to be paid to the way the sample is chosen. It is reasonable to assume that any subset of sand or coffee grains is representative of the rest of the sand on that particular beach or the rest of the coffee grains in the jar.

The issue of representativeness becomes very significant, though, when the population is composed of dissimilar units. The selection or sampling procedure has to be well thought out to make sure that the selected sample is representative of the dissimilarities present in the population.

Surveys and censuses concerning households and individuals deal with a mass made up of dissimilar units. It is clear that no matter how carefully the sampling procedure is applied, it is impossible to obtain a sample of dissimilar items which is completely representative of the total population from which the sample was drawn. Total representativeness of a sample of dissimilar units is achieved only when all elements of the population are studied (100% representation). Technically, 100% of the population is by definition not a sample. Because a sample is only a portion of the population, sampling errors will occur, no matter how carefully the sampling procedure is executed.

To facilitate an easier grasp of sampling procedures, here are a few basic definitions of terms frequently used in survey sampling. This and the following chapters on survey sampling draw heavily on material presented in Kish,[1] Yates,[2] and Cochran.[3]

Census. This is defined as the enumeration of a population in terms of basically simple data. By this definition, interviewing is not necessary. It is only necessary to count or to use extremely simple questionnaires. A census may be applied to an entire population or a sample.

Survey. A survey is an investigation which involves the collection of highly detailed information. When it is a survey of people, it will frequently be necessary to collect the information by interview, for which trained interviewers are required. When it is a survey of things, such as vehicles, it will often necessitate the use of trained investigators or sophisticated equipment to collect the information. A survey may be applied to an entire population or a sample.

Population. The population is the aggregate of the elements which are the subject of the census or survey. In this definition, the sample is a representative subset of the population. The population includes both sampled and nonsampled elements. A sample is required to be able to make statements about the population.

Sample Bias. This is the error introduced by "bad" sampling. It may be due to such factors as the use of an incorrect sampling frame, insufficient control of sampling, or excessive nonresponse. There is not always bias in a sample, and ideally should not be any.

Random Sampling Error. This is the error introduced by the fact that a sample is used to represent a population. There is always a sampling error in any sample. Its size, though, will depend on many things, including the method of sampling.

Sampling Frame. A sampling frame is a basic list or reference which unambiguously defines every element or unit in the population from which the sample is to be taken. The existence of a sampling frame is essential to the process of sampling.

Estimation. This is the calculation of population values from sample values. It involves the use of expansion factors, determined from the sampling rate, and may also utilize supplementary information about the population or from another sample.

Sampling Theory. Sampling theory concerns the means by which to obtain representativeness from a small part of the whole population. It permits computation of the degree to which errors exist in representing the whole.

These definitions are probably best illustrated by an example. This example, adopted from Stopher and Meyburg,[4] addresses the issue of how to select a good sample and illustrates some of the pitfalls the analyst can encounter when overlooking basic rules of sampling theory. To illustrate what sampling means, consider table 2–1, which is a record of the heights of 100 men. Certain specific figures can be determined from this table which describe the population of 100. First, the mean (average) height is 5'7.66" and the median height is 5'8". A frequency table can be constructed, showing the frequency with which each height occurs, as shown in table 2–2.

Now suppose one wishes to choose a sample of 10 individuals, rather than measuring all 100 men. In any sample of 10, the mean and median heights should closely approximate those of the whole population, and the height distribution should be similar to table 2–2. One could expect, in a sample of 10, the distribution shown in table 2–3. If the first 10 were chosen, the mean height would be 5'9.3", the median height 5'10", and the distribution as shown in table 2–4. It is clear that there is an overrepresentation of the taller men in this sample. The median is higher, not lower, than the mean, and there are too many individuals at the top end of table 2–4 and not enough at the bottom end.

Suppose one chose the first five and the last five observations (i.e., numbers 1–5 and 96–100). Now the mean height is 5'9.2" and the median height is 5'10". These results are almost identical to the previous ones, but quite different from the true values. The distribution, shown in table 2–5, is similarly biased.

This example serves to demonstrate two things. First, it clarifies what is meant by choosing a sample; and second, it shows that sampling to yield a representative sample is not a simple process of either haphazard or systematic selection. In fact, selecting a sample from dissimilar objects requires the use of specific sampling rules which serve to guarantee, within certain known limits, a representative sample. These known limits are termed sampling error, and this error can be determined for any of a number of different sampling procedures.

Table 2-1
Height Observations for One Hundred Individuals

Obs. No.	Height	Obs. No.	Height	Obs. No.	Height
1	5'10"	34	5' 8"	67	5'10"
2	5' 9"	35	5' 7"	68	5' 3"
3	6' 0"	36	5'11"	69	5' 1"
4	6' 0"	37	5' 9"	70	6' 1"
5	5'11"	38	5' 4"	71	5' 8"
6	5'10"	39	5' 8"	72	5' 5"
7	5' 7"	40	6' 1"	73	5' 0"
8	5' 5"	41	5' 0"	74	5' 2"
9	6' 1"	42	5' 6"	75	5' 6"
10	5' 4"	43	5' 7"	76	5' 3"
11	5' 8"	44	5'11"	77	5' 1"
12	5' 9"	45	6' 0"	78	5'11"
13	5' 3"	46	6' 1"	79	6' 0"
14	5' 7"	47	5' 3"	80	6' 3"
15	6' 2"	48	5' 5"	81	4'10"
16	6' 0"	49	5' 8"	82	5' 4"
17	5'11"	50	5'11"	83	5' 8"
18	5' 8"	51	5' 9"	84	5' 9"
19	5' 7"	52	5' 2"	85	5'11"
20	5' 1"	53	5'10"	86	5' 7"
21	6' 0"	54	5'10"	87	5' 5"
22	5' 9"	55	4'11"	88	6' 1"
23	5' 4"	56	5' 7"	89	5' 2"
24	5' 6"	57	5' 8"	90	5' 0"
25	5'10"	58	6' 0"	91	5' 9"
26	5' 5"	59	6' 2"	92	5' 7"
27	5' 6"	60	5' 9"	93	5' 3"
28	5' 3"	61	5' 7"	94	5' 4"
29	6' 1"	62	5' 8"	95	6' 0"
30	5'10"	63	5' 5"	96	5'11"
31	6' 0"	64	6' 0"	97	5' 8"
32	6' 3"	65	5' 6"	98	5'10"
33	5' 5"	66	5' 9"	99	5' 3"
				100	5' 6"

Sampling Bias and Sampling Error

The basic principle of all sampling procedures is random selection. In this context, random does not mean haphazard. It is specifically defined by a number of statistical measures. In simple terms, selecting at random means that every member of the population from which the sample is selected has an equal probability of being chosen. To assist in choosing random samples, there are tables of random numbers and most digital computers have programs which can generate random numbers.

To illustrate the use of simple random sampling, the earlier example can be used again. Using the RAND table,[5] page 263, line 13104 gives the

Table 2–2
Frequency Distribution of Heights

Frequency	Height
1	4′10″
1	4′11″
3	5′ 0″
3	5′ 1″
3	5′ 2″
7	5′ 3″
5	5′ 4″
7	5′ 5″
6	5′ 6″
9	5′ 7″
10	5′ 8″
9	5′ 9″
8	5′10″
8	5′11″
10	6′ 0″
6	6′ 1″
2	6′ 2″
2	6′ 3″

Table 2–3
Frequency Distribution of Heights for Ten
Representative Individuals

Frequency	Height
1	4′10″–5′ 2″
1	5′ 3″–5′ 4″
1	5′ 5″–5′ 6″
2	5′ 7″–5′ 8″
2	5′ 9″–5′10″
2	5′11″–6′ 0″
1	6′ 1″–6′ 3″

Table 2–4
Frequency Distribution of Heights for the First Ten Individuals

Frequency	Height
0	4′10″–5′ 2″
1	5′ 3″–5′ 4″
1	5′ 5″–5′ 6″
1	5′ 7″–5′ 8″
3	5′ 9″–5′10″
3	5′11″–6′ 0″
1	6′ 1″–6′ 3″

Table 2–5
Frequency Distribution for the First and Last Five Individuals

Frequency	Height
0	4′10″–5′ 2″
1	5′ 3″–5′ 4″
1	5′ 5″–5′ 6″
1	5′ 7″–5′ 8″
3	5′ 9″–5′10″
4	5′11″–6′ 0″
0	6′ 1″–6′ 3″

following 10 observations: numbers 65 8 62 50 96 77 36 95 54 93. Using this selection, the sample mean is 5′7.8″ (compared with 5′7.66″ true mean) and the sample median height is 5′8″ (the same as that of the total population). The distribution of values is shown in table 2–6. Though the distribution is not the same as that shown in table 2–3, it is more even than that obtained before (tables 2–4 and 2–5).

Provided there is no ordering of members of the population in a list, choosing every tenth member for a 10% sample would give similar results. Since table 2–1 presents a random ordering of individuals, an approximately random sample could be chosen. If, in the above example, the population had been ordered by increasing height (table 2–1), selecting every tenth member (say, numbers 2, 12, 22, etc.) would not produce a random sample.

Many variations on simple random sampling have been devised, designed to produce specific known characteristics. These procedures are discussed in detail in subsequent chapters.

A few comments concerning sampling bias are in order here, since this type of inaccuracy, unlike the sampling error, can be avoided through conscientious adherence to the rules of random sampling.

If the analyst is interested in averages, or means, measurements of a

Table 2–6
Frequency Distribution for a Random Sample of Ten Individuals

Frequency	Height
1	4′10″–5′ 2″
1	5′ 3″–5′ 4″
2	5′ 5″–5′ 6″
1	5′ 7″–5′ 8″
1	5′ 9″–5′10″
4	5′11″–6′ 0″
0	6′ 1″–6′ 3″

sample might be taken which he or she considers to be representative of averages in the population. However, the sample will be dependent on two things: the accuracy of the analyst's knowledge of the averages, and his or her subjective assessment of how near any member of the population is to the average. Bias will be introduced in both cases.

In another instance, the analyst may be interested in a representative sample of the whole population, but might deliberately try to choose members of the population who represent the extremes and the middle range of the population value. Again, subjective selection of members of the population will yield bias.

Bias can also arise by selecting sampling units on the basis of an attribute of the population measures in which the analyst is interested. An example of such an error would be sampling from the telephone directory, as a sampling frame, when the proportion of households having a telephone is one of the measures the analyst is interested in.

Another type of bias is particularly difficult to detect and no one may be conscious of its existence, including the perpetrator. While it can enter in a number of ways, it most commonly enters when an investigator is setting up the sample. The investigator may select the sample on the basis of a random sample table, but discard certain selections on some pretext that they do not fulfill the survey requirements. The pretext, however, may be consciously or unconsciously related to the results he or she hopes to obtain.

A fifth type of bias is a very common one, but can be eradicated easily by careful field control. It frequently happens in household interviewing, for instance, that the interviewer gets no reply when calling on a selected house. Either immediately or after one or two callbacks, the interviewer may select the next house to make up the sample to the right number. Such a practice will lead to an overrepresentation of households with families who tend to be home all day, against an underrepresentation of households with one or two people who are out at work or school all day.

The final type of bias that can be identified may arise in a similar type of situation to the one just illustrated. It occurs in the case where, for example, no reply was received from a particular household, and another household was not substituted. This is also the type of bias which occurs most frequently in postal questionnaires, where replies are most likely to come from people who have an interest in the objectives of the survey or who possess other characteristics which make them unrepresentative of the whole population.

In summary, it must be concluded that to avoid the introduction of sampling bias, it is essential that a truly random selection process is ensured or a selection process is used that is random subject to restrictions that will not introduce a bias into the sample.

The surest way of obtaining a truly random sample is to use random-number tables. Another safeguard against a biased selection process is to assure that all sample selection be performed away from the field, and none of it be done by the investigators while they are actually interviewing or measuring the sample.

Apart from in the sample selection process, bias may also occur at other stages in the survey task. One common source of bias is inaccuracy of measurement. For instance, if a survey of households is carried out in which each person interviewed is requested to disclose the income group into which the household falls, there is a tendency for misrepresentation by household members.

Another source of bias is estimation due to faulty methods of analyzing the survey results. A common source of such a bias occurs in estimating ratios. A simple example can best illustrate the point: data have been collected on household incomes, split into three groups: under $12,500, $12,500–$25,000, and over $25,000. The mean incomes for the groups have been determined as $9,000, $16,000, and $28,000, respectively. One hundred people were found in the lowest group, 60 in the middle, and 30 in the top. To calculate the mean income, the weighted mean should be calculated:

$$\frac{100 \times 9,000 + 60 \times 16,000 + 30 \times 28,000}{100 + 60 + 30} = \$14,200 \qquad (2.1)$$

If the average income would have been calculated from the group means, the result would have been $17,667. This would constitute an estimating bias of 24.5%.

The sources and implications of sampling bias have been discussed at some length. Before leaving this topic, it is appropriate to indicate under what circumstances a bias might be tolerable and permissible. If the analyst wants to determine, from a series of surveys, the changes that are occurring in a particular attribute, a constant bias from survey to survey will not affect the results. This bias must be constant over time and not related to the attribute whose change is being measured. Clearly, if the objective is to measure changes in car ownership and the bias consists of an overly large number of high-income households, the rate of change of car ownership in the sample is likely to be much smaller than in the population at large. This is because income is related to car ownership, but car ownership tends to reach a saturation level at higher incomes.

Again, if the survey is intended to produce data to allow comparisons between groups within the population, a constant bias in the data irrespective of the groups will be of little or no account. Similarly, in such a situation a bias, in terms of measuring numbers of units making up the groups which are not in proportion to their occurrence in the population, will be unimportant if the concern is to compare averages, or means, of the groups. (If the samples are very small, this bias could be important in giving less confidence than desired in the group means.)

Besides sample bias, there is another type of inaccuracy which can and does enter the sampling process, namely, the random-sampling error. While sampling bias arises from faulty sampling methods, errors in estimation, and inaccurate measurement, the random-sampling error is due to chance differences between those members of the population who are surveyed and those who are not. Clearly, if the entire population is sampled, there is no random-sampling error. In fact, it can be shown that random-sampling error is approximately proportional to $1/\sqrt{(n)}$, where n is the number of units surveyed.

Once it has been established that there is no important bias in the sample, the next concern must be with the random-sampling error. When carrying out a survey, it would be desirable at some stage to determine the level of accuracy which survey results are required to attain. The sample must be chosen in such a way as to ensure that the random-sampling error is sufficiently small to allow the achievement of the desired level of accuracy. Chapter 4 addresses this issue in substantial detail.

Notes

1. Leslie Kish, *Survey Sampling,* New York: John Wiley & Sons, 1965.

2. Frank S. Yates, *Sampling Methods for Censuses and Surveys,* London: Charles Griffin and Co., 1971.

3. William G. Cochran, *Sampling Techniques,* 2d ed., New York: John Wiley & Sons, 1966.

4. Peter R. Stopher and Arnim H. Meyburg, *Urban Transportation Modeling and Planning,* Lexington, Mass.: Lexington Books, D.C. Heath and Co., 1975.

5. RAND Corporation, *A Million Random Digits,* New York: Free Press Books, 1955.

3

Design of Sampling Procedures

Random Sampling

A primary consideration that underlies survey-sampling procedures is the determination of sample size to achieve a desirable degree of accuracy in the population parameters. As discussed in the preceding chapter, all things being equal, random-sampling error is approximately proportional to the inverse of the square root of the number of units in the sample. There are other means, though, by which random-sampling error can be affected. One such contributory factor to random-sampling error is the variability per unit. If restrictions on the fully random sample are imposed which do not introduce bias but allow the reduction of that part of the variance of the units which contributes to the sampling error, the size of the random-sampling error and, consequently, the size of the sample required for a given level of accuracy can often be reduced substantially.

The simplest form of such restrictions is *stratification*. This involves stratifying, or dividing, the population into blocks of units in such a way that the units within each stratum are as similar as possible. Each stratum is then sampled at random. If the same fraction of the population of each stratum is sampled, the final sample will clearly represent the correct proportions of each stratum within the whole population, and differences between the different strata will have been eliminated from the sampling error.

A further refinement of this method is the use of a *variable sampling fraction*. The form of this restriction on simple random sampling is that the strata are sampled at different rates in such a way that the strata which are of more importance, or are more variable, are sampled more intensively. This naturally requires that the survey results are weighted correctly before being combined to yield a total result for the entire population.

Another form of restricted sampling is the use of *multistage sampling*. In this form of sampling, the population is divided into a number of first-stage sampling units, which are in turn subdivided into second-stage units, and so on. For instance, in the case of a travel survey of the entire United States, as a first stage the United States is divided into states and a sample is drawn from the total population of states. Each selected state is then subdivided into counties and the counties are sampled. Within the selected counties, the population is subdivided by households, and a sample of households is drawn for the third stage. A fourth stage could be added in which households are subdivided into individuals and people are sampled from each selected household.

The final form of restricted sampling is a controversial one, namely, *cluster sampling*. There is considerable disagreement about the merits and demerits of cluster sampling, which is discussed in more detail later on. Basically, cluster sampling consists of grouping the total population into clusters, where each cluster can be considered to be a natural unit of the population. The clusters are then sampled at random, and the units within the cluster are either selected in total or sampled at a very high rate.

One further method that can be used to improve the accuracy of the survey results is that of the use of *supplementary information*. Information is used which is not derived from the sample, or is derived from a far more extended sample than the one which will supply the principal material. This can be explained more easily by using a simple example. Suppose that the number of private cars in the United States is to be determined, that a random sample of households throughout the states has been taken, and that the total number of cars owned in the sample of households determined. The total number of cars could then be found by multiplying the sample total by the reciprocal of the sampling rate. Alternatively, if the total number of households in the United States is known, the total number of cars may be determined by multiplying the average number of cars per household from the sample by the total number of households in the United States. Provided the number of households in the United States is known with sufficient accuracy, the second estimation method will be the more accurate of the two.

The Sampling Frame

A basic definition of sampling frame was given in chapter 2. The relationship between a sampling frame and the sampling method is discussed next. First, consider a sampling unit. This is the unit on which all the survey measurements are based. It may be a household, an individual, an SMSA (standard metropolitan statistical area), a freeway ramp, lengths of freeway, a census tract, and so on. The units may be natural units, such as individuals, or aggregates of natural units, such as households or SMSAs; or they may be artificial units, such as traffic zones. Whatever these units may be, it must be possible to define them clearly and unambiguously.

To give such definition, it is necessary to have a sampling frame, which is a basic list or reference which unambiguously defines every unit of the population from which the sample is taken. So if households are to be sampled in a specific city, a list is required of every household in that city, defined by its street address. Such a list would comprise the sampling frame. If traffic zones were sampled, a list of zones with their boundaries clearly defined on a map is necessary, so that they also can be identified unambiguously on the ground.

The sampling frame must be treated somewhat cautiously, however. If a list of households compiled for city tax purposes is used, there may be no clear distinction between single households and apartment buildings. In other words, households will be identified on the basis of the building structure, not on the aggregation of individuals comprising a household. If such a list were used for sampling households, a serious bias would be introduced into the sample due to the fact that a much higher proportion of apartment-dwelling households would appear in the sample than would be present in the total population. Again, if the sampling frame is a map of the area from which the sample is to be drawn, showing the individual buildings on each street, a sample drawn from this identification when a household sample is required would suffer from the same bias as the previous example.

A final example of the bad selection of a sampling frame should suffice to indicate clearly how important is an accurate, complete, and unambiguous sampling frame. Suppose a sample of households is desired and the sample is selected from the telephone directory. Unfortunately, this unacceptable procedure is used quite frequently. The pitfalls are obvious: even in the United States in 1979, not everyone has a telephone. And not all those who have a telephone have listed numbers. Then again, some professional people, such as doctors, may wish to be listed only under the address of their practice, which might be located in an area covered by a different directory, while some people may be listed more than once, by home and office.

These examples should illustrate the fact that a clear and unambiguous definition of every unit or element in the population to be sampled is required for the sampling frame. Without this, serious and possibly unmeasured bias will certainly be introduced into the survey sample.

Sampling Units

The aggregate of all the sampling units is termed the population of sampling units. If the sampling units themselves are aggregates of some other unit (e.g., households are aggregates of the individuals), a further population is involved which must be distinguished from the population of sampling units.

There are no rules governing the size and uniformity of sampling units. Sampling units may be individuals, households, countries, or

SMSAs. Clearly, these cover a wide range of unit size and also contain wide variations in size for each unit. A household may comprise one individual or 10, or even more. However, everything to do with the sampling procedure is simpler if the sampling units are approximately the same size. Also, the smaller the sampling units, the more accurate will be the results when a given proportion of the population is sampled.

Pure random sampling requires the selection of units from the total population in such a way that each unit has an equal probability of being selected. Such an equal-probability selection necessarily demands a completely rigorous method of sampling. The best method of obtaining such a sample is to use random-number tables. Examples of widely used random-number tables are those developed by Kendall and Smith[1] and by the RAND Corporation.[2]

Before illustrating this method by some examples, two further points about the method of selection should be made. Since the sample is to be selected on the basis of random numbers, it appears to be necessary to assign numbers to each unit in the population to be able to select them. However, this is not absolutely true. If a large number of units is to be sampled, the following procedure can be used. Suppose the population can be subdivided into p groups, with x_q members in the qth group ($q=1,...,p$). Successive subtotals of the population can be obtained by groups, $X_1, X_2, ... , X_p$, where $X_1 = x_1, X_2 = x_1 + x_2, X_3 = x_1 + x_2 + x_3$, etc. Random-number tables are then used to select the sample. The group is identified in which each selection falls. Only those units need to be numbered that fall into the groups which contribute units to the sample.

The second point is the use of random-number tables. In most applications of concern in sampling, once a unit is selected for the sample it ceases to be a candidate for further selection. In other words, once a household is selected for interviewing, even though its number may come up again, the household is not going to be interviewed twice. This is called *sampling without replacement* and means that once a particular number has been selected, it *cannot* be considered for selection again, and sampling continues on the basis only of the population members which are so far unselected.

When nonhuman populations are the subject of a survey, as well as sometimes when human populations are the subject but are only observed, *sampling with replacement* may be carried out. In this case, the selection of a particular number does not remove it from the pool of future drawings. Thus, one number and hence a given member of the population may be selected more than once to make up any given sample. In this context, it is useful to note that random sampling is strictly defined as equal-probability sampling. That is, each member of the population has an equal probability of being selected. If the population is rather small, sampling without replacement could violate equal-probability sampling in that the removal of sampled members from the population means that the last members to be drawn have a higher probability of being chosen than did the first members. Of course, the probabilities on each drawing are equal for all members of the population that have not yet been selected. In such a case, sampling with replacement is therefore more correct than sampling without replacement.

In random-number tables digits are usually arranged in pairs. However, digits may be combined into threes (for selection of numbers from 0 to 999), into fours (for selection of numbers from 0 to 9999), and so on. The tables may be used across or down. Two examples are presented to illustrate the basic methodology of random sampling.

Example 1

Select a sample of 15 from 3211 units: using Kendall and Smith's tables,[3] it will be found that the sample comprises 2977, 2194, 1237, 1661, 2595, 318, 588, 2012, 1045, 2202, 837, 579, 657, 626, 674. To obtain this sample, 40 four-digit numbers had to be examined and 25 rejected because they were greater than 3211. This is somewhat undesirable, and one of a number of methods can be applied to get around it. For example, units 3301–6600 may be considered to represent 1–3300, and similarly 6601–9900 to represent 1–3300 also. If this approach is followed, the sample comprises 1166, 2240, 2061, 3070, 1275, 2977, 2194, 1237, 11, 2042, 881, 2071, 1661, 2613, 2. In selecting this sample no numbers had to be rejected.

Another method would be to use a random-number generator of a computer, where it is possible to define the range within which random numbers are to be generated. One may also allocate multiple numbers to each member of the population and use random-number tables in the normal fashion. In this case, it is important that each member of the population be allocated the same number of reference numbers, otherwise equal-probability rules are violated. Note also that in the example above, the procedure is only partially more efficient since, as described, the numbers 3212–3300, 6512–6600, and 9812–9999 are all unused. This will distort the sampling somewhat.

Table 3–1 summarizes the distribution of the two selected samples over the whole range of the population. In neither case are the samples distributed evenly over the whole population, and random selection will always give samples which deviate from an even distribution (5, 5, and 5 in this case). The aggregate deviation of the first sample is 6, while that for

Table 3–1
Example of Sampling Distribution

	First Sample	Second Sample
1–1070	8	3
1071–2140	3	7
2141–3211	4	5
	15	15

the second is 4. It can be shown that the first method of selection generally yields a greater aggregate deviation than the second method. Hence the second method is clearly preferable.

Example 2

Select 15 houses from 12 streets in a census tract: these streets contain 18, 36, 15, 24, 19, 16, 57, 39, 27, 6, 25 and 36 houses, respectively. This can be done by successive totaling of the houses. Thus the following successive subtotals are found: 18, 54, 69, 93, 112, 128, 185, 224, 251, 257, 282, 318.

Using random-number tables, the following sample is obtained: 116, 186, 57, 267, 277, 31, 33, 261, 90, 110, 191, 226, 266, 317, 148.

Numbers 31 & 33	are in the 2d street
Number 57	is in the 3d street
Number 90	is in the 4th street
Number 110	is in the 5th street
Number 116	is in the 6th street
Number 148	is in the 7th street
Numbers 186 & 191	are in the 8th street
Number 226	is in the 9th street
Numbers 261, 266, 267, & 277	are in the 11th street
Number 317	is in the 12th street

Therefore, houses in 10 streets have to be numbered totaling 294 houses. A smaller sample would clearly have given far greater gains. If only 10 houses were sampled, only 7 streets would have to be numbered, totaling 174 houses; and if only 6 houses were sampled, 5 streets would have to be numbered, totaling 131 houses. This demonstrates the gains that can be achieved with this method.

Next, the determination of the required sample size in a fully random sample and general comparisons of errors between purely random sampling and other types of sampling are discussed. Two factors contribute principally to random-sampling error: the size of the sample and the variance of the units in the population. It remains to be determined what is the sample size necessary to obtain a prespecified level of accuracy in terms of the variance of the units in the population.

First, the standard error (s.e.) of an estimate from a sample is determined. This standard error is a measure of the average random-sampling error and also indicates the frequency with which errors of certain magnitudes will occur.

In a fully random sample from a large population, the standard error of estimate of the proportion of units having a particular attribute is as follows: if p is the proportion of units in the whole population having this attribute, and $q = 1 - p$ is the proportion not having this attribute, the standard error of the proportion of the units of this type in the random sample of n units is

$$\text{s.e. of } \rho = \sqrt{\left(\frac{pq}{n}\right)} \tag{3.1}$$

where ρ is the sample estimate of p. (If p, q, and ρ are in percentages instead of proportions, the expression is unchanged.)

If 30% of the units of the population are of the specific type, the standard error of the percentage of units in a sample of 50 is

$$\text{s.e. of } \rho = \sqrt{\left(\frac{30 \times 70}{50}\right)} = 6.48 \tag{3.2}$$

Thus, there is a 67% chance of obtaining estimates between 23.5% and 36.5%; and a 95% chance of obtaining estimates between 17% and 43%. Alternatively, the percentage standard error (p.s.e.) of estimate can be used. The percentage standard error, when p and q are measured in percentages, is

$$\text{p.s.e.} = 100 \sqrt{\left(\frac{q}{np}\right)} \tag{3.3}$$

Using the same details as before results in equations (3.4) and (3.5).

$$\text{p.s.e.} = 100 \sqrt{\left(\frac{70}{50 \times 30}\right)} \tag{3.4}$$

$$\text{p.s.e.} = 21.6\% \tag{3.5}$$

If a total population of 10,000 units is considered, 3000 will be of the specific type under investigation here. The estimate from the sample of 50 of the number of units of this type in the whole population would have a 0.67 probability of being between 2350 and 3650, and a 0.95 probability of being between 1700 and 4300. Clearly, the equation can be rewritten to determine the sample size.

$$n = \frac{pq}{(\text{required s.e. of } \rho)^2} \qquad (3.6)$$

$$n = \frac{10{,}000 \, q}{p(\text{required p.s.e. of } \rho)^2} \qquad (3.7)$$

If, in the earlier example, one wished to specify a percentage standard error of 10%, the result would be

$$n = \frac{10{,}000 \times 70}{30 \times 100} = \frac{700}{3} \qquad (3.8)$$

Thus, 2.33% of the population of 10,000 units would have to be sampled to reduce the standard error of estimate to 10%. These equations apply to the sampling units obtained by random sampling, and for which the proportion having particular attributes are to be estimated. If the sampling units are households and one wanted to estimate attributes of individuals, the standard errors of proportions of individuals determined from the sample will be larger (often very much larger) than those given by putting n as the number of individuals.

These equations apply to attributes which either exist or do not. Similar, also simple, equations can be derived for an attribute which is measured continuously, such as income. In this case, the percentage standard error and the sample size are

$$\text{p.s.e.} = \frac{\% \text{ s.d. of a unit}}{\sqrt{(n)}} \qquad (3.9)$$

$$n = \frac{(\% \text{ s.d. of a unit})^2}{(\text{required p.s.e.})^2} \qquad (3.10)$$

To calculate either of these, the unbiased estimate of the standard deviation, s, would be used.

$$s = \frac{1}{N-1} \sum_i (X_i - \bar{X})^2 \qquad (3.11)$$

It is obvious from this that the calculations of standard errors and of required sample sizes for simple random samples are very simple and can be performed rapidly. However, as soon as sampling other than simple random sampling is involved, this is no longer true. If the approximate order of the relationship between simple random-sampling standard errors and the standard errors of the more involved sampling techniques is known, a calculation of the sample size that would be required for a random sample may be a helpful guide.

The reader will remember that the idea of other restricted random samples was introduced as a means to reduce the amount of variability of the units that contributes to the random-sampling error. As a general rule, all the other forms of sampling can be expected to have smaller standard errors, and therefore require a smaller sample size for a given required standard error of estimate. In such cases, if only one type of sampling unit is considered, the reduction to the required sample size will be proportional to the fraction of the total variability of the units which is removed by the sampling method. In the case of sampling units of different types or sizes, the situation rapidly becomes more complicated.

Yates[4] sets out the following seven indicative rules which may be useful in this context.

1. The use of stratification, variable sampling fractions, or supplementary information can generally be expected to improve the accuracy of the survey results. In consequence, the sample size for a random sample can be considered to be the upper bound to the sample size required in any reasonable form of sampling using the same sampling units.

2. Although stratification will generally increase accuracy, it will only do so substantially when there are highly significant differences between the strata. Increases in accuracy will generally be greater for quantitative characteristics than for qualitative ones.

3. Again, the use of a variable sampling fraction will generally increase accuracy significantly only when there are large differences between strata and when the sampling units vary greatly in size. However, fractions which improve accuracy for quantitative characteristics will often worsen accuracy for qualitative characteristics.

4. The use of reliable supplementary information will generally increase the accuracy substantially, and can often serve as an alternative to stratifying the population.

5. In stratification, it is necessary to have at least one unit in each stratum. So for small samples, stratification may be undesirable or ineffective, since for a given sample size the strata will have to be fewer and larger. However, in large samples, more detailed stratification is possible with a consequent rapid increase in accuracy.

6. If sampling units of type A consist of aggregates of type B sampling units (e.g., households and individuals), the use of sampling units of type A in place of sampling units of type B will usually result in lower accuracy for a given sample size.

7. In multistage sampling, the number of units required in the final-stage sample will be more than if single-stage sampling had been used on the final-stage units.

Simple Stratified Sampling

This section focuses on some of the more complex versions of sampling. Consider, first, stratified sampling. Stratification is the division of the population to be sampled into a number of blocks, each of which is to be sampled separately. Assume that sampling takes place at a uniform rate over all the strata. The division into strata is, of course, carried out before sampling. The various strata may be of any size at all, and there is no requirement that each stratum should be of the same size, or even order of size, in terms of the number of units in each.

When properly used, stratification can increase the accuracy of population estimates from the sample. Accuracy is improved if the strata are so chosen to maximize between-strata variance, while minimizing within-strata variance. This reduces sampling error in the way discussed below.

Uses of Stratification

Effectively, sampling is performed at random from a set of subpopulations. In so doing, the calculation of sampling errors can be considered to be carried out for each subpopulation independently. Thus, the between-strata variances never enter the calculations of sampling errors or the estimation of population values. A very simple stratification can illustrate this point qualitatively. Suppose a survey of students is carried out and the population comprises, say, 3000 students, 2000 of whom are men and 1000 are women. The survey is designed to find out the courses these students are taking and their career ambitions. Obviously, one can expect some basic differences between the men and women. First, consider a purely random sample of 10% of the entire population. On the basis of the previous statements about sampling error, the probability can be calculated that a certain number of men will be observed in the sample.

The standard error of ρ, the proportion of men in the sample, is

$$\sqrt{\left(\frac{33.33 \times 66.67}{300}\right)} = 2.73 \tag{3.12}$$

There is a 95% probability that between 61% and 72% of the sample will be men, and consequently that between 28% and 39% will be women.

Since one expected the men and women to give different results in the survey, the sampling-error calculations and estimation of population values must both take into account the difference between the number of women included in the sample and the number that could have been expected to be included. This is even more crucial if in fact the proportion of men to women in the whole population is not known. If, however, the population is stratified into men and women, and 10% from each population is sampled at random, assuming the previous figures, the first 100 women and the first 200 men would have been selected from the population. Then the sampling errors and estimations of population values do not have to take into account the proportions of men and women in the sample compared with those to be expected in the total population.

From this example, a serious drawback to stratification becomes obvious immediately. If statements are required about the total student population on the basis of the sample, correct stratification can only be achieved if one has a priori knowledge of the numbers of men and women in the total population. Clearly, if in the above example there were actually 1500 men and 1500 women in the population and stratification took place as suggested, choosing 100 women and 200 men by random sampling, using these survey results to estimate total population values would lead to more erroneous results than purely random sampling would give. On the basis of this observation, it can be stated that proper use of stratification to reduce random-sampling error requires a priori knowledge of the population with respect to the characteristics on which stratification is based.

Stratification can also achieve a second purpose, namely, to ensure that each subdivision of the population in which one is interested is adequately represented in the sample. Thus, in the earlier example about 3000 students, there are 150 foreign students of whom 50 are women and 100 are men. If a random sample was taken over the unstratified population, the proportion of foreign students of either sex that would be in the sample can be calculated.

$$\text{s.e.} = \sqrt{\left(\frac{5 \times 95}{300}\right)} = 1.26 \tag{3.13}$$

Therefore, there is a 95% probability of obtaining between 2.5% and 7.5% of the sample as foreign students. Within each sex, the error becomes even larger in proportion to the percentage required. Clearly, only 1.67% of the entire population are female foreign students, and the standard error of this is

$$\text{s.e.} = \sqrt{\left(\frac{1.67 \times 98.33}{300}\right)} = 0.74 \tag{3.14}$$

So there is a 95% probability that only 0.23% of the sample will be female foreign students (300 in the sample gives an expected number of 0.69 female foreign students) or as many as 3.11% will be female foreign students. In other words, one would expect 5 female foreign students in the sample, but random sampling would generate anything from none to 10 in 95% of the samples. Stratified sampling can be used to ensure that 5 female foreign students, 10 male foreign students, 95 American female students, and 190 American male students are obtained by stratifying the population.

A third advantage, or use, of stratified sampling occurs when it may be desirable to use different surveying or sampling methods within each stratum. Such a situation might occur in a survey carried out to determine use of a rail-transit facility. One may consider three strata in relation to this: people living within walking distance (e.g., up to two blocks from the station), people on a bus line to a station, and people who have to use a car to reach the rail-transit facility. One may then decide to sample the first group, using purely random sampling. The second group might also be selected by using random sampling or by using cluster sampling at various distances along the bus line. The last group might also be selected by using cluster sampling. Similarly, the first two groups, or strata, are more likely to be transit-oriented, and one may feel confident in using a postal questionnaire on these people. The third stratum is more likely to be auto-oriented, and one may consider it necessary to use interviewers for this part of the survey.

Methods of Selection

The preceding section illustrated the basic uses of stratified sampling; now the discussion focuses on the methodology of selection of a stratified sample. Several alternatives in the structure of the sampling frame are considered.

1. In the first case, it is assumed that each sampling unit is defined in the sampling frame by the classification desirable to use for stratification. For instance, if stratification is by sex, the list of individuals may have Mr., Miss, Mrs., and Ms. designated. In such a case, the selection procedure is similar to the one used for simple random sampling. As soon as the random sampling has generated all the females required for the sample, any further ones selected are rejected in the same way that any number previously selected is to be rejected. Alternatively, one may reconstruct the sampling frame into separate frames for each stratum and select each stratum's sample at random in the usual way.

2. Assume that the sampling units are not conveniently classified in

this manner, but it is known how many units are required in each stratum of the sample. In this case, units are selected at random from the entire population but each unit is classified by its stratum when it is selected. Selections for any stratum will be rejected once its quota has been filled.

3. Finally, if the sampling frame does not classify the units, and it is not known how many units should be selected within each stratum, it will be necessary to enumerate the entire population before sampling. While doing this, it would be a simpler matter to classify the units of the population, so the first selection method could be used.

Now consider a small matter of definition of terms in stratified sampling. Suppose a population of 10,000 units is to be sampled at the rate of 5% over each of eight strata. For simplicity, suppose each stratum is the same size, that is, 1250 units. A 5% sample would thus require the selection of 62.5 units from each stratum. If units are indivisible (e.g., individuals), one clearly cannot select 5% from each stratum. Therefore, say, 63 units will be selected from each stratum. This gives a sampling rate of 5.04%. This situation arises quite frequently, so it becomes convenient to talk in terms of exact sampling fractions and working sampling fractions. The exact sampling fraction is obtained from sampling a whole number of units from a stratum, and the working sampling fraction is used as the selection basis. In this example, the exact sampling fraction is 5.04%, and the working sampling fraction is 5%.

In calculations of population values and sampling errors, it is usual to use the working sampling fraction. In most cases this will lead to only very slight inaccuracies, which will usually be of a significantly lower order than most of the errors in the sample. Obviously, in the case of very small samples from small populations, this error could be troublesome. This type of problem is also less likely to arise when the strata are all the same size.

One feature of stratified sampling vis-à-vis simple random sampling should be noted. If the number of units in the whole population in each stratum is known, and a sufficiently large simple random sample is taken to ensure that each stratum is adequately represented in the sample, adjustment of the survey results so that each stratum is represented in the correct proportion would yield the same accuracy as the stratified sample. One of the keys here is "sufficiently large . . . sample." The size of sample that may be required to do this will often make stratification a much more economical proposition. Also, calculations in this case for the stratified sample will be simpler than those for the simple random sample adjusted for stratum sizes.

There are cases, however, where the simple random sample is the better alternative. For example, one may wish to stratify the population by car ownership (no cars, one car, two cars, three or more cars). Though the number of cars owned by the whole population and the distribution of car ownership may be known, it may be impossible to select households on the basis of car ownership since prior knowledge of a household's car ownership is not likely to be available.

Multiple Stratification and Variable Sampling Fractions

Two extensions of simple stratification are multiple stratification and variable sampling fractions. Multiple stratification is the term applied to a stratification of a population on the basis of more than one characteristic. In the example of a survey of students, the population was first stratified by sex. It was then proposed to further stratify by ethnic origin (in that case, American or non-American), which is in fact an example of multiple stratification. As with simple stratification, different levels of a priori knowledge of the population to be sampled are assumed.

Consider the same example again. Suppose stratification is performed by both sex and ethnic origin. In the first case, suppose the number of each sex in each ethnic group is not known. This situation is called multiple stratification without control of substrata, and is basically undesirable, because it causes considerable theoretical and practical difficulties, and the sampling error calculation is awkward.

Yates[5] gives details of a method that can be used to choose the sample in this type of situation. It is basically a converging iterative solution for the contents of each substratum. This method is illustrated below, using Yates'[6] data. In this situation, one must know how many units to select in each stratum of the two-way classification. Then the conventional random-sampling process is performed until every stratum has at least as many units as are required.

In this example, one wants to select a sample of 1000 units from a population classified into two sets of four strata for which the substrata totals are unknown. The correct strata totals for the sample are known to be 120, 280, 350, 250 for each set. After a sample of 1125 units has been drawn, the numbers are as shown in table 3–2.

From the table, it is found that there is an excess of units sampled in every stratum except B_1. Each figure in the table is then reduced by deducting the excesses in proportion to the number of units in the substrata of each row. The excesses are:

$B_1 = 0$ $A_1 = 9$
$B_2 = 37$ $A_2 = 37$
$B_3 = 58$ $A_3 = 26$
$B_4 = 30$ $A_4 = 53$

Table 3–2
Example of Two-way Stratification without Control

Strata B	Strata A				Total	Required
	1	2	3	4		
1	37	40	35	8	120	120
2	39	140	82	56	317	280
3	45	97	173	93	408	350
4	8	40	86	146	280	250
Total	129	317	376	303	1125	1000
Required	120	280	350	250	1000	—

So the following values are deducted: A_1B_1, A_2B_1, A_3B_1, A_4B_1 all 0.

$A_1B_2 = 5$	$A_2B_2 = 16$	$A_3B_2 = 10$	$A_4B_2 = 6$
$A_1B_3 = 6$	$A_2B_3 = 14$	$A_3B_3 = 25$	$A_4B_3 = 13$
$A_1B_4 = 1$	$A_2B_4 = 4$	$A_3B_4 = 9$	$A_4B_4 = 16$

These are then totaled and show deficits in each row of $A_1 = 12$, $A_2 = 34$, $A_3 = 44$, $A_4 = 35$. The required deficits are 9, 37, 26, 53, which means that these deficits are wrong by $+3$, -3, $+18$, -18, respectively. Again, these excesses are distributed with the signs reversed in the same way as before but on the columns. The results are shown in table 3–3. Now rows B_3 and B_4 have to be rebalanced by the same process; the results are given in table 3–4. Finally, an adjustment is made to columns A_3 and A_4 to eradicate the imbalances. Now the substrata totals are given in table 3–5.

It is obvious that this process is tedious, somewhat arbitrary toward the end, and a little inaccurate (note that the answers disagree slightly with Yates'[7]). Also, the process cannot be guaranteed to converge.

In the more usual situation, in which the required sample size in each substratum is known, the selection process is, of course, the usual simple one that was dealt with under ordinary stratification. Multiple stratification serves basically the same purposes as simple stratification in reducing much of the variance of the samples and in allowing concentrated

Table 3–3
Calculation of Numbers to Be Rejected

	A_1	A_2	A_3	A_4	
B_1	0	0	0	0	0
B_2	-1	$+2$	-4	$+3$	0
B_3	-2	$+1$	-9	$+6$	-4
B_4	0	0	-5	$+9$	$+4$

Table 3–4
Results of Rebalancing Rows B_3 and B_4

	A_1	A_2	A_3	A_4	
B_3	0	+1	+2	+1	+4
B_4	0	−1	−1(−2)	−2(−1)	−4
	0	0	+1	−1	

Table 3–5
Results of Rebalanced Two-way Stratification

	A_1	A_2	A_3	A_4	Total
B_1	37	40	35	8	120
B_2	35	122	76	47	280
B_3	41	81	155	73	350
B_4	7	37	84	122	250
Total	120	280	350	250	1000

identification of every part of the population which is of interest. This last function leads into the next aspect of stratification discussed here.

The method is called stratification with variable sampling fractions (sometimes called disproportionate sampling). As its name implies, this method of stratification involves the use, not of a constant sampling fraction over the whole population, but a separate fraction for each stratum. In this method one begins to move toward the idea of optimization in sampling.

Several times the desirability was mentioned of reducing within-stratum variance while increasing between-strata variances as a means of introducing some increase of accuracy of the sample. Clearly, it is not always likely that one would be able to stratify in such a way as to be able to minimize within-strata variances of the sampling units. Under these circumstances, one can achieve considerable gains in accuracy by sampling at a greater rate from the more variable strata. How much more should be sampled from the more variable strata? Obviously, the optimum is to use sampling fractions, or rates, which are directly proportional to the within-strata standard deviations of the units. If each stratum is denoted by subscripts 1, 2, ..., n and the sampling fractions as $f_p(p=1,2,...,n)$, equation 3.15 defines the ideal sampling fractions relative to each other.

$$\frac{f_1}{\sigma_1} = \frac{f_2}{\sigma_2} = \frac{f_3}{\sigma_3} = \dots = \frac{f_n}{\sigma_n} \qquad (3.15)$$

In practice, it is possible to obtain a ratio which demands a sampling fraction greater than 1. In such a case, one would in fact just sample the entire stratum for which this occurred.

Another situation in which this method can be used to give considerable gains is that where the stratification is based on size of units. Often, the characteristics of interest vary proportionately with the size of the units, such that the within-strata standard deviations are roughly proportional to the mean sizes of the units in the various size groups. Then the sampling fractions should be chosen in proportion to the mean sizes. If the characteristics of the units are highly correlated with size, the range within each size group is often a good estimate of the within-strata standard deviations. In this case, the sampling fraction can be chosen to be proportional to the ranges.

These rules must of course be adjusted to some extent if the strata represent study domains. In such situations, the sampling fractions must always be big enough to give adequate representation of each stratum.

Another use for this method of stratification occurs when the strata are selected in such a way that as well as being a stratification into some type of natural grouping, the strata also contain units for which one characteristic of interest varies with the natural grouping. This point may be best illustrated by a real-world example.

Several years ago, the Greater London Council had the task of carrying out a number of economic evaluations of the urban freeway plan by recosting a sample of freeway sections. The freeway plan consisted of three rings of freeways and a system of interconnecting radial freeways. Because of the structure of London, these various rings and radial sections could be divided by cost, using their geographical pattern. Thus, if the strata were chosen to comprise the box, the inner radials, the C-ring, the outer radials, and the D-ring, the five strata were also highly correlated with land and construction costs. Since the evaluations that were performed concerned estimates of cost savings or added expenditure for various alternatives, the overall costs were the items for which greatest accuracy was required. For this reason, it was decided that the greatest sampling errors could be tolerated on the cheapest sections, and the smallest sampling errors would be required on the most expensive sections. The sampling fractions, therefore, were chosen in proportion to the mean section costs on each stratum, with the qualification that a certain size of sample (25% in fact) was required for even the most lightly sampled stratum, to ensure sufficient detail in the analysis and that the overall sample rate should not exceed about 33.33%. (This was set so that real gains would be achieved by sampling rather than evaluating the entire network.)

It can in fact be shown that over a range of sampling fractions around the optimum, the accuracy of the sample is somewhat insensitive. Since the analyst will frequently have various reasons for not adopting an optimal sampling fraction (due to cost, requirement of adequate representation of all strata, disproportionately arrayed standard deviations of the several study characteristics with respect to stratification, etc.), it is worth noting that any sampling fraction which is near-optimal will give almost as high gains in accuracy as an optimal fraction. Furthermore, since optimal fractions would rarely be used in practice, the authors do not adopt the term "optimal allocation" for this sampling method (in agreement with Yates,[8] but not with Kish[9]). Kish points out that sampling fractions between one-half and twice the optimum fraction will produce within 10% of the optimum variance. This allows considerable flexibility where additional restrictions have to be imposed.

Some guidelines can be set up for the use of stratification with variable sampling fractions. The method should be used only when both the between-strata and the within-strata variances are substantial, because the gain in accuracy obtained otherwise is unlikely to be worth the additional work and cost involved in calculations and sampling procedure over those of simple stratified sampling. One should attempt to avoid awkward sampling fractions since these cause unnecessary complications in calculating and sampling. Since it was shown that the variance is relatively insensitive to the sampling fraction near the optimum, serious losses in accuracy will not be incurred by taking the nearest simple fraction for the sampling rate. As few different sampling rates as possible should be used. As is seen later, the calculation of population values and sampling errors involves fairly lengthy calculations for each sampling fraction. Kish[10] suggests an approach using two rates only—a low rate for most of the sampling units and a separate stratum of large units which can be selected with certainty (i.e., a sampling fraction of unity).

Miscellaneous Sampling Schemes

Systematic Sampling

So far, a great deal of time was spent emphasizing the necessity of using a random sample. However, the use of a sampling frame was mentioned which comprises some sort of list of the sampling units. A frequently used method of sampling from such a list (particularly if it is very long) is to take every nth entry in the list. This is called systematic sampling.

A frequently employed gimmick in this method of sampling is to choose the first entry by generating a random number between 1 and n. This procedure is erroneously thought to "randomize" the selection process. In fact, all it does is to reduce the likelihood of a bias due to selecting the starting point without the aid of random number tables.

Such lists for systematic sampling will vary considerably in their approximation to a random source. A list of houses arranged by blocks will tend to be biased, giving houses in densely developed blocks a far greater probability of being included in the sample than those in less dense blocks. On the other hand, a list of individuals, arranged alphabetically, is the nearest approach to a random order of individuals. Even these lists will have biases. For instance, in Britain the majority of Scotsmen have surnames which begin with Mac or Mc. If one took every nth entry, one would not have a fully random sample. Similarly, the sampling interval must be smaller, preferably much smaller, than the natural divisions of the lists. For instance, the sampling interval for an alphabetic list should be much smaller than the minimum number of units or individuals under any letter of the alphabet.

In systematic sampling, it is not possible to estimate a sampling error per se. However, an approximate sampling error can be obtained by considering a sample stratified in the major subdivision of the list (e.g., the initial alphabetical letter). If the sample is assumed to be fully random, the tendency is to overestimate the error, and this inaccuracy will increase as neighboring entries in the list increase in similarity.

It should be clear that systematic sampling is likely to be much easier and simpler to carry out than random sampling. However, to use it, the survey designer must be aware how crucial is the form of the basic list from which the sample is taken. Two basic properties of the list must be checked for, because their presence will invalidate a systematic sample by introducing a significant bias. The first is a trend of a characteristic of interest. Kish[11] gives an excellent example of this. Suppose a bank has a list of the mortgages it has financed over the last 15 years, arranged in chronological order of when they were advanced. The analyst is interested in knowing both the amount advanced and the current indebtedness of each mortgagee. Because housing costs have been rising, the amount loaned will increase from the earliest to the latest. Likewise, the current debt of each borrower will increase toward the present. If the list consists of 10,000 people and every 100th is selected (an average of 670 mortgages would have been advanced each year), one would be likely to obtain quite different means for debts and loans according to where in the first 100 mortgages the systematic selection was commenced.

The other undesirable property of a list is some periodic pattern in it, particularly if the sampling interval should turn out to be a direct multiple of the periodic pattern. For instance, if a list of the addresses of houses on an estate is available which has developed in a regular grid pattern with a consistent density of 10 houses per block, a sampling interval which is 10 or a multiple of 10 will select a house in the identical position in each block. The estate developer may have laid out the blocks in such a way that houses of different layout and size are repeated in the same order in each block. In this case, the sample will appear to be extremely consistent and have a low variance. However, the real population may have a very high variance that will be completely unmeasured by the sample.

Multistage Sampling

Another type of sampling is multistage sampling, sometimes termed sub-sampling. Multistage sampling can be considered as a compromise between fully random sampling and cluster sampling. The method comprises random selection of a hierarchy of units. The first-stage units are the largest aggregates to be used and the final-stage ones are the smallest breakdown of units. An example of this type of sampling might occur in a resurvey of an urban area on the basis of census tracts. Assume such a widespread random sample of census tracts is not needed, but a higher accuracy at household and individual levels is desired. Consider that the census tracts are first aggregated into districts, each comprising about 10 census tracts. Selection from these districts is done at random. Now only the census-tract numbers in the selected districts need be known to avoid listing all the tract-district relationships. Tracts within the selected districts are then selected at random, ready for a household survey. Therefore, only house-hold lists for the selected tracts are needed. The final stage of the selection process will involve the selection of a sample of households.

As pointed out earlier, the accuracy of this final sample is less than that which would have been obtained, had a single-phase selection of households been carried out. The method has considerable advantages, however, where no sampling frame exists initially. By using this form of selection, lists are prepared only for a small proportion of the total population of final-stage units. It should also be noted that there is no necessity for the sampling procedures at each stage to be the same. In the last example, one could have taken an entirely random sample of districts, a stratified sample of tracts within each district, based on quadrants of the districts (to ensure geographical representation of the entire district) and a stratified, variable-sampling-fraction selection of households to obtain better representation of single households against multiple households.

Moving-Observer Selection

The moving-observer method is particularly pertinent to transportation studies; it is used when the subjects of the survey are in motion. Consider, for the moment, that the objective is to count people who are walking about. The concern is to estimate the number of people on one sidewalk on one block in a downtown area.

The method that would appear to be applicable here would be to demarcate a number of short sections of sidewalk and put observers in the center of each section. These observers would then count the number of people in their own areas. This method is in fact extremely difficult to carry out and highly inaccurate, particularly when pedestrian flows are very heavy.

A similar technique for determining flow of vehicles on a street, particularly where different classes of vehicles have to be separately recorded, becomes an even greater problem. Also, the number of people or vehicles counted will depend on the speed of movement, which adds another complication.

All these difficulties can be overcome by using the moving-observer technique. Return to the simple case of pedestrians on a block in a downtown area. One observer is required who walks the length of the block and records the number of people he or she passes and the number who pass him or her. The observer then walks back again at the same speed counting the same. Subtracting the number of people who overtake him or her from those the observer passes, and averaging the results of these from the two traversals, results in an estimate of the average number of people on the block during the time of the counts.

This method can be applied to vehicle flow. The velocity of the vehicle becomes important in the estimation of the total number of vehicles in a section. Simultaneously, the method can yield estimates of the mean vehicle speed for the flow or for each vehicle type in the flow. The formulae for calculating speeds and flows are not covered here. This method is relevant since it permits a survey to be conducted under circumstances where the types of sampling considered so far are not applicable.

Cluster Sampling

Basically, cluster sampling comprises grouping of sampling units on a spatial or geographical basis. These clusters are sampled at random, and the units within them are either selected with certainty or selected by random sampling, using a higher than usual sampling fraction. One may, for instance, sample households on the basis of blocks, where the blocks are the clusters of households. One might then select blocks at random, and survey all the households, or 75% of them, within each block selected.

The basic advantage of cluster sampling is that it is cheaper to carry out and more easily controlled than simple random sampling. However, it suffers from two basic disadvantages: natural clusters tend to be fairly homogeneous, while between-cluster variances are high and frequently not measured by the survey; and statistical analysis of cluster sampling is more complicated. However, if adequate information is readily available about the clusters, cluster sampling can combine a number of the advantages of multistage and stratified sampling while still being cheaper than either of these. However, multistage or stratified sampling are each more accurate and decrease the sampling error quite markedly.

In cluster sampling it is common practice to stratify the clusters. Stratification has even greater advantages in cluster sampling than in unit sampling. The way in which stratification reduces sampling errors substantially by reducing within-strata variances has already been discussed. Since one of the major drawbacks to cluster sampling is the high variance between clusters and the irregular within-cluster variances, it follows that this stratification will reduce considerably the high sampling errors of cluster sampling.

Cluster sampling could effectively be applied to a problem used in an earlier section, where households were surveyed to determine their use of a rapid transit facility. In that example, stratification into three basic strata was suggested according to the location of the household relative to the rail facility. Although a completely random sample of households in the smallest stratum would probably be both simple and relatively cheap, such a sampling method in the two larger strata would almost certainly not be, and considerable gains could be made by taking cluster samples within those two cluster strata. Similarly, the clusters would have one added advantage in this type of situation. In that example, the two larger strata comprised people who were more than walking distance from the rail facility and who were either close to a bus line, or who would have to use auto to reach the rail line. In this situation, distance from the rail line would be an important factor. Each cluster (if these were a block, for example) would represent a particular distance from the rail station and from a bus line, but with a reasonable variation in household and individual characteristics. Clearly, for a given number of households to be included, the cost of the survey would be lower using cluster sampling, and the usefulness of the survey data would be enhanced.

Choice-based Sampling

There are various circumstances in which the above sampling techniques may not offer the most appropriate procedure because of the cost of sampling or the difficulty of finding the sample. Under certain circumstances, a biased method of sampling, called choice-based sampling, may be suitable.

The term choice-based is derived from the fact that the sample is drawn on the basis of the outcome of the decision process under study. The sampling method has been used in transportation[12] and marketing, where a sample is drawn from a choice process, such as a purchasing or travel decision. Some examples of such processes would include roadside interviews of car drivers, on-board interviews of transit or airline passengers, interviews of housewives purchasing a given product in a supermarket.

Choice-based sampling is conducted by identifying the decision group of interest and then choosing individuals from this group at random. It is important to ensure as near a random sample within the choice group as possible, since the only bias permitted is the lack of representation of those who do not choose the option under study. No other biases are permitted or acceptable. As discussed for other sampling techniques, any survey method may be used with choice-based sampling. Most commonly, the survey method is either an interview conducted during or immediately after the choice process, or a handout, mail-back survey to be administered by the recipients of the survey.

In choice-based sampling, the survey results can only be used to describe the choice subpopulation studied. Provided the sample is drawn at random within that subpopulation, population values and sampling errors, for that subpopulation, can be calculated as described in chapter 4. However, it is important to note that without supplementary information the data cannot be used to describe any general properties of the total population.

An example of the restriction of data to the subpopulation can be given from the following situation. Suppose a survey is carried out of television viewers to determine what programs they watch on a specific evening. Such a survey, if conducted by a proper random sampling of that evening's viewers, will provide reliable information on the proportion of viewers tuned to different programs on that evening. However, since the survey gained no information on those who were not watching television that evening or on those who do not have TV, the information could not be applied to the total population of the region or nation. To make up such an application, it would be necessary to have supplementary data giving the number of people who were not watching on that evening and the number who do not own televisions.

An important use of choice-based sampling is for the calibration and use of a particular type of choice model, called a logit model. This application is described in detail in chapter 15.

Notes

1. Maurice G. Kendall and B. Babington Smith, *Tables of Random Sampling Numbers,* Cambridge, England: Cambridge University Press, 1939.

2. RAND Corporation, *A Million Random Digits,* New York: Free Press Books, 1955.

3. Kendall and Smith, *Random Sampling Numbers.*

4. Frank S. Yates, *Sampling Methods for Censuses and Surveys,* London: Charles Griffin and Co., 1971, p. 98.

5. Yates, *Sampling Methods.*

6. Ibid, pp. 26–27.

7. Ibid.

8. Ibid.

9. Leslie Kish, *Survey Sampling,* New York: John Wiley & Sons, 1965.

10. Ibid.

11. Ibid, p. 120.

12. Charles F. Manski and Steven R. Lerman, "The Estimation of Choice Probabilities from Choice-based Samples," *Econometrica,* 1977.

4

Population Estimates and Sampling Errors

Introduction

Estimating Population Values

The ultimate purpose of survey sampling is to determine the characteristics of the population from which samples were drawn by any one or more of the techniques discussed in chapter 3. The determination of these population values is to be performed within known error limits. This chapter illustrates the mathematical process by which population values are obtained from sample values. It also discusses the determination of the size of sampling errors and therefore errors in the population estimates. The authors lean heavily on Yates' work.[1]

The preceding chapters discussed how to draw a sample which can be considered representative of the local population. In doing this, it was implied that the primary interest of the analyst is to be able to make statements about the whole population. Therefore, the necessary tools need to be developed to use the sample results as estimates of the population values of the characteristics that were measured.

Information derived from sample surveys is subject to error. Strict adherence to prescribed sampling procedures allows one to calculate the size of sampling errors, thereby providing some feel for the degree of accuracy of the population estimates. This chapter is concerned with the development of procedures for calculating both population values and associated sampling errors for all the sampling schemes described in chapter 3.

Mention was made briefly of the use of supplementary information as a means of increasing the accuracy of the data obtained from a survey. Supplementary information was defined as consisting of data produced from some other source, including other surveys and censuses, the addition of which could be used to remove lack of information in the survey under study. Much of this chapter is concerned with the use of such supplementary data, particularly in terms of its optimum use in estimating population values and its effects on the calculation of sampling errors for those values.

First, consider the choice of values to estimate from the survey data. As a simple example, the arithmetic mean of a random sample is an estimate of the mean of the population. Sampling errors will prevent it from being an exact estimate of the population mean. However, one may not necessarily be interested in the arithmetic mean of a population. Rather, one may wish to estimate the median, the geometric mean, or the average (i.e., the mean of the lowest and highest values). Other estimates that can be made with supplementary data may also be of interest. The question arises as to which one of all these possible values to estimate should be used to make statements about the population.

Yates[2] suggests three criteria in deciding which estimates to choose. These are freedom from bias, accuracy, and computational convenience. The mean of a random sample from a normally distributed population (with respect to the characteristic being measured) is free from bias, of maximum accuracy (if no supplementary information is available), and computationally simple. No matter what the population distribution is, the arithmetic mean remains a bias-free estimate for the total population, although it will probably no longer be the most accurate estimate that could be obtained. It also has the advantage that sampling errors can be assessed easily and do not depend very greatly on the form of the distribution of population values.

Like Yates, this discussion only considers what appear, on the basis of these criteria, to be the most useful estimates for each type of sampling. This treatment assumes a freedom from bias for the estimates, and does not calculate bias where it is known to exist.

A few statements about convenient notation are in order here. As a general rule, lower-case letters are used for sample values and upper-case for population values. A hat ($\hat{}$) over an upper-case letter represents the estimate of a population value. The letters Y,y are used to refer to the quantitative variate under consideration, and X,x to a supplementary quantitative variate, such as the size of unit. The subscript i is used to denote a specific stratum. Other important symbols are:

A	= total area covered by the sampling grid
b	= estimated regression coefficient
f	= working sample fraction
θ	= exact sampling fraction
g	= working raising factor $(= 1/f)$
γ	= exact raising factor $(= 1/\theta)$
m	= number of strata
m^i	= value for an individual
n_0	= total number of sampling points
P,p	= proportion of units with a given attribute
R,r	= ratio $Y/X, y/x$
T	= sum of squares of deviations of the ys from the values given by the ratio line
U,u	= number of units with a given attribute

z = deviation of sampling unit from true population

$\overset{n}{\Sigma}$ = summation over the sample

$\underset{i}{\Sigma}$ = summation over the strata

There are five basic rules of estimation which apply to all forms of sampling (Yates[3]).

1. To estimate *totals* for a population of a variate, multiply all the sample values by their raising factors.

2. To estimate the number of *sampling units* in the population, multiply the number of sampling units in each part of the sample by their raising factors.

3. To estimate the population *mean* of a variate, divide the estimated population total by the estimated number of sampling units in the population.

4. To estimate the *proportion* of units in the total population possessing a given attribute, proceed as rule 2, except score 1 for each unit possessing the characteristic and 0 for those not possessing it. Divide the estimate by the estimated number of units in the population.

5. To estimate the *ratio* of two quantitative variates, estimate the totals of the two variates by rule 1, and take the ratio of these totals.

These rules are applied later in the chapter for the estimation of a number of population values.

Estimating Sampling Errors

Initially, a simple random sample is used to develop the basic principles of the estimation of sampling errors. Some matters of elementary statistics are included which are relevant and will help to clarify the procedure. It is assumed, in this discussion, that the sample is drawn from a large population.

Suppose a sample of one unit is taken from a population to estimate the mean, $\bar{Y}$, of the population. Each unit in the population has a deviation from the true population which can be denoted by z_r, and z_r is defined by equation (4.1).

$$z_r = y_r - \bar{Y} \tag{4.1}$$

where r denotes the rth member of the population.

The summation of the deviations over the entire population equals zero. This, in terms of sampling, represents an assumption of no bias. Hence the mean deviation is also zero (if account is taken of the sign). Two measures of the magnitude of error are possible: the mean absolute deviation and the mean square deviation. The latter is usually adopted in statistics and is the one used in calculating the sampling errors in survey sampling.

Now in the sample of one unit, the standard error of estimate of the population mean obtained from this one unit is the standard deviation of the unit. In a sample of two units, r and s, the actual error of estimate is

$$\hat{\bar{y}} - \bar{Y} = \bar{y} - \bar{Y} = (z_r + z_s)/2 \qquad (4.2)$$

The standard error of estimate is the square root of the average value of equation 4.3.

$$(z_r + z_s)^2/4 = (z_r^2 + z_s^2 + 2z_r z_s)/4 \qquad (4.3)$$

All of this may seem to be very elementary and simple, but two observations on its implications are important. Normally, one would assume that the average value of $z_r z_s$ is zero, in which case, since the average values z_r^2 and z_s^2 are both σ^2, the average value of the above expression is $\sigma^2/2$. Therefore the standard error of estimate is $\sigma/\sqrt{2}$.

This argument does not depend on any assumption about the distributions of the zs. However, it is assumed that the sample is selected in such a way that each unit is selected randomly and independently. Thus, if there were some tendency to select a second unit with a similar deviation to the first, the average value of $z_r z_s$ would not be zero.

If n units are sampled, the argument can be extended to show that the variance and the standard error are

$$V(\hat{\bar{y}}) = \frac{1}{n} V(y) \qquad (4.4)$$

$$\text{s.e.}(\hat{\bar{y}}) = \frac{1}{\sqrt{n}} \sigma \qquad (4.5)$$

Similarly, the standard error for the population estimate can be shown to be

$$\text{s.e.}(\hat{Y}) = \text{s.e.}(gn\hat{\bar{y}}) = gn[\text{s.e.}(\hat{\bar{y}})] = g\sigma\sqrt{n} \qquad (4.6)$$

One may also be interested in the sampling variance of $\overset{n}{\Sigma}(y)$ and the standard error, which are

$$V[\overset{n}{\Sigma}(y)] = nV(y) \qquad (4.7)$$

$$\text{s.e.}[\overset{n}{\Sigma}(y)] = \sigma\sqrt{n} \qquad (4.8)$$

As discussed earlier, one may wish to express standard errors in percentages. In fact, if a standard error is expressed as a percentage of the population value of the estimated quantity, this has the advantage of being independent of the units of the quantity. If the standard deviation of a single unit is expressed as a percentage of the mean value of a single unit, this is called the coefficient of variation and denoted $\sigma\%$. For large populations, the standard error can then be expressed

$$\text{s.e.}\%(\hat{\bar{y}}) = \text{s.e.}\%[\overset{n}{\Sigma}(y)] = \text{s.e.}\%(\hat{Y}) = (\sigma\%)/\sqrt{n} \qquad (4.9)$$

The proof of the unbiased estimate of σ, denoted s, is briefly presented here (see also chapter 13). Equation 4.10 states a well-known identity.

$$\overset{n}{\Sigma}(y-\bar{y})^2 = \overset{n}{\Sigma}(y-\bar{Y})^2 - n(\bar{y}-\bar{Y})^2 \qquad (4.10)$$

The average value of the first term is $n\sigma^2$, and the average value of the second term is σ^2, since $(\bar{y}-\bar{Y})$ is the error in the estimate of the mean. Therefore, on average, equations 4.11 and 4.12 hold true.

$$\overset{n}{\Sigma}(y-\bar{y})^2 = n\sigma^2 - \sigma^2 = \sigma^2(n-1) \qquad (4.11)$$

$$\sigma^2 = \frac{1}{n-1}\overset{n}{\Sigma}(y-\bar{y})^2 \qquad (4.12)$$

Thus an estimate of σ^2 is

$$s^2 = \frac{1}{n-1}\overset{n}{\Sigma}(y-\bar{y})^2 \qquad (4.13)$$

This is the estimate that is used in the equations developed for the standard errors of sample estimates. The theory considered so far really applies only to sampling from an infinite population. In a finite population σ^2 is

$$\sigma^2 = \frac{1}{N-1}\overset{n}{\Sigma}(y-\bar{Y})^2 \qquad (4.14)$$

where N = the number of units in the total population. The equation for s^2 still stands as before. The equations for the standard errors of estimation of population estimates now require to be modified by a factor $\sqrt{(1-f)}$, or preferably $\sqrt{(1-\theta)}$.

$$\text{s.e.}(\bar{y}) = \sigma\sqrt{\left(\frac{1-f}{n}\right)} \qquad (4.15)$$

The derivation of this expression is not developed here. The algebra used to derive the standard error can be extended to prove this. The necessity of the term $(1-f)$ in sampling errors is obvious. If f equals 1, the sampling error is zero, as it should be. This factor should not be used when tests are being carried out between the means of two sampled populations, for instance, to determine whether they are subject to the same causal factors.

Error Theory

Before proceeding further to develop sampling error equations, it is advisable to look at the theory of errors on which these equations are based. This theory could be called the normal law of error. Basically, the law states that although the distributions of the deviations of individual sampling units may vary widely, over a wide range of such distributions the errors to which estimates are subject are distributed approximately normally. The larger the sample, the more closely is this law followed.

In the event that this law can be applied to standard errors, the concept of confidence limits can also be used. Essentially, this means that probabilities, based on the normal distribution, can be attached to errors of greater than certain magnitudes. For example, one can have 34.1% confidence that deviations greater than one standard deviation from the mean will occur (or that 68.3% of estimates will be within one standard deviation of the mean), and so forth. Inaccuracies in the error estimate can be allowed for by using the t distribution instead of the normal distribution.

If the normal law of error does not apply, estimates of accuracy will be overestimates. There does not appear to be a theory of sampling errors developed for nonnormal errors, and the treatment of these is left largely as a matter of choice for the individual. Two basic problems exist here: if the sample size is small, it will frequently occur that the lower limit of confidence will become negative, where positive values are the only possible ones. In effect, this skews the distribution. If the mean or total is large and the sample is also large, the skewness introduced by this will not usually be significant. The second problem is that if the errors do not obey the normal error law at all, there is no convenient way to deal with error estimates. If the true distribution of errors is known, this could be used, but this is unlikely to be the case. It is advisable to assume the normal law holds and to adapt the conclusions if a lack of adherence to this law appears to be significant.

In a random sample of n units from a normal population with standard deviation σ, the standard error of s is

$$\text{s.e.}(s) = \frac{\sigma}{\sqrt{[2(n-1)]}} \tag{4.16}$$

The use of equation 4.16 can be illustrated by using the data of chapter 2 (tables 2–1 and 2–6). The standard deviation of the 100 values of table 2–1 is 3.97. Assuming this is the true value, σ, the standard error of the estimate of σ, s, from the data of table 2–6 can be calculated. Using the actual sample of 10 (as listed in the text of chapter 2), the value of s is found to be 3.85. Evaluating equation 4.16 produces

$$\text{s.e.}(s) = \frac{3.97}{\sqrt{2 \times 9}} = \pm 0.936 \tag{4.17}$$

Applying the normal law of error, the confidence limits on s at 95% would be 3.85 ± 1.835 (since there is a 95% probability that the value will lie within 1.96 standard deviations from the estimate). It can be seen that the actual estimate obtained is very much closer to the true value than the 95% confidence limit.

The key issue from the foregoing is that the sampling errors are a function of the variance or standard deviation of the values measured by the survey. To develop the appropriate estimates of variances, a number of properties of variances need to be considered. Suppose a linear function is formed, $\hat{L}$, of the estimates $\hat{y}_1, \hat{y}_2, \hat{y}_3,...,\hat{y}_m$, where the estimates have sampling variance, $V(\hat{y}_1)$, $V(\hat{y}_2)$, $V(\hat{y}_3)$. $\hat{L}$ is defined as

$$\hat{L} = l_1\hat{y}_1 + l_2\hat{y}_2 + l_3\hat{y}_3 + \ ... \ + l_m\hat{y}_m \tag{4.18}$$

where l_m = multipliers whose values are not subject to sampling. The sampling variance of $\hat{L}$ is

$$V(\hat{L}) = l_1^2 V(\hat{y}_1) + l_2^2 V(\hat{y}_2) + l_3^2 V(\hat{y}_3) + \ ... \ + l_m^2 V(\hat{y}_m) \tag{4.19}$$

This property depends on the fact that $V(\hat{y}_1)$, $V(\hat{y}_2)$, ...,$V(\hat{y}_m)$ are independent. The sampling errors will be independent if the estimates are derived from independent sets of values. Estimates from samples of different populations, different strata, or different samples from a large population fulfill this requirement. Estimates derived from two variates belonging to the same population are not generally independent.

On the basis of the above property, the variance and standard error of $l y_1$ are

$$V(l\hat{y}_1) = l^2 V(\hat{y}_1) \tag{4.20}$$

$$\text{s.e.}(l\hat{y}_1) = l\,\text{s.e.}(\hat{y}_1) \tag{4.21}$$

The standard error of the difference of two independent estimates of a variate is the square root of the sum of the squares of the standard errors of the two estimates.

$$V(\hat{y}_1 - \hat{y}_2) = V(\hat{y}_1) + V(\hat{y}_2) \tag{4.22}$$

$$\text{s.e.}(\hat{y}_1 - \hat{y}_2) = \sqrt{([\text{s.e.}(\hat{y}_1)]^2 + [\text{s.e.}(\hat{y}_2)]^2)} \tag{4.23}$$

The standard error of the sum of a number of independent estimates is

$$V(\hat{y}_1 + \hat{y}_2 + \hat{y}_3 + ...) = V(\hat{y}_1) + V(\hat{y}_2) + V(\hat{y}_3) + ... \qquad (4.24)$$

$$\text{s.e.}(\hat{y}_1 + \hat{y}_2 + \hat{y}_3 + ...) = \sqrt{[\{\text{s.e.}(\hat{y}_1)\}^2 + \{\text{s.e.}(\hat{y}_2)\}^2 + ...]} \qquad (4.25)$$

The standard error of estimate of the mean of a large population can be derived from this. Similarly, one can derive the standard error of a weighted mean, $\hat{\bar{y}}_w$,

$$\text{s.e.}\hat{\bar{y}}_w = \frac{w_1\hat{y}_1 + w_2\hat{y}_2 + ...}{w_1 + w_2 + ...} \qquad (4.26)$$

where the ws are weights.

Provided the ys are independent and their variances known, the variance of $\hat{\bar{y}}_w$ can be calculated. Equations 4.27 through 4.30 show two common cases. Equation 4.28 shows the variance on the basis of the definition of $\hat{\bar{y}}_w$ provided in equation 4.26, and where the individual variances are assumed to be equal, as shown in equation 4.27.

$$V(\hat{y}_1) = V(\hat{y}_2) = ... = V(\hat{y}) \qquad (4.27)$$

$$V(\hat{\bar{y}}_w) = \frac{\overset{n}{\Sigma}(w^2)}{[\overset{n}{\Sigma}(w)]^2} V(\hat{y}) \qquad (4.28)$$

In the second case, the variance of the variate is expressed as a function of λ, where λ is assumed to be a constant.

$$V(y_1) = \lambda/w_1 \qquad (4.29)$$

The variance is

$$V(\hat{\bar{y}}_w) = \frac{\overset{n}{\Sigma}(w)}{[\overset{n}{\Sigma}(w)]^2} \lambda = \frac{\lambda}{\overset{n}{\Sigma}(w)} \qquad (4.30)$$

This is one form of weighted mean, used when combining estimates from a number of independent sources. Another form of weighted mean is that in which the weights, w, are supplementary information, in which case the estimate of the weighted mean, $\hat{\bar{y}}_w$, is

$$\hat{\bar{y}}_w = \frac{\overset{n}{\Sigma}(wy)}{\overset{n}{\Sigma}(w)} \qquad (4.31)$$

For a random sample, and when the relationship of the variances of the ys and ws is not known, an unbiased estimate of $V(\hat{\bar{y}}_w)$ is required. This is obtained by weighting the deviations of y from $\hat{\bar{y}}_w$.

$$T = \overset{n}{\Sigma}w^2 (y - \hat{\bar{y}}_w)^2 \qquad (4.32)$$

Then the variance of the deviations is

$$s_t^2 = \frac{T}{(n-1)} \tag{4.33}$$

Hence, the variance of the estimate of $\hat{\bar{y}}_w$ is

$$V(\hat{\bar{y}}_w) = \frac{(1-f)ns_t^2}{\frac{n}{[\Sigma(w)]^2}} \tag{4.34}$$

The approximate estimation procedures for the standard errors of the product and ratio of two estimates whose sampling errors are independent are

$$V(\hat{y}_1\hat{y}_2) = \hat{y}_2^2 V(y_1) + \hat{y}_1^2 V(\hat{y}_2) \tag{4.35}$$

$$V\left(\frac{\hat{y}_1}{\hat{y}_2}\right) = \left(\frac{\hat{y}_1}{\hat{y}_2}\right)^2 \left(\frac{V(\hat{y}_1)}{\hat{y}_1^2} + \frac{V(\hat{y}_2)}{\hat{y}_2^2}\right) \tag{4.36}$$

These are approximations which are satisfactory only if the $V(\hat{y}_1)$ and $V(\hat{y}_2)$ are small relative to $\hat{y}_1^2$ and $\hat{y}_2^2$. Likewise, if the sampling errors of $\hat{y}_1$ and $\hat{y}_2$ are not independent, the covariance terms omitted from the above expressions must be included. This also applies to the previous expressions for $V(\hat{L})$, $V(\hat{y}_1-\hat{y}_2)$, and $V(\hat{y}_1+\hat{y}_2)$. Under these circumstances the equations should read

$$V(\hat{L}) = l_1^2 V(\hat{y}_1) + l_2^2 V(\hat{y}_2) + l_3^2 V(\hat{y}_3) + \dots$$
$$+ 2l_1l_2 \, \text{cov}(\hat{y}_1\hat{y}_2) + 2l_1l_3 \, \text{cov}(\hat{y}_1\hat{y}_3)$$
$$+ 2l_2l_3 \, \text{cov}(\hat{y}_2\hat{y}_3) + \dots \tag{4.37}$$

$$V(\hat{y}_1-\hat{y}_2) = V(\hat{y}_1) + V(\hat{y}_2) - 2\text{cov}(\hat{y}_1\hat{y}_2) \tag{4.38}$$

$$V(\hat{y}_1\hat{y}_2) = \hat{y}_2^2 V(\hat{y}_1) + \hat{y}_1^2 V(\hat{y}_2) + 2\hat{y}_1\hat{y}_2 \, \text{cov}(\hat{y}_1\hat{y}_2) \tag{4.39}$$

$$V(\hat{y}_1/\hat{y}_2) = \left(\frac{\hat{y}_1}{\hat{y}_2}\right)^2 \left(\frac{V(\hat{y}_1)}{\hat{y}_1^2} + \frac{V(\hat{y}_2)}{\hat{y}_2^2} - \frac{2\text{cov}(\hat{y}_1\hat{y}_2)}{\hat{y}_1\hat{y}_2}\right) \tag{4.40}$$

It should also be noted that b, the regression coefficient, and r, the correlation coefficient, can be expressed in variances and covariances.

$$b = \frac{\widehat{\text{cov}}(xy)}{\hat{V}(x)} \tag{4.41}$$

$$r = \frac{\widehat{\text{cov}}(xy)}{\sqrt{[\hat{V}(x)\hat{V}(y)]}} \tag{4.42}$$

Simple Random Samples

The estimation of both population values and sampling errors for simple random samples is performed by applying some very simple and straightforward equations. In this chapter, four population estimates[4] are considered, based upon Yates' rules, namely:

1. The population mean, $\bar{Y}$;
2. The population total, Y;
3. A ratio of two values, R; and
4. A proportion of two units in a given category, P.

A simple random sample is drawn at a sampling rate of f from a large population. The number of units in the sample is n and in the population is N. A variate, y, such as number of children in the household, is measured. The population total is

$$\hat{Y} = g\overset{n}{\Sigma}(y) = N\hat{\bar{y}} \qquad (4.43)$$

If N is known and differs from $\hat{N}$, equation 4.44 would be used, instead of equation 4.43.

$$\hat{Y} = \gamma\overset{n}{\Sigma}(y) = N\hat{\bar{y}} \qquad (4.44)$$

The estimate of the population size is

$$\hat{N} = gn \qquad (4.45)$$

If N is known, θ, the true sampling rate, is n/N and γ, the true raising factor, is N/n. The estimate of the mean population value is

$$\hat{\bar{y}} = \bar{y} = \frac{1}{n}\overset{n}{\Sigma}(y) \qquad (4.46)$$

The sampling errors for these estimates were used to illustrate the theory of error, but are repeated here for convenience. If N is assumed to be infinite, the standard errors of the total and mean are

$$\text{s.e.}(\hat{Y}) = g\sigma\sqrt{n} \qquad (4.47)$$

$$\text{s.e.}(\hat{\bar{y}}) = \sigma/\sqrt{n} \qquad (4.48)$$

If N is not infinite, but is known or estimated from equation 4.45, the sampling errors are

$$\text{s.e.}(\hat{Y}) = g\sigma\sqrt{n(1-f)} \qquad (4.49)$$

$$\text{s.e.}(\hat{\bar{y}}) = \sigma\sqrt{(1-f)/n} \qquad (4.50)$$

The use of these equations can be illustrated again with the data from chapter 2 on heights. Using the random sample of 10, the total height of the 10 individuals is 678 inches. Applying equation 4.44, since N is known exactly, the value of $\hat{Y}$ is found to be 6780 inches. (The true value, from table 2–1, is 6766 inches.) Applying equation 4.49, since N is small, the sampling error of $\hat{Y}$ is found to be ± 115.5, using $s = 3.85$ as the estimate of σ. Thus, for these data, the total of heights, estimated from the sample, is 6780 inches, with a standard deviation of ± 115.5. Using the normal law of error, one could state that there would be 95% confidence that the true value of Y would lie between 6553.6 and 7006.4 inches.

Similarly, equations 4.46 and 4.50 can be used to show that the estimated mean of height, $\hat{\bar{y}}$, is 67.8 inches with a sampling error of ± 1.155, yielding a 95% estimate of 65.54 to 70.06 inches.

The next measures of interest are those relating to the proportion of units in a particular category. Equation 4.51 shows the estimate of the proportion of units in the sample within a particular category or possessing a particular characteristic.

$$\hat{p} = \frac{u}{n} \qquad (4.51)$$

The number possessing the characteristic or being in the category is

$$\hat{U} = gu = \hat{N}\hat{p} \qquad (4.52)$$

If N is known, equation 4.52 can be rewritten

$$\hat{U} = \gamma u = N\hat{p} \qquad (4.53)$$

Sampling errors are also to be estimated for the quantities of equations 4.51, 4.52, and 4.53, where these are often referred to as *qualitative variates*. The basis of this procedure is the same as that for quantitative variates—estimation of the variance of individual sampling units and then derivation of sampling error in terms of this variance. The variance of an attribute, however, is proportional to the number of units possessing the attribute. The variance of the individual sampling units therefore need not be estimated.

If in a random sample from a large population P is the proportion with the attribute and Q is $1 - P$, the variance and the standard error are

$$V(\hat{p}) = \frac{PQ}{n} \qquad (4.54)$$

$$\text{s.e.}(\hat{p}) = \sqrt{\left(\frac{PQ}{n}\right)} \qquad (4.55)$$

If the population is finite and is sampled at rate f, these expressions change to

$$\text{s.e.}(\hat{p}) = \sqrt{PQ(1-f)/n} \qquad (4.56)$$

$$\text{s.e.}(\hat{U}) = g\sqrt{[nPQ(1-f)]} = \hat{N}[\text{s.e.}(\hat{p})] \qquad (4.57)$$

The standard error of the percentage of the population possessing an attribute varies with $100\,\hat{p}$. It is a maximum at $P = 0.5$ and a minimum of zero at $P = 0$ or 1.

From these equations, one might ask, if P and Q are known, why is a survey needed? Normally, P and Q will not be known and these equations cannot be used. With some loss of accuracy, one may substitute the estimates $\hat{p}$ and $\hat{q}$ for P and Q. In large samples, the loss of accuracy will not be great. However, these equations only apply under specific circumstances:

1. The units whose attributes are under consideration must be the sampling units;
2. The sample must be random; and
3. In stratified sampling the equations apply to each stratum separately.

In other forms of sampling, for example, multistage, and any form of sampling with supplementary information, the variance no longer depends on P and Q alone. When this is the case, the variance of individual units must be calculated. This can be done by assigning values 1 and 0 to units according to whether or not they possess the attribute, and then proceeding as for quantitative variates. The following simple examples serve to illustrate the use of these equations.

Example 1

In a household survey of a town, a random sample of all households was taken with a sampling fraction of 1/30; 632 households out of 7962 in the sample did not have a car available. What is the estimated number and percentage of non-car-owning households in the town, and the error associated with each? Equations 4.58 and 4.59 provide the estimated number and percentage, using equations 4.51 and 4.52.

% of non-car-owning households

$$= 100\,\hat{p} = 100 \times \frac{632}{7962} = 7.94\% \qquad (4.58)$$

total number of non-car-owning households

$$= \hat{U} = 30 \times 632 = 18,960 \qquad (4.59)$$

Using equations 4.56 and 4.57, the standard errors of $100\,\hat{p}$ and $\hat{U}$ are found to be ± 0.298 and ± 711.6, respectively. Applying the normal law of error, one would state there is 95% confidence that the percentage of non-car-owning households is between 7.35 and 8.52, and the number of

Table 4–1
Average Grades for a Sample of Twenty Students

1.9	2.8	2.9	1.7	2.2
2.6	1.8	2.0	2.2	2.5
3.0	2.7	1.8	2.3	2.8
2.5	2.6	1.7	2.7	1.9

such households in the population is between 17,565 and 20,355, also with 95% confidence.

Example 2

Table 4–1 represents grade averages for 20 students sampled at random from a list of all students in a university, using a sampling fraction of 1/1000. Estimate the mean grade average for the university, and the total of average grades for all students. These estimates are

$$\hat{N} = 1000 \times 20 = 20,000 \qquad (4.60)$$

$$\overset{n}{\Sigma}(y) = 46.6 \qquad (4.61)$$

$$\hat{\bar{y}} = \frac{1}{20} \times 46.6 = 2.33 \qquad (4.62)$$

$$\hat{Y} = 1000 \times 46.6 = 46,600 \qquad (4.63)$$

Hence the mean grade average is 2.33, and the total of average grades is 46,600. Assume it is known that there are in fact 20,110 students in the university. A more accurate result for $\hat{Y}$ is

$$\hat{Y} = 20,110 \times 2.33 = 46,856.3 \qquad (4.64)$$

From table 4–1, the estimated standard deviation of y, s, is found to be ± 0.433. Hence the sampling errors of $\hat{\bar{y}}$ and $\hat{Y}$ are ± 0.097 and $\pm 1,935.5$, respectively, and can be used to establish confidence limits on the values of equations 4.62 and 4.63.

Stratified Samples

Uniform Sampling Fraction

For estimating population values, the same equations as developed for simple random samples hold true, except that if the N_i are known and these differ from $\hat{N}_i$, more accurate results can be obtained from

$$\hat{Y} = \sum_i (N_i \, \hat{\bar{y}}_i) \tag{4.65}$$

$$\hat{U} = \sum_i (N_i \, \hat{p}_i) \tag{4.66}$$

Otherwise, the total and mean are

$$\hat{Y} = g \sum_i \overset{n_i}{\sum} (y) = g \sum_i n_i \hat{\bar{y}}_i \tag{4.67}$$

$$\hat{\bar{y}} = \frac{1}{n} \sum_i \overset{n_i}{\sum} (y) \tag{4.68}$$

The proportion and number possessing a particular attribute are

$$\hat{p} = \frac{1}{n} \sum_i u_i \tag{4.69}$$

$$\hat{U} = \sum_i (\hat{N}_i \hat{p}_i) = g \sum_i u_i \tag{4.70}$$

Similarly, the sampling errors are obtained from the equations developed for a simple random sample. The variance and standard error of the estimated mean are

$$V(\hat{\bar{y}}) = (1-f) \sum_i (n_i \sigma_i^2)/n^2 \tag{4.71}$$

$$\text{s.e.}(\hat{\bar{y}}) = \sqrt{\frac{(1-f) \sum_i (n_i \sigma_i^2)}{n^2}} \tag{4.72}$$

These equations assume that the within-strata variances, σ_i^2 are unequal. If, however, it can be assumed that these variances are equal, the standard error is

$$\text{s.e.}(\hat{\bar{y}}) = \sqrt{\frac{(1-f)\sigma_1^2 n}{n^2}} \tag{4.73}$$

where σ_1^2 = within-stratum variance (equal in all strata)
and $\quad n = \sum_i n_i$, the total number of observations.

Equation 4.73 can be simplified to yield

$$\text{s.e.}(\hat{\bar{y}}) = \sigma_1 \sqrt{\left(\frac{(1-f)}{n}\right)} \tag{4.74}$$

For a simple random sample, equation 4.75 represents the standard error.

$$\text{s.e.}(\hat{\bar{y}}) = \sigma \sqrt{\left(\frac{(1-f)}{n}\right)} \tag{4.75}$$

Equation 4.75 is the same as the equation for the stratified sample except for the use of σ_1 in place of σ in the latter. This value σ_1 is the common within-strata standard deviation. The most accurate estimate of σ_1 is obtained by weighting the estimates of σ_i^2 for each stratum by the number of degrees of freedom on which they are each based. This is the same as adding the sums of squares of the deviations from the strata means and dividing by the sum of the associated degrees of freedom. For t strata, the associated degrees of freedom are $n - t$.

It may be more convenient, however, to estimate σ_1 from an analysis of variance. The total sum of squares is equal to the sum of squares of the strata means about the overall mean (between strata) plus the sum of squares within the strata;

$$\sum_i \overset{n_i}{\Sigma} (y - \bar{y}_i)^2 + \sum_i n_i (\bar{y}_i - \bar{y})^2 = \overset{n}{\Sigma} (y - \bar{y})^2 \qquad (4.76)$$

where $\sum_i \overset{n_i}{\Sigma} (y - \bar{y}_i)^2 = $ within-strata sum of squares

and $\sum_i n_i (\bar{y}_i - \bar{y})^2 = $ between-strata sum of squares.

It is useful to rewrite the between-strata sum of squares by expanding the squared term, which produces

$$\sum_i n_i (\bar{y}_i - \bar{y})^2 = \sum_i \bar{y}_i \overset{n_i}{\Sigma} (y) - \bar{y} \overset{n}{\Sigma} (y) \qquad (4.77)$$

Using the identity in equation 4.76 and the expression in equation 4.77, analysis of variance can be used to set out the computations, as shown in table 4–2. The most convenient form of computation for purposes of this discussion is to calculate the between strata and total sums of squares and calculate the difference. The mean squares are, of course, the sums of squares divided by their degrees of freedom. Only s_1^2 is of interest here, which is obtained by dividing the difference in the sums of squares by $(n - t)$.

Table 4–2
Analysis of Variance for Pooled Estimate of Error

Source	d.f.	Sum of Squares	Mean Square
Between strata	$t - 1$	$\sum_i \bar{y}_i \overset{n_i}{\Sigma} (y) - \bar{y} \overset{n}{\Sigma} (y)$	A
Within strata	$n - t$	$\sum_i \overset{n_i}{\Sigma} (y - \bar{y}_i)^2$	$B = s_1^2$
Total	$n - 1$	$\overset{n}{\Sigma} (y^2) - \bar{y} \overset{n}{\Sigma} (y)$	$C = s^2$

If the within-strata variances are unequal, as will generally be the case in practice, one may still use this approach provided it can be ascertained that the errors are due to the assumption of equal variances. If all the sampling fractions are equal, equation 4.78 represents the variance estimate (from equation 4.71).

$$\hat{V}(\hat{\bar{y}}) = \frac{(1-f)}{n} \frac{\sum_i (n_i s_i^2)}{n} \tag{4.78}$$

The second factor, $[\sum_i (n_i s_i^2)]/n$, is a weighted mean of the s_i^2 using n_i as the weights. In this estimate using an assumption of equal variances, s_1^2 is the weighted mean of the s_i^2, with weights proportional to (n_i-1). Therefore, unless the n_i are very small, no significant inaccuracy will be introduced with the assumption of equal variances. However, estimates of the sampling errors of means or totals for a single stratum will not be accurate, and would be highly misleading.

Variable Sampling Fraction

For a stratified sample with variable sampling fraction, the estimating equations are slightly different. First, the population estimate is

$$\hat{N} = \sum_i (g_i n_i) \tag{4.79}$$

The population total and population mean are

$$\hat{Y} = \sum_i [g_i \overset{n_i}{\sum}(y)] \tag{4.80}$$

$$\hat{\bar{y}} = \hat{Y}/\hat{N} \tag{4.81}$$

The number and proportion possessing a particular attribute are

$$\hat{U} = \sum_i (g_i u_i) \tag{4.82}$$

$$\hat{p} = \hat{U}/\hat{N} \tag{4.83}$$

Again, slightly more accurate results can be obtained for equations 4.80 and 4.82 using the same equations for the constant sampling fraction, when the N_i are known and are different from $\hat{N}_i$.

In considering sampling errors for this case, it is first assumed that there are unequal variances within the strata. It should be noted that the total variance of the sampling units over the whole population does not have to be calculated, but only the variances of the sampling units within each stratum.

If the sample sizes within each stratum are large enough to estimate within-strata variances per sampling unit, sampling variances of the population estimates for each stratum can be estimated separately. The sampling variances for the total population estimates can be estimated by using equation 4.19 for $V(\hat{L})$. This method is valid for all forms of stratification, regardless of the inequality of within-stratum variance per sampling unit. In this case, the sample estimate might be that of the mean, computed from

$$\hat{\bar{y}} = \sum_i [g_i \overset{n_i}{\underset{}{\Sigma}}(y)]/\hat{N} \tag{4.84}$$

If the variance within the ith stratum is denoted σ_i^2, the variance of y is

$$V[\overset{n_i}{\underset{}{\Sigma}}(y)] = n_i \sigma_i^2 (1-f_i) \tag{4.85}$$

Using equations 4.84 and 4.85, the variance of $\hat{\bar{y}}$ is

$$V(\hat{\bar{y}}) = \sum_i [g_i^2 n_i \sigma_i^2 (1-f_i)]/\hat{N}^2 \tag{4.86}$$

When σ^2 is not known, the estimated variance is

$$\hat{V}(\hat{\bar{y}}) = \sum_i g_i^2 n_i s_i^2 (1-f_i)/\hat{N}^2 \tag{4.87}$$

When the sampling fraction varies among the strata, the pooled estimating method, described for the case of a uniform sampling fraction, cannot be used. The weights in this case should be proportional to $g_i^2 n_i (1-f_i)$, as can be seen from equation 4.86, developed earlier. Therefore, the pooled estimate of variance will not be applicable and would be highly misleading if used.

Sampling errors for the other estimates can be obtained by following the same principles described above. For example, the sampling error of $\hat{Y}$ is

$$s.e.(\hat{Y}) = \sqrt{[\sum_i g_i^2 n_i \, \sigma_i^2 (1-f_i)]} \tag{4.88}$$

Noncoincident Study Domains and Strata

In survey sampling the strata often do not represent study domains, but the study domains cut across strata. For example, in a study of urban recreation-choice behavior, the survey may stratify on socioeconomic characteristics as a convenient and useful stratification base. However, study domains may include such characteristics as participation in a particular activity, such as golf, swimming, or bicycling. Clearly, for a household survey, stratification could not be done on participation, since this will be unknown until analysis of the survey. When study domains are defined in this manner, the population estimates and sampling errors are needed for the study domains and not for the original strata.

When stratified sampling with a uniform sampling fraction is used, one may consider the sampling to have been done by domain and calculate population and domain values and sampling errors by the same equations as for the normal uniform-sampling case. When a variable sampling fraction is used, however, it is necessary to compute the values of the sampling errors differently. In this case, the estimated within-stratum variance of units in the study domain which are in the ith stratum are denoted by $s_i'^2$, and the proportion of units in the ith stratum not in the study domain by q_i'; thus q_i' is

$$q_i' = (n_i - n_i')/n_i \qquad (4.89)$$

The estimated variances of the total in the study domain, $\hat{Y}'$, and the mean, $\hat{\bar{y}}'$, are

$$\hat{V}(\hat{Y}') = \sum_i g_i^2 n_i (1-f_i)[n_i' q_i' \bar{y}_i'^2 + (n_i'-1)s_i'^2]/(n_i-1) \qquad (4.90)$$

$$\hat{N}'^2 \hat{V}(\hat{\bar{y}}') = \sum_i g_i^2 n_i (1-f_i)[n_i' q_i'(\bar{y}_i'-\bar{y}')^2 + (n_i'-1)s_i'^2]/(n_i-1) \qquad (4.91)$$

This analysis assumes that a study domain cuts across all strata. Where a study domain does not appear in a stratum, the above equations will still be appropriate; n_i' will become zero and q_i will become 1 for that stratum.

Ratio and Regression Methods for Supplementary Data

So far, these equations are simple and straightforward. A somewhat more complex problem is the estimation of population values when supplementary information is available. Supplementary information on a quantitative characteristic can be used as the basis of stratification or to adjust an unstratified sample by stratification after selection. To do this, the values of the supplementary variate must be known for every unit of the population. Alternatively, it can be used directly, without stratification, using one of two methods: the ratio method or the regression method. In both methods, only the total or mean of the supplementary variate for the total population and the values for each of the sampling units in the selected sample have to be known.

The ratio method consists of estimating the ratio Y/X for the population from the sample, and multiplying this estimated ratio by the total X for the population to give $\hat{Y}$, where X is the variable on which supplementary data exist. This process of estimation must be carried out in such a way as to avoid bias. The regression method requires an estimation of the regression coefficient b of the relationship between y and x. This coefficient is then used to adjust the sample results for any discrepancy between the mean size of units in the sample and that for the population.

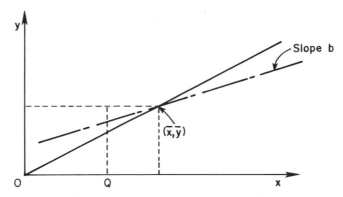

Figure 4–1. Graphical Representation of the Ratio and Regression Methods

It is useful to compare the differences between these two methods. In a graphical representation (figure 4–1) of these two methods, the ratio method is the straight line through the origin and the sample mean, while the regression method is the least-squares estimate (see chapter 7) through the sample mean, in a plot of y against x.

The adjusted estimate of the population is given by the intersection of the ordinate at Q with the ratio line or the regression line. Therefore, the adjusted estimate of $\hat{\bar{y}}$ is $(\bar{y}/\bar{x})\bar{X}$ from the ratio method and $\bar{y} + b(\bar{X} - \bar{x})$ for the regression method. If the supplementary variate is size of unit, the regression line should pass through the origin, but its curvature may cause the straight regression line to intersect the y-axis some distance from the origin. The gain in accuracy of the regression method is usually small, and the added complexity of computation is not usually worthwhile. Since the concepts of the regression method are needed in later considerations of sampling errors, the estimation equations for this method are derived here.

If multistage sampling is carried out, the population mean of x, $\bar{X}$, can be estimated from observation at, say, the first stage, while y is observed at the second stage. Under these circumstances, the equations developed below will still hold true. However, if the sampling is single stage, the estimate of $\bar{X}, \hat{\bar{X}}$, for the population will give no gain in the accuracy of the population estimate $\hat{\bar{Y}}$. In other words, $\bar{X}$ must be known or must be estimated from a larger sample than is used to obtain y.

Estimation by the Ratio Method

In developing the necessary equations for these estimates, some other uses for the supplementary variate become evident. The ratio method is considered first for each of the sampling methods.

Random Sample. Equations 4.92, 4.93, and 4.94 show the estimating equations for the population mean, population total, and ratio, respectively.

$$\hat{\bar{y}} = \frac{\overset{n}{\Sigma}(y)\bar{X}}{\overset{n}{\Sigma}(x)} = \frac{\bar{y}}{\bar{x}}\,\bar{X} \tag{4.92}$$

$$\hat{Y} = \frac{\overset{n}{\Sigma}(y)}{\overset{n}{\Sigma}(x)}\,X \tag{4.93}$$

$$\hat{R} = \frac{\overset{n}{\Sigma}(y)}{\overset{n}{\Sigma}(x)} \tag{4.94}$$

The sampling errors for this method are based on the calculation, first, of the variance of the ratio $(\hat{R})$, since this ratio features in each of the estimates for $\hat{Y}$ and $\hat{y}$, as shown in equations 4.92 and 4.93. To calculate the variance of the ratio, $\hat{R}$, the correlation between the values of x and y for the same sampling unit has to be taken into account.

$$V(\hat{R}) = \frac{(1-f)}{n}\,\hat{R}^2\left[\frac{V(y)}{\bar{y}^2} - \frac{2\,\text{cov}(xy)}{\bar{x}\bar{y}} + \frac{V(x)}{\bar{x}^2}\right] \tag{4.95}$$

This is only approximate, but will usually be sufficiently accurate. Equation 4.95 can be rewritten in a more convenient way. Equation 4.96 defines T which is the sum of squares of the deviations of the ys from values given by the ratio line.

$$T = \overset{n}{\Sigma}(y-\hat{R}x)^2 \tag{4.96}$$

The values of y and x may be written as deviations from their respective means of $\bar{y}$ and $\bar{x}$.

$$T = \overset{n}{\Sigma}[(y-\bar{y}) -\hat{R}(x-\bar{x})]^2 \tag{4.97}$$

Expanding the square of equation 4.97 produces

$$T = \overset{n}{\Sigma}(y-\bar{y})^2 - 2\hat{R}\overset{n}{\Sigma}(y-\bar{y})(x-\bar{x}) + \hat{R}^2\overset{n}{\Sigma}(x-\bar{x}) \tag{4.98}$$

The estimated mean square deviation from the true ratio line is defined as s_t^2.

$$s_t^2 = T/(n-1) \tag{4.99}$$

The ratio $\hat{R}$ and the variance of the ratio $\hat{V}(\hat{R})$ are

$$\hat{R} = \bar{y}/\bar{x} \tag{4.100}$$

$$\hat{V}(\hat{R}) = \frac{(1-f)}{n}\,\frac{s_t^2}{\bar{x}^2} \tag{4.101}$$

Equations 4.102 through 4.104 express the variances for the random-sample-ratio method in a more convenient form.

$$\hat{V}(\hat{R}) = \frac{(1-f)}{\frac{n}{[\Sigma(x)]^2}} \, ns_t^2 \qquad (4.102)$$

$$\hat{V}(\hat{Y}) = \frac{X^2}{\frac{n}{[\Sigma(x)]^2}} \, (1-f)ns_t^2 \qquad (4.103)$$

$$\hat{V}(\hat{y}) = \frac{\bar{X}^2}{\bar{x}^2} \frac{(1-f)}{n} \, s_t^2 \qquad (4.104)$$

The reader will recall the estimates of variances for a random sample without supplementary information.

$$\hat{V}(\hat{Y}) = (1-f)s^2 \qquad (4.105)$$

$$\hat{V}(\hat{\bar{y}}) = \frac{(1-f)}{n} s^2 \qquad (4.106)$$

The difference between the two methods lies in the factor $\bar{X}^2/\bar{x}^2$ and in the fact that deviations of y are from the ratio line instead of the mean, $\bar{y}$.

A frequent problem in the use of supplementary information is that it is out of date or otherwise subject to error. If values of x are known for all units of the population, the calculation of $\bar{x}$ for the selected units can be assessed for error by using all the data, provided the original frame is complete. If the original frame is not complete, the sampling has to be split into two parts: that covering units included in the original frame, for which one can use either the ratio or regression method; and that covering units not included in the original frame, for which one would use the appropriate estimation method without supplementary information.

A further property of this estimation procedure is worth noting. If the variance of r, where r is the fixed ratio of y/x, for fixed x is constant over the whole range of values of x, and if r shows no trend over this range, $\hat{V}(\hat{R})$ can be estimated.

$$\hat{V}(r) = \sum_{}^{n}(r-\bar{R})^2/(n-1) \qquad (4.107)$$

$$\hat{V}(\hat{R}) = \frac{\sum_{}^{n}(x^2)}{\frac{n}{[\Sigma(x)]^2}} \, V(r) \qquad (4.108)$$

Integer Values of the Supplementary Variate. The following issue is particularly pertinent to household surveys. This problem arises from the situation where the supplementary variate x can only take small integer values. Consider the situation in which a survey of households was carried out, but data of individuals are required. If the household values are the ys, these can be considered as the aggregate values for all individuals in the household. The values for an individual are denoted by m^i and $[m^i]$ is defined as the total for individuals in the household (where there is more than one in a household). Then $[m^i]$ is equal to y, and $\hat{R}$ is equal to $\bar{m}^i$. If the size of the household is denoted by suffixes, such that n_1 equals the number of one-individual households, and n_2 equals the number of two-individual households, the calculation of T can be simplified.

$$T = \overset{n_1}{\Sigma}(y^2) + \overset{n_2}{\Sigma}(y^2) + \ldots - 2\hat{R}[\overset{n_1}{\Sigma}(y) + 2\overset{n_2}{\Sigma}(y) + \ldots] + \hat{R}^2(n_1 + 4n_2 + \ldots)$$

(4.109)

Replacing y with m^i generates

$$T = \overset{n_1}{\Sigma}(m^i)^2 + \overset{n_2}{\Sigma}(m^i)^2 + \ldots$$
$$- 2\bar{m}^i[\overset{n_1}{\Sigma}(m^i) + 2\overset{n_2}{\Sigma}(m^i) + \ldots]$$
$$+ \bar{m}^{i2}(n_1 + 4n_2 + \ldots)$$

(4.110)

Collecting terms and simplifying leads to an expression for T.

$$T = \overset{n_1}{\Sigma}(m^i - \bar{m}^i_1)^2 = \overset{n_2}{\Sigma}(m^i - 2\bar{m}^i_2)^2 \ldots$$
$$+ n_1(\bar{m}^i_1 - \bar{m}^i)^2 + 4n_2(\bar{m}^i_2 - \bar{m}^i)^2 + \ldots$$

(4.111)

These expressions pertain to the case of a simple random sample. Similar transformations can be made for the other sampling methods. Each of the above formulations may be used, according to which is the most convenient form for computation in each situation.

Stratified Sample with Uniform Sampling Fraction. In this case two alternative situations exist. In the first instance, it is assumed that the ratio R is the same for each stratum, so random-sample equations apply. Alternatively, the ratio may take different values in each stratum. Then each stratum is treated separately, using random-sample equations, and total estimates are obtained by summing the estimates for the separate strata and dividing the total by N or $\hat{N}$. The population estimate is

$$\hat{Y} = \Sigma_i \left[\frac{\overset{n_i}{\Sigma}(y)}{\overset{n_i}{\Sigma}(x)} X_i \right]$$

(4.112)

One may choose which alternative is to be assumed on the basis of three criteria: simplicity, size of strata, and variance of the ratio between the strata. Briefly, the assumption of a constant ratio is simpler and is more accurate when the strata populations are small and when the ratio variance is small. If the strata populations are large and the ratio variance is also large, the second assumption will give better estimates.

In estimating sampling errors, it is again necessary to consider two situations, that is, the same or different ratios in the strata. In the first alternative with the same ratio for all strata, if there are m strata, T and s_t^2 may be defined as

$$T = \sum_i \sum^{n_i} [(y - \bar{y}_i) - \hat{R}(x - \bar{x}_i)]^2 \tag{4.113}$$

$$T = \sum_i \sum^{n_i} (y - \bar{y}_i)^2 - 2\hat{R} \sum_i \sum^{n_i} (y - \bar{y}_i)(x - \bar{x}_i) + \hat{R}^2 \sum_i \sum^{n_i} (x - \bar{x}_i)^2 \tag{4.114}$$

$$s_t^2 = T/(n - m) \tag{4.115}$$

These are the sums of squares and products within the strata, and the derivation of the error variances then follows as for the simple random sample. When there are different ratios in different strata, $\hat{R}$ is replaced with the separate $\hat{R}_i$, which then remain within the summation signs in the formulation of T; thus T and s_t^2 are

$$T = \sum_i \sum^{n_i} (y - \bar{y}_i)^2 - 2\sum_i R_i \sum^{n_i} (x - \bar{x}_i)(y - \bar{y}_i) + \sum_i R_i^2 \sum^{n_i} (x - \bar{x}_i)^2 \tag{4.116}$$

$$s_t^2 = T/(n - m) \tag{4.117}$$

It may be convenient to calculate the sums of squares and products for each stratum separately, which also may be desirable to apply to the error variance calculations for $\hat{Y}$.

If the separate strata population totals, X_i, are not known, but only the population total over all strata, X, the calculations must be carried out as for the case of the same ratio for all strata, even though one may be certain that the ratio varies across the strata.

The following agriculture example was adapted from Yates.[5] The calculations are based on the ratio method for a stratified sample with uniform sampling fraction and variable ratio (table 4–3). Using this method results in a total estimated wheat acreage of 43,010 acres. The difference between this estimate and two others that could have been obtained is shown below. In the first instance, assume that a constant ratio over all strata can be used.

$$\hat{R} = \frac{\sum^n (y)}{\sum^n (x)} = \frac{2,301}{15,114} = 0.15224 \tag{4.118}$$

Table 4–3
Example of Ratio Method for Stratified Sample with Uniform Sampling Fraction

District No.	No.	Wheat n_i $\Sigma(y)$	Crops & Grass n_i $\Sigma(x)$	Ratio r_i	District Crops & Grass X_i	Estimated District Wheat $r_i Y_i$	T_i
1	15	141	1,935	.0729	22,932	1,670	5,107.6
2	8	259	1,385	.1870	43,591	8,150	1,550.7
3	40	763	4,851	.1573	57,263	9,010	7,964.0
4	24	837	4,034	.2075	73,946	15,340	20,566.6
5 & 7	14	127	882	.1440	40,905	5,890	3,737.1
6	24	174	2,027	.0858	34,437	2,950	1,080.9
	125	2,301	15,114		273,074	43,010	40,006.9

Source: Yates, F., 1971, p. 161.

The new acreage estimate is

$$\hat{Y} = \hat{R}X = 0.1521 \times 273,074 = 41,580 \text{ acres} \qquad (4.119)$$

From these data, T is 40,006.90, which is calculated by determining T_i for each district in table 4–3 and summing. The standard error of $\hat{Y}$ is given by equation 4.120, which is the square root of equation 4.103.

$$\text{s.e.}(\hat{Y}) = \frac{X}{\frac{n}{\Sigma(x)}} \sqrt{[(1-f)ns_i^2]} \qquad (4.120)$$

Substituting the appropriate values into this equation yields

$$\text{s.e.}(\hat{Y}) = \frac{273,074}{15,114} \sqrt{\left[\left(1-\frac{1}{20}\right)125 \frac{40,006.9}{119}\right]}$$

$$= \pm 3610 \qquad (4.121)$$

Hence, the 95% confidence limits on $\hat{Y}$ are that the number of acres of wheat lie between 34,504 and 48,656.

Another alternative is the estimate that would have been obtained without supplementary information. Using a stratified sample, Yates[6] gives the result to this example as 41,100 acres. By not stratifying the sample, an estimated wheat acreage of 46,020 acres was obtained. However, the stratification differs from that used in the illustrations here of the ratio method. The actual total wheat acreage is given as 44,676 acres (Yates[7]). Since insufficient data are provided to determine the sampling fractions by district, the alternative stratification cannot be used here. It is clear that the first method gave the more accurate results.

Stratified Sample with Variable Sampling Fraction. The same two alternative assumptions can be made about the ratio. If it is assumed to be the same for all strata, the estimates for the ratio, the sample, and the population are

$$\hat{R} = \frac{\sum_i [g_i \overset{n_i}{\underset{}{\Sigma}}(y)]}{\sum_i [g_i \overset{n_i}{\underset{}{\Sigma}}(x)]} \tag{4.122}$$

$$\hat{\bar{y}} = \hat{R}\,\bar{X} \tag{4.123}$$

$$\hat{Y} = \hat{R}\,X \tag{4.124}$$

If the ratio is different in each stratum, the calculations are the same as for a stratified sample with uniform sampling fraction. The sampling fraction does not enter the calculations.

For calculating the sampling errors, a constant ratio over all the strata necessitates only one change in the calculations, that is, T_i is calculated using the common value of $\hat{R}$. On the other hand, if a different ratio exists in each stratum, the variance of the estimated total is

$$\hat{V}(\hat{Y}) = \sum_i \left[\frac{x_i^2}{[\Sigma(x)]^2} (1-f_i)n_i s_{ti}^2 \right] \tag{4.125}$$

$$\hat{V}(\hat{Y}) = \sum_i [g_i^2(1-f_i)n_i s_{ti}^2] \tag{4.126}$$

$$\hat{V}(\hat{R}) = \hat{V}(\hat{Y})/X^2 \tag{4.127}$$

where s_{ti}^2 are estimated separately for each stratum, as s_{ti}^2 for a random sample. These values are obtained by dividing T_i by (n_i-1).

Estimation by the Regression Method

Random Sample. The equation of the regression line is

$$y_i = \bar{y} + b(x_i - \bar{x}) \tag{4.128}$$

where b is defined as

$$b = \frac{\overset{n}{\underset{}{\Sigma}}(y-\bar{y})(x-\bar{x})}{\overset{n}{\underset{}{\Sigma}}(x-\bar{x})^2} \tag{4.129}$$

The population estimates $\hat{\bar{y}}$ and $\hat{Y}$ are

$$\hat{\bar{y}} = \bar{y} + b\,(\bar{X}-\bar{x}) \tag{4.130}$$

$$\hat{Y} = N\hat{\bar{y}} \tag{4.131}$$

If N is not known exactly, it must be estimated from the sample. All values of b will give unbiased estimates, so any value b_0, which appears to be appropriate to the data being studied, may be used. If b_0 is assumed to equal one, this is equivalent to using the differences $y - x$. (The first equation becomes $y - x = \bar{y} - \bar{x} = $ constant.) The regression method gives the value of b which best fits the data being studied, but requires somewhat more computation. Regressions can also be used in the same way as ratios in two-phase sampling.

The use of the regression method can be illustrated by applying the method to the Yates[8] data used earlier; the results are

$$b = 0.19316 \qquad (4.132)$$

$$\hat{\bar{y}} = 16.185 \qquad (4.133)$$

$$\hat{Y} = 40,400 \text{ acres} \qquad (4.134)$$

The correct value of Y is 44,676 acres, and the ratio estimate was 43,010. Therefore the ratio method gives, in this case, a more accurate result for less computational effort.

The estimation of the sampling error for the regression method follows similar lines to that for the ratio method. A sum of squares of deviations, from the regression line, is defined as

$$T = \overset{n}{\Sigma}(y - \hat{y})^2 \qquad (4.135)$$

Substituting for $\hat{y}$, in terms of the regression estimate, equation 4.135 can be rewritten

$$T = \overset{n}{\Sigma}(y - \bar{y})^2 - b\overset{n}{\Sigma}(x - \bar{x})(y - \bar{y}) \qquad (4.136)$$

Substituting for b in equation 4.136 produces

$$T = \overset{n}{\Sigma}(y - \bar{y})^2 - \frac{[\overset{n}{\Sigma}(x - \bar{x})(y - \bar{y})]^2}{\Sigma(x - \bar{x})^2} \qquad (4.137)$$

Hence the variance is

$$\hat{s}^2 = \frac{T}{(n-2)} \qquad (4.138)$$

where $\hat{s}$ and $\hat{y}$ refer to estimates of s and y from the regression line. The use of $(n-2)$ as the divisor of T is to take account of the additional lost degree of freedom as a result of the calculation of b. Then neglecting errors in b, the variance can be expressed as

$$\hat{V}(\hat{\bar{y}}) = \frac{(1-f)}{n}\hat{s}^2 \qquad (4.139)$$

A term is introduced to allow for the errors in b.

$$\hat{V}(b) = \frac{\hat{s}^2}{\overset{n}{\Sigma}(x-\bar{x})^2} \tag{4.140}$$

Then the error variance of a standardized value y_0 of y for the value x_0 of x is

$$\hat{V}(\hat{y}_0) = \left[\frac{1}{n} + \frac{(x_0-\bar{x})^2}{\overset{n}{\Sigma}(x-\bar{x})^2}\right]\hat{s}^2 \tag{4.141}$$

The $(1-f)$ correction has been omitted from the equation since standardized values are generally used only for comparative purposes. This use demands that such a correction not be made. A similar adjustment can be made to $\hat{V}(\hat{y})$, but the error variance of b will usually be of a much lower order than the rest of the error variance of y, so its inclusion will not normally be worthwhile.

The use of an estimate of b, b_0, where b_0 is an arbitrary value assigned to the regression, was also discussed earlier. If this term is used, the expressions for T and $\hat{s}^2$ develop.

$$T = \overset{n}{\Sigma}(y-\bar{y})^2 - 2b_0\overset{n}{\Sigma}(x-\bar{x})(y-\bar{y}) + b_0^2\overset{n}{\Sigma}(x-\bar{x})^2 \tag{4.142}$$

$$\hat{s}^2 = \frac{T}{(n-1)} \tag{4.143}$$

Here, b_0 is not calculated from the data, so only one degree of freedom is lost because the regression line passes through the mean point $(\bar{x},\bar{y})$. No allowance has to be made for errors in b_0, but this method should not be used for standardization unless it is known that b_0 approximates b sufficiently closely so the error $(b_0-b)(x_0-\bar{x})$ introduced into the standardization is small enough to be neglected.

Stratified Sample with Uniform Sampling. With the same regression coefficient for each stratum, the equations for estimating population values for a random sample hold, except that b is defined as

$$b = \frac{\Sigma[\overset{n_i}{\underset{i}{\Sigma}}(y-\bar{y}_i)(x-\bar{x}_i)]}{\Sigma[\overset{n_i}{\underset{i}{\Sigma}}(x-\bar{x}_i)^2]} \tag{4.144}$$

The method applicable for cases with varying regression coefficients is the same as that for ratios, where the estimation is performed for each stratum separately, using the equations for a random sample.

For the case of an identical regression coefficient for all strata, the expressions for sampling errors are also analogous to those shown in the instance of the random-sample case. Equations 4.145 through 4.147 show these computations for the regression method.

$$T = \sum_i \overset{n_i}{\Sigma}(y-\bar{y}_i)^2 - b\sum_i \overset{n_i}{\Sigma}(y-\bar{y}_i)(x-\bar{x}_i) \tag{4.145}$$

$$\hat{s}^2 = T/(n-m-1) \tag{4.146}$$

$$V(b) = \hat{s}^2 \sum_i \overset{n_i}{\Sigma}(x-\bar{x}_i)^2 \tag{4.147}$$

Similarly, the equation for T can be rewritten as in equation 4.148 if an arbitrary b_0 is used.

$$T = \sum_i \overset{n_i}{\Sigma}(y-\bar{y}_i)^2 - 2b_0\sum_i \overset{n_i}{\Sigma}(y-\bar{y}_i)(x-\bar{x}_i)$$
$$+ b_0^2 \sum_i \overset{n_i}{\Sigma}(x-\bar{x}_i)^2 \tag{4.148}$$

Where different regression coefficients exist for different strata, b_i is defined

$$b_i = \frac{\overset{n_i}{\Sigma}(y-\bar{y}_i)(x-\bar{x}_i)}{\overset{n_i}{\Sigma}(x-\bar{x}_i)^2} \tag{4.149}$$

For a pooled estimate of error, the variance for b_i can be derived.

$$T = \sum_i \overset{n_i}{\Sigma}(y-\bar{y}_i)^2 - \sum_i b_i \overset{n_i}{\Sigma}(y-\bar{y}_i)(x-\bar{x}_i) \tag{4.150}$$

$$\hat{s}^2 = T/(n-2m) \tag{4.151}$$

$$V(b_i) = \hat{s}^2/\overset{n_i}{\Sigma}(x-\bar{x}_i)^2 \tag{4.152}$$

Stratified Sample with Variable Sampling Fraction. Where the regression coefficient is constant for all strata, the estimate of $\hat{\bar{y}}$ is

$$\hat{\bar{y}} = \bar{y}_w + b(\bar{X}-\bar{x}_w) \tag{4.153}$$

where $\bar{x}_w$ and $\bar{y}_w$ are the estimates of $\bar{X}$ and $\bar{Y}$ that would be obtained with no supplementary information on x. In this case the regression coefficient is

$$b = \frac{\sum_i [\lambda_i \overset{n_i}{\Sigma}(y-\bar{y}_i)(x-\bar{x}_i)]}{\sum_i [\lambda_i \overset{n_i}{\Sigma}(x-\bar{x}_i)^2]} \tag{4.154}$$

where λ_i are numerical weighting coefficients for the strata and $\bar{y}_w$ and $\bar{x}_w$ are

$$\bar{y}_w = \frac{\sum_i [g_i \overset{n_i}{\underset{}{\Sigma}}(y)]}{\sum_i (g_i n_i)} \tag{4.155}$$

$$\bar{x}_w = \frac{\sum_i [g_i \overset{n_i}{\underset{}{\Sigma}}(x)]}{\sum_i (g_i n_i)} \tag{4.156}$$

Almost any value of λ_i will give unbiased estimates, though optimal values can be obtained. If the regressions within strata are truly linear, with the values of b being identical in all strata (as assumed), λ_i should be chosen as being inversely proportional to the within-strata variances. If the coefficients do vary (although it is assumed they do not), the error of the assumption will be minimized if λ_i are chosen proportional to g_i^2. In fact, it is not usually worthwhile to determine which are the best estimates of λ_i, because the gain in accuracy is slight. If all the within-strata variances are similar, then λ_i can be taken as unity; and if the sampling fractions have been chosen to minimize sampling error, λ_i can be made equal to g_i. These values make for simple computations of only marginally less accuracy than the best possible estimates.

For stratified samples with variable sampling fractions and with varying regression coefficients, the computational procedure is identical to that for the ratio method for a constant sampling fraction.

For sampling errors, the computations are similar to those for the ratio method. When a constant regression coefficient is assumed over all the strata, values of T_i are calculated using the common value of b. Similarly, if different regression coefficients are assumed for each stratum, s_{ti}^2 are estimated separately as though each stratum were a separate simple random sample.

Calibration of Eye Estimates

The regression method may be used for another purpose, namely, for calibrating eye estimates. In this instance, it is assumed that y can be measured only on a nonrandom sample, which is a subsample of a random sample in which eye estimates could be made of the variate, x. Eye estimates are meant to denote some form of subjective measurement, while measurements of y are objective. This situation might arise, for instance, in a study of the effect of uniformity of street lighting and accidents. In such a study, a random sample of streets may be taken and the uniformity of lighting (mean illumination divided by minimum illumination) may be measured on a nonrandom subsample, while making subjective estimates for the entire sample. The eye estimates cannot be used as supplementary information, because any bias in the subsample which contains both subjective and objective measurements will give a biased value of b. Instead, the regression of x on y is calculated so the regression equation becomes

$$\hat{\bar{y}} = \bar{y} + \frac{1}{b'} (\bar{x}_1 - \bar{x}) \qquad (4.157)$$

where b' = the regression coefficient of x on y

$$(\sum_{}^{n}(x-\bar{x})(y-\bar{y})/\sum_{}^{n}(y-\bar{y})^2),$$

$\bar{y}$ and $\bar{x}$ = actual means for the subsample

and $\bar{x}_1$ = the mean for the entire sample.

In this procedure, it must be assumed that the subsample is effectively random for units having a given value of y, even though the subsample is not random. Also, the eye estimates must be reasonably accurate, that is, they must have a relatively small variance about the regression line, which must have a significant slope and be a good estimate of the data. If the true line is curved to any great extent, the use of a linear regression will lead to biased estimates.

The sampling variance of $\hat{\bar{y}}$ can be split into three parts: that due to the error of b'; that due to the sampling variance of $\bar{x}_1$ arising from the main sampling process; and that due to the variance about the regression line. The first can usually be neglected since it is typically very small. It is given by $(\bar{x}_1-\bar{x})^2\hat{V}(b')/b'^4$, so unless b' is very small and $V(b')$ is large, this will usually be a very small component of the errors.

The variance of $\bar{x}_1$ is calculated for the values of x in the manner appropriate to the method of sampling used. The contribution to the error variance of $\hat{\bar{y}}$ is $\hat{V}(\bar{x}_1)/b'^2$, or more accurately $\hat{V}(\bar{x}_1)[\hat{V}(x) - \hat{V}_l(x)]/\hat{V}(x)$, where $\hat{V}_l(x)$ is the residual variance of x about the regression line, and $\hat{V}(x)$ is the part of the variance of x that contributes to $\hat{V}(\bar{x}_1)$.

The variance about the regression that contributes to $\hat{V}(\hat{\bar{y}})$ is calculated from $V_l(x)$. If n_1 and n are the numbers of units in the original sample and the subsample of eye estimates, the required value is given by $(n_1-n)\hat{V}_l(x)/b'^2nn_1$, if the sample units are all given equal weight. If the weights are unequal, this becomes

$$\frac{V_l(x)}{b'^2} \left(\frac{\sum\limits_{}^{n_1-n}(a)}{\sum\limits_{}^{n_1}(a)} \right)^2 \left[\frac{\sum\limits_{}^{n}(a^2)}{[\sum\limits_{}^{n}(a)]^2} + \frac{\sum\limits_{}^{n_1-n}(a^2)}{[\sum\limits_{}^{n_1-n}(a)]^2} \right] \qquad (4.158)$$

where the as are the weights.

Sampling with Probabilities Proportional to Size of Unit

This is one of a variety of situations that can arise in surveys and may be a feature of multistage sampling,[9] although Yates[10] seems to consider it a sampling method in its own right. The technique of sampling units with a probability proportional to the size of the unit appears to be due originally to Hansen and Hurwitz.[11] Typically, this method may be useful when units at a primary stage are grouped in some way that makes surveying easier. For example, households may be listed by county or school district, where such units provide considerable advantages for surveying as the primary units. Suppose a statewide survey is to be carried out on employment. Primary selection may be made at the level of the county. As Cochran suggests,[12] stratification may also be applied to the counties, for example, grouping them into agricultural, mining, and urban counties. If the counties are chosen with equal probabilities, the estimated mean over all primary units will be biased. Sampling with the probability being proportional to the size of the unit removes the bias.

In the most usual situation, either the size, x, of all the units of the sample or X, the size for all units in the population, is known. Here, x acts as a supplementary variate and the ratio method is usually appropriate. Because the probability of selection is proportional to x, raising factors proportional to $1/x$ must be introduced into the estimating equations for $\hat{R}$, $\hat{Y}$, and $\hat{\bar{y}}$.

$$\hat{R} = \frac{1}{n} \sum^{n} \left(\frac{y}{x} \right) \qquad (4.159)$$

$$\hat{Y} = \hat{R}X \qquad (4.160)$$

$$\hat{\bar{y}} = \hat{R}\bar{X} \qquad (4.161)$$

In the second situation, total size, X, of the population is not known. Here, X must be estimated from the sample as well as Y and R. Sample selection has to be carried out by some such method as randomly or systematically locating points on a map, and points not falling in the units under consideration must be taken into account. If n_0 is the total number of sampling points and A the total area covered by the sampling grid, the estimating equations for the ratio and the population variates are

$$\hat{R} = r \qquad (4.162)$$

$$\hat{X} = An/n_0 \qquad (4.163)$$

$$\hat{Y} = \hat{R}\hat{X} = rAn/n_0 \qquad (4.164)$$

If A is not known, the density of points per unit area, d, may be used which leads to

$$\hat{A} = n_0/d \qquad (4.165)$$

$$\hat{X} = n/d \qquad (4.166)$$

When this form of sampling is conducted within a stratified sampling process, the size of all units must be known and the population estimates are

$$\hat{R}_i = \frac{1}{n_i}\sum^{n_i}\left(\frac{y}{x}\right) = r_i \qquad (4.167)$$

$$\hat{Y} = \sum_i [\hat{R}_i X_i] \qquad (4.168)$$

$$\hat{R} = \hat{Y}/X \qquad (4.169)$$

To estimate the sampling errors for this method, the estimated variance of the ratio r has to be determined.

$$s_r^2 = \sum^n (r - \bar{r})^2/(n-1) \qquad (4.170)$$

$$\hat{V}(\hat{R}) = \frac{1}{n} s_r^2 \qquad (4.171)$$

If the population size is known, the variances are

$$\hat{V}(\hat{\bar{y}}) = \bar{X}^2 s_r^2/n \qquad (4.172)$$

$$\hat{V}(\hat{Y}) = X^2 s_r^2/n \qquad (4.173)$$

Otherwise, $\hat{V}(\hat{X})$ has to be determined first,

$$\hat{V}(\hat{X}) = \frac{\hat{X}^2}{n}\frac{n_0-n}{n_0} \qquad (4.174)$$

where n_0 = the total number of sampling points.

Then the variance of the population estimate $\hat{Y}$ is

$$\hat{V}(\hat{Y}) = \hat{X}^2\hat{V}(\hat{R}) + \hat{R}^2\hat{V}(\hat{X}) \qquad (4.175)$$

Substituting equation 4.174 into equation 4.175 results in the expression for the variance of the population estimate $\bar{Y}$.

$$\hat{V}(\hat{Y}) = \frac{\hat{X}^2}{n}\left(s_r^2 + \frac{n_0-n}{n_0}\bar{r}^2\right) \qquad (4.176)$$

If A is not known, as is estimated from equation 4.175, $\ddot{V}(\hat{X})$ is

$$\ddot{V}(\hat{X}) = n\frac{(n_0-n)}{d^2 n_0} \qquad (4.177)$$

The other equations are derived as before. Two examples, again adapted from Yates,[13] serve to illustrate this method.

Example 1

A survey is being carried out to estimate the area and yield of a crop. The survey is conducted by locating points systematically at a density of one per four square miles, and the yields per acre of the fields in which the points fall and which carry the crop are determined by harvesting small areas; 8317 points are obtained yielding 529 sample locations of fields carrying the crop. The arithmetic mean of the yields per acre of the selected fields is 0.785 tons per acre. The object is to estimate the total area and yield of the crop.

The density of survey location d is

$$d = 1/4 \text{ sq. mi.} = 1/2560 \text{ per acre} \qquad (4.178)$$

The estimates for total area, $\hat{X}$, and yield, $\hat{Y}$, are

$$\hat{X} = 529 \times 2560 = 1,354,000 \text{ acres} \qquad (4.179)$$

$$\hat{Y} = 0.785 \times 1,354,000 = 1,063,000 \text{ tons} \qquad (4.180)$$

Yates provides no data on the individual survey points and does not give a value of s_r^2. Assuming a reasonable value for this of 0.039, equation 4.176 is used to estimate the variance of $\hat{Y}$.

$$V(\hat{Y}) = \frac{(1,354,000)^2}{529} \left[0.039 + \left(\frac{8317-529}{529} \right) (0.785)^2 \right]$$

$$= 3.15749 \times 10^{10} \qquad (4.181)$$

The sampling error of $\hat{Y}$ is therefore 177,693. Applying the normal law of error, the estimate of equation 4.181 may be restated as follows: the estimated crop yield is (with 95% confidence) between 714,722 and 1,411,278 tons.

Example 2

Assume a two-phase survey was carried out and a further 24,938 points for type of crop only were surveyed; 1673 fields carrying the crop of interest were found, and the additional sample gives a density of one point per square mile. The estimate can now be revised by using the two sets of observations as the first-phase sample, and the original set as the second phase.

The total number of survey points, n_0', and the number of fields carrying the crop of interest, n', are

$$n_0' = 24,938 + 8317 = 33,255 \qquad (4.182)$$

$$n' = 1673 + 529 = 2202 \qquad (4.183)$$

Given that phase-one density, d, equals 1/640 per acre, the total number of acres, $\hat{X}$, and of tons yielded, $\hat{Y}$, are 1,409,000 acres and 1,106,000 tons, respectively.

Multistage and Systematic Sampling

Multistage Sampling

Estimating equations for both situations with and without supplementary information are determined. When estimating values for multistage sampling, one may take estimates at each stage, using the appropriate estimation techniques. However, it is frequently more desirable to treat multistage sampling as a single process for estimation purposes. The combined raising factor for the single process is given by the product of the raising factors for each separate stage, as shown in equation 4.184 for a two-stage process in which g' and g'' are the raising factors for the first and the second stages of sampling.

$$g = g' g'' \tag{4.184}$$

When there is no supplementary information, the general equation for $\hat{Y}$ is

$$\hat{Y} = \overset{n}{\Sigma}(gy) \tag{4.185}$$

where $\overset{n}{\Sigma}$ is the summation over all units. Similar equations can be derived for the other population values, for example, for $\hat{N}$.

When supplementary information is available, four basic situations exist: the ratio method with the same ratio for the whole population, ratio method with different ratios for different parts of the population, regression method, and sampling with the probability proportional to the size of the unit. Basically, each situation requires just a simple application of one of the estimating methods already considered.

In estimating sampling errors, two possible situations exist. In the first case, if the sampling fraction at the first stage is small, the total sampling error can be calculated by estimating the unit values of the first-stage units from the results of the second and subsequent stages. The additional variance for each stage beyond the first is automatically included in this method.

If the sampling fraction for the first stage (f') is not small, the correction factor ($1-f'$) may not be neglected. The increase in the sampling error variance will be equal to f' times the variance resulting from the second and subsequent stages. For instance, in two-stage sampling where n' first-stage units are selected, and n'' second-stage units, with similar connotations for s'^2 and s''^2, f' and f'', the variance is

$$\hat{V}(\hat{\bar{y}}) = \frac{1-f'}{n'}s'^2 + f'\frac{(1-f'')}{n'n''}s''^2 \qquad (4.186)$$

Systematic Sampling

The estimation techniques discussed previously can be applied to systematic sampling. Yates[14] points out an improvement in accuracy of systematic sampling by making "end corrections." These end corrections are weights applied to the sampling points at the ends of the sample list. Consider a systematic sample from a line AB, as shown in figure 4–2.

Consider P_2, P_3, ... to be sampling points representing the intervals Q_1Q_2, Q_2Q_3, ... where the Ps are sampling points and the Qs are midpoints between the Ps. However, in systematic sampling $Q_1Q_2 = Q_2Q_3 = Q_3Q_4 = ...$, so the Ps represent measures from equal intervals, except P_1 and P_6; P_1 represents a measure from the interval AQ_1 and P_6 from Q_5B. If each P is weighted by the length it represents, equal weights are obtained, Q_1Q_2, say, for each of P_2, P_3, P_4 and P_5. Weights for P_1 and P_6 are AQ_1 and Q_5B, respectively. Hence the end corrections are AQ_1/Q_1Q_2 and Q_5B/Q_1Q_2.

Cochran[15] also details a number of other estimation methods that can be used for systematic samples, based on alternative ways of looking at the sampling procedure.

In systematic sampling there is no completely valid method of calculating the sampling error, since the sampling method is not random and stratified. The best approximation is to stratify the material arbitrarily and calculate the sampling errors as for a random stratified sample. In the event that the sample is chosen from a one-dimensional frame, stratification can be performed so that each stratum contains two observations. The variance is then estimated from the differences between the members of each of the pairs. Each difference contributes one degree of freedom, so if there are n' pairs, with differences d, the error variance per unit is

$$s^2 = \frac{1}{2}\overset{n}{\Sigma}(d^2)/n' \qquad (4.187)$$

Further refinements are detailed in both Yates and Cochran,[16] as is an extension to a two-dimensional case.

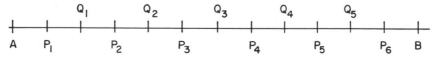

Figure 4–2. Sampling Points and Intervals in Line AB

Sampling on Successive Occasions

Consider now the problems of estimation associated with sampling on successive occasions. In the first instance, consider only the problem associated with sampling on two successive occasions. If the second sample is totally independent of the first or is identical, no problems exist. Each occasion can be treated as though it were a single unrepeated sample, and estimation can be achieved by using the appropriate equation. However, if the second sample is a subsample of the first or if there is a partial replacement of the sample on the second occasion, more elaborate techniques are needed.

If the second sample is a subsample of the first, changes can be estimated most simply by considering the subsample alone on both occasions and identifying the changes between the units. The population mean and total for the second occasion may be estimated by adding the change between the first and second to the mean and total determined from the first survey. A more accurate method would, however, comprise use of the data of the first survey as supplementary data and use of the regression method for estimating the new mean and total. This in turn yields the most accurate measure of change over the time period. Very briefly, this method works as follows.

Denote the values from the first survey by x and those for the second as y, the values belonging to both samples as x', and y', using a double-prime for those confined to the first survey, x'', leads to

$$\hat{\bar{x}} = \lambda \bar{x}' + \mu \bar{x}'' \tag{4.188}$$

where λ = the proportion of units in the first sample that are sampled on the second occasion and $\mu = 1 - \lambda$

The corresponding expression for the second survey is

$$\hat{\bar{y}} = \bar{y}' + b(\hat{\bar{x}} - \bar{x}') \tag{4.189}$$

$$\hat{\bar{y}} = \bar{y}' + \mu b(\bar{x}'' - \bar{x}') \tag{4.190}$$

where $\bar{x}$ = the overall estimate for the first survey
and $\hat{\bar{y}}$ = the adjusted estimate for the second survey.

The change in the population mean is obviously $\hat{\bar{x}} - \hat{\bar{y}}$.

$$\hat{\bar{x}} - \hat{\bar{y}} = \lambda \bar{x}' + \mu \bar{x}'' - \bar{y}' - \mu b(\bar{x}'' - \bar{x}') \tag{4.191}$$

Collecting terms and rearranging equation 4.191 leads to expressions of the change in the population mean, shown as

$$\hat{\bar{x}} - \hat{\bar{y}} = \mu \bar{x}''(1-b) + \bar{x}' - \bar{x}'\mu(1-b) - \bar{y}' \tag{4.192}$$

$$\hat{\bar{x}} - \hat{\bar{y}} = \bar{x}' - \bar{y}' - \mu(1-b)(\bar{x}' - \bar{x}'') \tag{4.193}$$

Here, b is estimated from values of the units which are included in both surveys. The estimated change described earlier is

$$\hat{\bar{x}} - \hat{\bar{y}} = \bar{x}' - \bar{y}' \qquad (4.194)$$

Hence the gain in accuracy of this method comes from the term $-\mu(1-b)(\bar{x}'-\bar{x}'')$. If b tends to unity, which will occur as the change between the occasions becomes very small compared with differences between members of the population, this term goes to zero. Hence it is used in situations where the change over the period of time is greater than the variance of the units.

If in the second sample partial replacement occurs between the two surveys, in effect three populations are involved: one surveyed only on the first occasion (x''), one surveyed on both occasions (x',y'), and one surveyed only on the second occasion (y''). Likewise, one can obtain two estimates of $\hat{\bar{y}}$, $\hat{\bar{y}}_1$, and $\hat{\bar{y}}_2$: from the proportion of the population (λ) common to both surveys and the proportion included in the second survey only, respectively.

Then $\hat{\bar{y}}_2$ equals $\bar{y}''$, and one wishes to find $\hat{\bar{y}}_w$, the weighted mean of $\hat{\bar{y}}_1$ and $\hat{\bar{y}}_2$. The weights for these two means will be $\lambda/(1-\mu^2r^2)$ and $\mu(1-\mu r^2)/(1-\mu^2r^2)$, where r is the correlation coefficient between the values of the units for the two successive occasions. For a pair of random samples the correlation r is

$$r = \frac{\overset{n}{\Sigma}(x'-\bar{x}')(y'-\bar{y}')}{\sqrt{[\overset{n}{\Sigma}(x'-\bar{x}')^2\overset{n}{\Sigma}(y'-\bar{y}')^2]}} \qquad (4.195)$$

The sums of squares and cross products have to be modified for more complex sampling methods, similar to the modification of b in these methods. This results in the expression for the weighted mean of $\hat{\bar{y}}_w$.

$$\hat{\bar{y}}_w = \frac{\lambda}{1-\mu^2r^2}\,\hat{\bar{y}}_1 + \frac{\mu(1-\mu r^2)}{1-\mu^2r^2}\,\hat{\bar{y}}_2 \qquad (4.196)$$

Using the regression method on the common population provides the estimate of $\hat{\bar{y}}_1$.

$$\hat{\bar{y}}_1 = \bar{y}' + b(\hat{\bar{x}}-\bar{x}') \qquad (4.197)$$

Remembering that $\hat{\bar{y}}_2$ is equal to $\bar{y}''$ and μ is equal to $1-\lambda$, equation 4.196 can be rewritten

$$\hat{\bar{y}}_w = \frac{\lambda}{1-\mu^2r^2}\,[\hat{\bar{y}}_1+b(\hat{\bar{x}}-\bar{x}')] + \left(1 - \frac{\lambda}{1-\mu^2r^2}\right)\bar{y}'' \qquad (4.198)$$

A further modification is necessary if the two samples are not the same size,

$$\hat{\bar{y}}_w = \frac{n'[\bar{y}'+b(\hat{\bar{x}}-\bar{x}')] + n''(1-\mu r^2)\bar{y}''}{n' + n''(1-\mu r^2)} \qquad (4.199)$$

where n' = number included in both samples
and n'' = number in second sample only.

An estimate of the change between the two occasions can also be determined, using the weighted mean of the two estimates of change $y' - \bar{x}'$ and $\bar{y}'' - \bar{x}''$.

$$\hat{\bar{y}}_w - \hat{\bar{x}}_w = \frac{\lambda}{1-\mu r} (\bar{y}'-\bar{x}') + \frac{\mu(1-r)}{1-\mu r} (\bar{y}''-\bar{x}'') \qquad (4.200)$$

The sampling errors for two occasions with subsampling on the second occasion are derived from the computation of the variances.

$$\hat{V}(\hat{\bar{y}}) = [\hat{V}(y)-\mu b^2 \, \hat{V}(x)]/\lambda n \qquad (4.201)$$

$$\hat{V}(\hat{\bar{y}}) = (1-\mu r^2)\hat{V}(y)/\lambda n \qquad (4.202)$$

$$\hat{V}(\hat{\bar{y}}-\hat{\bar{x}}) = [\hat{V}(y)+(\lambda-2\lambda b-\mu b^2)\hat{V}(x)]/\lambda n \qquad (4.203)$$

For two occasions with partial replacement and using the notation employed before, the variance is

$$\hat{V}(\hat{\bar{y}}_w) = \frac{(1-\mu r^2)\hat{V}(y)}{n(1-\mu^2 r^2)} \qquad (4.204)$$

For the case of unequal numbers on the two occasions, the equation reads

$$\hat{V}(\hat{\bar{y}}_w) = \frac{(1-\mu r^2)\hat{V}(y)}{n'+n''(1-\mu r^2)} \qquad (4.205)$$

Likewise, the error variance of the change over the two occasions is

$$\hat{V} \text{ (change)} = \frac{(1-r)[\hat{V}(y)+\hat{V}(x)]}{n(1-\mu r)} \qquad (4.206)$$

If the surveying is extended to more than two successive occasions, some further modifications are necessary. If subsequent samples are subsamples of the first sample, there is no problem and a direct extension of the original calculations for two occasions can be made. However, if resampling occurs on each occasion with partial replacement, the problem is more complex. Although an exact general solution for this case cannot be derived, there are some approximate ones which will usually suffice.

In this type of sampling, the analyst is usually interested in the most accurate estimate of the population mean on each occasion, without revising estimates for previous occasions. So if $\hat{\bar{y}}_h$ is the most accurate estimate of the population mean on the hth occasion, using the results of the sampling up to and including the hth occasion, and $\hat{\bar{y}}_{h-1}$ is that for the $h-1$ occasion, subject to certain limitations, equation 4.207 holds.

$$\hat{\bar{y}}_h = (1-\phi)[\bar{y}'_h + r(\hat{\bar{y}}_{h-1}-\bar{y}'_{h-1})] + \phi\bar{y}''_h \qquad (4.207)$$

There are three limitations to equation 4.207:

1.   a given fraction of units is replaced on each occasion;
2.   the variances and correlations on successive occasions are constant; and
3.   the correlation between two successive occasions is r, between two apart r^2, between three apart r^3, etc.

Also, ϕ is the weighting factor for the most recent occasion and is a function of μ, the proportion of units that are not resampled, r, the correlation, and h, the number of occasions of successive sampling. As h becomes large, ϕ becomes a function of μ and r only. Thus, if μ units are replaced on each occasion, ϕ tends to a limiting value as h increases.

$$\phi = \frac{-(1-r^2) + \sqrt{[(1-r^2)[1-r^2(1-4\lambda\mu)]]}}{2\lambda r^2} \qquad (4.208)$$

Note that ϕ has already been determined for two successive occasions ($h=2$), as can be seen by examining equation 4.198. Hence in this case ϕ is

$$\phi = 1 - \frac{\lambda}{(1-\mu^2 r^2)} \qquad (4.209)$$

For more than two occasions, the limiting value of ϕ, equation 4.208, can be used.

Usually, an estimate $\hat{\bar{y}}_h - \hat{\bar{y}}_{h-1}$ is sufficient for the change in means between two successive occasions. If change is of particular interest, an equation similar to that derived for two successive occasions may be used. If the sample sizes tend to vary from occasion to occasion (a practical sampling scheme will rarely yield absolutely identical sampling sizes from a human population), ϕ must be replaced by ϕ' which is

$$\phi' = \frac{n''_h}{\mu n_h}\phi \qquad (4.210)$$

where n_h = the number of units in the sample on the hth occasion and n''_h = the number of units not included on the previous occasion.
     For calculating the sampling errors for sampling on successive occa-

sions, one can build on equation 4.208. From this, a limiting value of $\hat{V}(\hat{y}_h)$ can be defined.

$$\hat{V}(\hat{\bar{y}}_h) = \phi\hat{V}(y)/\mu n \tag{4.211}$$

The variance of the estimate of change is

$$\hat{V}(\hat{\bar{y}}_h - \hat{\bar{y}}_{h-1}) = 2\phi\hat{V}(y)[1 - r(1-\phi)]/\mu n \tag{4.212}$$

Some Closing Points

The preceding discussion covers the basic equations presented here for estimating population values and sampling error. There are, however, two or three further points which warrant brief attention before the topic of sampling errors is completed.

The use of pooled estimates of error in stratified sampling was mentioned earlier in the chapter. An important assumption here is that the within-strata variances are constant. If in fact the variances are not constant, error estimates based on pooling are not valid. However, the number of degrees of freedom in the different parts of the population may be too small for accurate determination of the error variances. In such cases, an alternative method is necessary.

The simplest and most convenient method is the error graph, in which error estimates are plotted against a characteristic believed to affect the magnitude of the error. A smooth curve is fitted to these points, and this represents the law for the variance from which error variances can be reestimated.

In large-scale surveys, it may not be necessary or desirable to calculate sampling errors from the entire survey results. A convenient method therefore is calculating the sampling error for certain parts or strata of the population. The parts to be used for the calculations should be selected randomly to avoid bias. Provided the survey is a large one, this process is unlikely to involve an excessive amount of work, and will produce a quite acceptable estimate of the errors of the entire sample.

One final matter is the effect of grouping and rounding-off. The additional variance per unit of the grouped data is $\sigma^2 + a^2/12$, where σ^2 is the true variance per unit and a is the grouping interval. Rounding-off is a form of grouping, and the same correction is necessary. The fractional loss of accuracy in the variance per unit is then $a^2/12\sigma^2$, while the loss of accuracy in the estimation of the variance of a sample from a normal distribution is $a^2/6\sigma^2$. These expressions are approximate and apply if the distribution is reasonably symmetrical and if a is not likely to be greater than σ. If these conditions are not satisfied, an additional bias is introduced. The best treatment of this is more careful selection of the sample.

Notes

1. Frank S. Yates, *Sampling Methods for Censuses and Surveys,* 3d ed., London: Charles Griffin and Co., 1971.

2. Yates, *Sampling Methods.*

3. Ibid, pp. 147–148.

4. Ibid; William G. Cochran, *Sampling Techniques,* 2d ed., New York: John Wiley & Sons, 1966, pp. 19–20.

5. Yates, *Sampling Methods,* p. 161.

6. Ibid, p. 153.

7. Ibid, p. 31.

8. Ibid, p. 162.

9. Cochran, *Sampling Techniques,* chapter 11.

10. Yates, *Sampling Methods,* p. 35.

11. M.H. Hansen and W.N. Hurwitz, "On the Theory of Sampling from Finite Populations," *Annals of Mathematical Statistics,* 1943, vol. 14, pp. 333–362.

12. Cochran, *Sampling Techniques,* p. 293.

13. Yates, *Sampling Methods,* pp. 168–169.

14. Ibid.

15. Cochran, *Sampling Techniques,* chapter 8.

16. Yates, *Sampling Methods,* pp. 229–233; Cochran, *Sampling Techniques,* chapter 8.

5 Efficiency

Definition

In survey sampling, the notion of efficiency is similar to that used in statistical estimation except that in the latter, efficiency and precision are used interchangeably. In survey sampling, small samples are often encountered, for which precision and efficiency will connote slightly different properties. Efficiency in survey sampling is not well covered in other texts.[1]

Two concepts relating to efficiency in survey sampling are discussed in this chapter: efficiency in terms of the method of sampling and size of sample necessary to give a required level of accuracy; and efficiency in terms of minimization of cost subject to a constraint of the requisite level of accuracy. A full treatment, such as that given by Yates,[2] is not provided here, but an attempt is made to provide the reader with a sufficient grasp of the notions of efficiency and related computations to permit him or her to handle more complex situations.

Sample-size Efficiency

In the preceding chapter the theoretical basis for the determination of required sample size was discussed. This basis resides in the calculation of sampling errors. Clearly, if a given accuracy is required, one can theoretically apply the sampling-error equations in reverse to determine the required sample size, n, or sampling fraction, f. It should be noted that for some sampling methods, it is not possible to obtain an algebraic solution for the sample size. For example, in stratified sampling with variable sampling fractions, one would have to know the relationship between sampling fractions. Alternatively, one would need to define a required accuracy for each stratum.

The major problem that arises in calculating sampling fractions and sample sizes is that these calculations require a knowledge of the standard deviations of relevant population values. However, these standard deviations are usually not known until after the survey has been carried out. It must be stated now that where no prior data have been collected on the relevant population values, a situation that might arise in developing countries, the most efficient sample size and sampling fraction *cannot* be determined. It will be necessary to use some rule of thumb or a "guestimate."

Here, the assumption is made that prior data have been collected, but not necessarily by the method to be used in the survey of current concern. Using such data, the issue is how to calculate either n, the sample size, or f, the sampling fraction, for a given level of accuracy.

A more useful notion is, perhaps, to calculate the relative precision of two sampling methods for any given sample size. Yates[3] defines the relative efficiency (which is equal to the relative precision only in large populations) as the reciprocal of the ratio of the sample sizes needed for a given level of accuracy with two alternative sampling methods. If the relative efficiency is greater than one, the method used for the numerator is more efficient than that used for the denominator.

The relative precision is defined as the reciprocal of the ratio of the sample variances of a population estimate for two methods, when the same sample size is used for both methods.

Small Samples

Before proceeding to a consideration of specific sampling methods, it should be noted that the equations derived in chapter 4 for sampling errors are only accurate for very large populations. For relatively small populations a correction factor should be applied, as shown below.

Calculate the number of units n_0 that would be required if the population were very large, and then compute the corresponding sampling fraction f_0,

$$f_0 = n_0/N \tag{5.1}$$

where N = total number of units in the population. It should be noted that f_0 can be larger than unity if the population is very small or the required accuracy is very high. The required sampling fraction is

$$f = \frac{f_0}{1 + f_0} \tag{5.2}$$

Obviously, one can then proceed to determine required sample sizes for any method of sampling by the same procedure as was used in the earlier discussions, that is, rewriting the sampling error equations in terms of n.

Some General Notions

Assuming that some past survey or census exists from which one can calculate the appropriate variances, a simple rule can be suggested for sample-size determination. Using the symbol $\sim$ to distinguish the values required for designing the new survey from those obtained in a past survey, a simple rule[4] is given by

$$\frac{\tilde{n}}{n} = \frac{V(\bar{y})\,(1-\tilde{f})}{\tilde{V}(\bar{y})\,(1-f)} \qquad (5.3)$$

The finite-population corrections, $(1-f)$, can be ignored if the population is very large. The value $\tilde{V}(\bar{y})$ is the desired variance, so that equation 5.3 can be rearranged to give $\tilde{n}$.

$$\tilde{n} = \frac{n\,V(\bar{y})(1-\tilde{f})}{\tilde{V}(\bar{y})(1-f)} \qquad (5.4)$$

This simple rule makes a number of restrictive assumptions, including that the past and planned surveys both use the same design, which is simple random sampling, or at least uniform sampling fractions for stratified sampling. It is useful to note from this that equation 5.5 is true under these assumptions.

$$\frac{\tilde{n}}{n} \propto \left[\frac{\text{s.e.}(\bar{y})}{\widetilde{\text{s.e.}}(\bar{y})}\right]^2 \qquad (5.5)$$

Thus the sample size is inversely proportional to the square of the standard error, from which it follows[5] that a reduction in the error by a factor k requires an increase in the sample size by a factor k^2.

A useful idea is to conduct a pilot survey (see chapter 6) that is large enough and appropriately sampled to provide the values of the population variances. This is rarely done in practice, frequently because of practical difficulties in sampling the appropriate population and because pilot surveys are often conducted too hurriedly and on too small a budget.

Kish[6] suggests a concept that may be useful called the *design effect*, ψ. The design effect is defined as the ratio of the sample variance to the variance from a simple random sample of the same size, n.

$$\psi = \frac{V(\bar{y})}{(1-f)\sigma_y^2/n} \qquad (5.6)$$

Rearranging equation 5.6 produces equation 5.7, which defines n.

$$n = \frac{(1-f)\sigma_y^2}{V(\bar{y})} \qquad (5.7)$$

As Kish[7] points out, this is most useful because ψ summarizes the effects of almost any complexity of sample design, including clustering, stratification, ratio and regression methods of using supplementary data, and multistage sampling. If a large sample is available, σ_y^2 can be estimated quite well by s_y^2, with slight error when ψ is not very close to one.

The most useful application of equation 5.7 follows from experience with sampling and the properties of sampling methods. Many survey designers have sufficient experience so they can estimate a value for ψ quite readily. In addition, it can be noted that ψ will generally be less than one for stratified sampling, since stratification is designed to reduce sampling variance for given sample size. Similarly, cluster sampling would give a value of ψ greater than one, showing the losses due to clustering. Hence an estimate can be made for ψ and used with information from a simple random sample.

Qualitative Data

The simplest method for use in this case is to estimate the percentage standard error of an available sample and calculate the required sample size.

$$\frac{\text{size of sample required}}{\text{size of sample available}} = \frac{(\text{actual p.s.e.})^2}{(\text{required p.s.e.})^2} \qquad (5.8)$$

The required sample size can be modified for a finite population by use of equation 5.2.

If a stratified sample with variable sampling fraction is to be used on a finite population, the correction of equation 5.2 will not be adequate. Correction need be made only if any of the correction factors, $(1-f_i)$, are sufficiently large to be of importance. In this case, a process of trial and error is probably best for calculating the sample sizes. This is done by ignoring the correction factors first, and computing the sample size and individual sampling fractions as though the population were infinite. Subsequently, the correction factors can be calculated and a new sample size computed. This procedure can be repeated until the sample size becomes sufficiently stable. Yates[8] suggests, however, that this method of sampling may be considered a rather unlikely one for quantitative data.

Comparing a Random Sample and a Stratified Sample with Uniform Sampling Fraction

The reader will recall that an analysis-of-variance table was set up for errors in stratified samples (table 4–2). The method used here is to construct the appropriate analysis of variance for the required sample according to the available data.

Suppose, for example, past data are available from a stratified sample with uniform sampling fraction, then an analysis of variance is undertaken to estimate s_1^2, the within-strata mean square, and s^2, the total mean square. The former gives an estimate of the error variance per unit for a stratified sample and the latter for a simple random sample. The relative efficiency can be determined by using s_1^2 and s^2, respectively, in equation 5.9.

$$n = \frac{(\text{p.s.e. of a unit})^2}{(\text{required p.s.e.})^2} \qquad (5.9)$$

It should be noted that the use of s_1^2 and s^2, as described here, involves a slight approximation.[9]

Similar calculations would be made if the past data were from a simple random sample, where it is necessary to know only how many units of the past sample fall in each stratum. In the case of data from a stratified sample with variable sampling fraction, s_1^2 and s^2 are determined a little differently. Here, the proportion of the population falling in each stratum of the past sampling procedure must be known. If this proportion is denoted h_i, then s_1^2 is determined as the pooled estimate across the strata.

$$s_1^2 = \sum_i h_i s_i^2 \qquad (5.10)$$

Similarly, s^2 is

$$s^2 = s_1^2 + \sum_i h_i \bar{y}_i^2 - \bar{y}^2 - \sum_i h_i(1-h_i)s_i^2/n_i \qquad (5.11)$$

The last term of equation 5.11 is a correction for the errors in the $\bar{y}_i$ terms, and is usually very small, unless the between-strata variances are small, the number of strata are large, and the number of units in each stratum is small.

To illustrate the calculations, suppose past data have been collected on household car ownership by a stratified sample with uniform sampling fraction. The stratification for a new stratified sample is to be by income group. Using the past data, the values of s_1^2 and s^2 are found to be 0.47 and 1.08, respectively. From this, the relative precision is 2.30. In other words, the stratified sample will be 2.30 times as precise for a given sample size. If the population is very large, for example, the households of a large metropolitan area, the relative efficiency will also be 2.30. If the population is small, then n must be determined for each method by using equation 5.2 and the relative efficiency will be somewhat less than 2.30.

Multiple Stratification

The issue in this case is whether further gains in precision can be obtained by further substratification of the main strata compared with using a single stratification scheme. The method for determining the required sample size for a desired level of precision is very similar to the one just described. The analysis of variance is constructed as follows:

$$\text{whole sample}(s^2)\begin{cases}\text{between main strata}\\\text{within main strata}(s_1^2)\begin{cases}\text{between substrata}\\\text{within substrata}(s_2^2)\end{cases}\end{cases}$$

The required relative precision is obtained by computing the ratio of the mean squares s_1^2 and s_2^2 within main strata and within substrata.

Stratified Sample with Variable Sampling Fraction

The sample sizes and sampling fractions within strata for this case can only be determined uniquely if the relationships among the n_i or f_i are specified a priori. If such relationships are not established, it is possible to estimate values of n_i and f_i by using an appropriate design strategy. It was discussed previously that the sampling fractions, f_i, in each stratum should be chosen proportional to the standard deviations σ_i for maximum accuracy, or proportional to the mean sizes of the size groups if the population is stratified by size. The sampling fractions can then be expressed as

$$f_i = c\lambda_i \tag{5.12}$$

where λ_i = the required proportions. The sample variance estimate can be expressed

$$N\hat{V}(\bar{y}) = \Sigma s_i^2 h_i(1-f_i)/f_i \tag{5.13}$$

where $h_i = N_i/N$ as before
 Substituting equation 5.12 into equation 5.13 yields

$$\frac{1}{c}\overset{i}{\Sigma}(s_i^2 h_i/\lambda_i) = N\hat{V}(\bar{y}) + \overset{i}{\Sigma}s_i^2 h_i \tag{5.14}$$

Rewrite equation 5.14 to isolate c.

$$c = \frac{\overset{i}{\Sigma}(s_i^2 h_i/\lambda_i)}{N\hat{V}(\bar{y}) + \overset{i}{\Sigma}s_i^2 h_i} \tag{5.15}$$

Multiplying c by the sum of the elements in all strata will provide the total number of elements required in the sample to achieve a prespecified level of accuracy in the population estimates. This procedure can lead to estimates of some f_i that are greater than one. If this occurs, these strata must be sampled at a value of f_i equal to one. However, this would fail to achieve the desired level of accuracy. Under these circumstances, c must be reestimated for those strata where f_i is less than one. This will increase c and permit the desired level of accuracy to be attained.

 Using the calculations here, the relative precision and relative efficiency can be found for a stratified sample with a variable sampling fraction against either a simple random sample or a stratified sample with uniform sampling fraction.

Supplementary Information

For the estimation of sample size in this case, the equations developed for sampling errors in this type of technique may be used where the appropriate variance is estimated by the methods used then. In general, $s_{\hat{r}}^2$ is estimated in place of s^2, but the methods are otherwise the same as already described in this chapter.

An issue of interest here is the gain in precision due to the use of supplementary data. In other words, one may wish to know if it is worthwhile to utilize supplementary information. If the regression method is used to calculate population values, the relative precision is dependent only on r, the correlation between y and x.

$$\text{rel. precision} = \frac{1}{1-r^2} \qquad (5.16)$$

This is again an approximate expression,[10] but is usually adequate for the purpose. If an arbitrary regression coefficient, b_0, is used, as discussed in chapter 4, the relative precision is

$$\text{rel. precision} = 1/[1-r^2 + r^2(1-b_0/b)^2] \qquad (5.17)$$

For the ratio method, the relative precision is

$$\text{rel. precision} = 1/[1-r^2 + r^2(1-\hat{r}/b)^2] \qquad (5.18)$$

where $\hat{r}$ = estimated ratio from the past survey.

Other Sampling Methods

Similar procedures can be applied to other sampling methods, such as two-phase sampling, sampling on successive occasions, and cluster sampling. Most of these methods add nothing new to the problems already addressed here. However, the reader will find treatments of other methods in Yates[11] and, for cluster sampling, in Kish.[12]

Cost Efficiency

The previous discussion considered the most efficient sampling size for a desired level of accuracy. It cannot be stated generally that the smallest sample will be the most efficient in terms of cost. If costs are to be minimized, they are a function of not only the sample size but also the sampling method. To calculate the exact total costs of a survey is tedious, difficult, and not usually necessary. However, estimates based on previous experience or on a pilot survey are best.

Supplementary Information

The potential use of supplementary information represents one of the more interesting issues for computing cost efficiency.[13] A number of costs are considered here:

c_s = cost per unit of obtaining supplementary information.
c_0 = marginal cost per unit if no supplementary information is collected.
C_{ac} = additional computation cost of using the supplementary information.
n_s = number of units with supplementary information.
n_0 = number of units without supplementary information.

The total cost of the sample of n_s units is

$$C_s = C_{ac} + n_s(c_0 + c_s) \qquad (5.19)$$

For n_0 units with no supplementary information, the cost is

$$C_0 = n_0 c_0 \qquad (5.20)$$

If the error variance is inversely proportional to the number of units in the sample, the two samples will be equally accurate when the number of units is in inverse ratio to the relative precision of the two methods. If a random sample is under consideration in each case and the regression method is used, where ρ equals the true correlation coefficient between the main variate and supplementary variate, equation 5.21 represents the variance for a random sample with no supplementary information.

$$\hat{V}(\hat{\bar{y}})_t = s_t^2 \frac{(1-f)}{n_0} \qquad (5.21)$$

The variance s_t^2 in equation 5.21 is defined as

$$s_t^2 = \frac{1}{(N-1)} \sum^{N}(y - \bar{y})^2 \qquad (5.22)$$

The corresponding expressions for a random sample with supplementary information (if N is large) are

$$\hat{V}(\hat{\bar{y}})_t = s_t^2 \frac{(1-f)}{n_s} \qquad (5.23)$$

$$s_l^2 = \frac{1}{N-1} \overset{N}{\Sigma}(y-\hat{y})^2 \tag{5.24}$$

$$s_l^2 = \frac{1}{N-1} \overset{N}{\Sigma}(y-\bar{y})^2 - \left[\frac{\{\overset{N}{\Sigma}(x-\bar{x})(y-\bar{y})\}^2}{\overset{N}{\Sigma}(x-\bar{x})^2} \right] \tag{5.25}$$

The true correlation coefficient ρ is

$$\rho = \frac{\overset{N}{\Sigma}(x-\bar{x})(y-\bar{y})}{\sqrt{[\overset{N}{\Sigma}(x-\bar{x})^2 \overset{N}{\Sigma}(y-\bar{y})^2]}} \tag{5.26}$$

Substituting the true correlation coefficient into equation 5.25 yields

$$s_l^2 = \frac{1}{N-1} \overset{N}{\Sigma}(y-\bar{y})^2 \left[1 - \frac{\{\overset{N}{\Sigma}(x-\bar{x})(y-\bar{y})\}^2}{\overset{N}{\Sigma}(x-\bar{x})^2 \overset{N}{\Sigma}(y-\bar{y})^2} \right] \tag{5.27}$$

By simplifying, equation 5.28 is derived.

$$s_l^2 = \frac{1}{N-1} \overset{N}{\Sigma}(y-\bar{y})^2[1-\rho^2] \tag{5.28}$$

Under the condition that the variances for a random sample with or without supplementary information are the same,

$$\hat{V}(\bar{y})_q = \hat{V}(\bar{y})_l \tag{5.29}$$

Substituting equations 5.28 and 5.22 into equations 5.23 and 5.21, respectively, permits the rewriting of equation 5.29.

$$\frac{(1-f)}{N-1} \frac{\overset{N}{\Sigma}(y-\bar{y})^2}{n_0} = \frac{(1-f)}{N-1} \frac{\overset{N}{\Sigma}(y-\bar{y})^2}{n_s}(1-\rho^2) \tag{5.30}$$

Equation 5.30 can be simplified to obtain

$$\frac{n_s}{n_0} = 1 - \rho^2 \tag{5.31}$$

Therefore, the use of supplementary information is more efficient if inequality 5.32 holds.

$$n_0 c_0 > C_{ac} + n_s(c_0+c_s) \tag{5.32}$$

Substituting for n_s, as defined in equation 5.31, provides

$$n_0 c_0 > C_{ac} + n_0(1-\rho^2)(c_0+c_s) \tag{5.33}$$

Reordering terms in equation 5.33 gives rise to inequality 5.34. Again, if this inequality holds, the use of supplementary information is more efficient from a cost standpoint.

$$n_0(c_0+c_s)\rho^2 > C_{ac} - n_0 c_s \tag{5.34}$$

If the cost of adjustment, C_{ac}, is negligible, equation 5.34 becomes equation 5.35, by dividing throughout by n_0 and rearranging.

$$\frac{c_s}{c_0} < \frac{\rho^2}{(1-\rho^2)} \tag{5.35}$$

Thus, under these conditions, the cost ratio is independent of n_0 and therefore of the accuracy required. If ρ equals 1/2, then c_s/c_0 equals 1/3 and the ratio of the total costs is

$$\frac{n_0 c_0}{n_s(c_0+c_s)} = \frac{c_0}{c_0 + c_s} = \frac{c_0}{\frac{4}{3}c_0} = \frac{3}{4} \tag{5.36}$$

Thus the ratio of the costs of not using supplementary information to those of using it is $3:4$ in this case. Suppose that c_0 is \$21 and c_s is \$7; ρ^2 is found to be 0.65 and the cost of using the supplementary information is \$2500. Based on required accuracy, a sample size without supplementary information is determined as 300 units. Using equation 5.34, one can determine if it is cost-efficient to use the supplementary information. The left side of the inequality is

$$n_0 (c_0+c_s)\rho^2 = 300(21+7)(0.65)^2 \tag{5.37}$$

This has the value 3549; the right side of the inequality is

$$C_{ac} - n_0 c_s = 2500 - 300 \times 7 = 400 \tag{5.38}$$

Hence the inequality holds and it is worthwhile to use the supplementary data.

Cost Minimization

If the sampling fractions are not fully determined by the required accuracy, one may determine them by cost minimization subject to the required accuracy. In this procedure costs have only been compared, not minimized. Consider a generalized case of a stratified sample with $n_1, n_2, \ldots, n_t$ units in each stratum. The total cost can be expressed as a linear function of these numbers, and a multiple K of this function can be added to the expression for the variance of the required estimate. The first differential of this with respect to each of $n_1, n_2, \ldots, n_t$ equated to zero will give the conditions for minimum cost. The following are examples of this procedure.

Variable Sampling Fraction. Denoting the marginal cost of taking an additional unit in the ith stratum as c_i, the total cost across all strata can be expressed.

$$C = \overset{i}{\Sigma}c_i n_i \tag{5.39}$$

The variance of the estimate is

$$V(\hat{Y}) = \overset{i}{\Sigma}\sigma_i^2(1-f_i)N_i^2/n_i \tag{5.40}$$

If a multiple K of the linear function of cost is added to this, the variance of the required estimate is

$$V(\hat{Y}) = \overset{i}{\Sigma}\sigma_i^2(1-f_i)N_i^2/n_i + K(\overset{i}{\Sigma}c_in_i-C) \tag{5.41}$$

Differentiating with respect to $n_1, n_2, \ldots, n_t$, the t equations of equation 5.42 are obtained.

$$-\sigma_i^2N_i^2/n_i^2 + Kc_i = 0 \qquad i=1,2,\ldots,t \tag{5.42}$$

Given that equation 5.42 holds, the relationships between sampling fractions are

$$n_i/N_i = f_i \tag{5.43}$$

$$\frac{f_1}{\sigma_1/\sqrt{c_i}} = \frac{f_2}{\sigma_2/\sqrt{c_2}} = \cdots = \frac{1}{\sqrt{K}} \tag{5.44}$$

To determine the actual values of f_i, one can proceed as follows. Rewriting equation 5.40 and using the relationship of equation 5.43 results in

$$V(\hat{Y}) = \overset{i}{\Sigma}\sigma_i^2N_i^2/n_i = \overset{i}{\Sigma}\sigma_i^2 f_i N_i^2/n_i \tag{5.45}$$

The optimal fractions, f_i, can be defined as shown in equation 5.46, according to the preceding discussion.

$$f_i = \frac{\sigma_i}{\sqrt{c_i}\ \sqrt{K}} \tag{5.46}$$

By substituting equations 5.43 and 5.46 into equation 5.45, a new equation for the variance of the required estimate is obtained.

$$V(\hat{Y}) = \overset{i}{\Sigma}\sigma_i^2\ \frac{N_i\sqrt{c_i}\ \sqrt{K}}{\sigma_i} - \overset{i}{\Sigma}\sigma_i^2N_i \tag{5.47}$$

The actual values of the sampling fractions necessary for a specified accuracy can then be obtained by solving for K. A trial solution may be required if any of the f_i equal one, since the terms for such a stratum must be omitted from the entire variance expression.

Two-phase sampling. For this example a fresh notation is used:

c_1 = cost per unit of first-phase information.
n_1 = number of first-phase units.
c_2 = cost per unit of second-phase information.
n_2 = number of second-phase units.

The total cost C is

$$C = n_1 c_1 + n_2 c_2 \tag{5.48}$$

In two-phase sampling, the variance is expressed as

$$V(\hat{\bar{y}}) = \frac{(1-f_1)}{n_1}\sigma_1^2 + \frac{1}{n_2}\left(1-\frac{n_2}{n_1}\right)\sigma_2^2 \tag{5.49}$$

By the same method as before, the variance can be rewritten

$$V(\hat{\bar{y}}) = \frac{1-f_1}{n_1}\sigma_1^2 + \frac{1}{n_2}\left(1-\frac{n_2}{n_1}\right)\sigma_2^2 + K(n_1 c_1 + n_2 c_2 - C) \tag{5.50}$$

Differentiating with respect to n_1 and n_2 and equating to zero results in

$$\frac{n_2^2}{n_1^2} = \frac{c_1}{c_2}\frac{\sigma_2^2}{\sigma_1^2 - \sigma_2^2} = \frac{c_1}{c_2}\frac{\eta^2}{1-\eta^2} \tag{5.51}$$

where $\eta = \sigma_2/\sigma_1$. The values of n_1 and n_2 can then be determined for a given accuracy by substituting for one in terms of the other in the expression for $V(\hat{\bar{y}})$.

Two-stage Sampling. Let c' be the cost per first-stage unit and c'' the additional cost per second-stage unit; C is then defined as

$$C = n'c' + nc'' \tag{5.52}$$

The error variance $V(\hat{\bar{y}})$ is

$$V(\hat{\bar{y}}) = U/n' + V/n + \text{constant} \tag{5.53}$$

where $U = \sigma_0'^2 - \sigma''^2/N$, and $V = \sigma''^2$.

It is then found that this relationship holds.

$$\frac{n^2}{n'^2} = n''^2 = \frac{Vc'}{Uc''} \tag{5.54}$$

Additional examples could be drawn from any of the other methods of sampling that have been considered. The preceding treatment should be sufficient, though, to indicate the method for minimizing cost. Further details may be found in other texts.[14]

Notes

1. For example, the concept does not appear at all in Cochran (*Sampling Techniques*) and is only obliquely treated by Kish (*Survey Sampling*) in his chapter on "The Economic Design of Surveys." Yates (*Sampling Methods*), however, devotes chapter 8 to efficiency.

2. Frank S. Yates, *Sampling Methods for Censuses and Surveys,* 3d ed., London: Charles Griffin and Co., 1971, chapter 8.

3. Yates, *Sampling Methods,* p. 247.

4. Leslie Kish, *Survey Sampling,* New York: John Wiley & Sons, 1965, p. 248.

5. Kish, *Survey Sampling,* p. 258.

6. Ibid, pp. 258–259.

7. Ibid.

8. Yates, *Sampling Methods,* p. 248.

9. Ibid, p. 250.

10. Ibid, p. 257.

11. Ibid, pp. 258–283.

12. Kish, *Survey Sampling,* pp. 254–255.

13. Yates, *Sampling Methods,* p. 284.

14. Kish, *Survey Sampling,* pp. 263–272, 279–282.

6

Survey Procedures

Purpose

The purpose of this chapter is to discuss the ways in which survey design and sampling interact. In particular, the decision process is examined that should precede the choice of a sampling technique and will tend to act as a major restraint on this choice. Problems associated with imperfect frames and nonresponse are also considered. The overriding concern here is with surveys of human populations. The sampling and surveying of nonhuman populations, such as road vehicles, is usually much simpler, although much of the material covered in definitions of survey aims and the design of the survey are partially applicable.

Seven broad categories of problems are considered:

1. design and use of pilot surveys;
2. definition of the population for the survey;
3. definition of the information to be collected;
4. choices of methods of collecting information;
5. choice of sampling frame and associated problems;
6. choice of sampling unit and sampling method, including problems of repeated surveys; and
7. problems of nonresponse.

Although each is considered, the problems are certainly not unrelated. However, an attempt is made to array them in the order in which one normally expects to meet them in a sequential design process. One exception to this occurs with item 1. Clearly, decisions on the survey population and the information to be collected have to be decided on for the pilot survey and, similarly, decisions have to be made about how to gather this information, how to sample, and what sampling unit to use before the pilot survey can be undertaken, even though these decisions may differ from those applied to the final survey.

Design and Use of Pilot Surveys

Pilot surveys are of use for four basic reasons. The importance of the reasons will tend to vary with the type of survey and the amount of knowledge previously gained about the population and the characteristics for the study. The four basic objectives of pilot surveys are

1. to provide the means of testing the survey forms and the sampling techniques, and for training the investigators;
2. to provide some outline information on variability of the population within the study domain;
3. to provide a basis for estimating the costs and times required for the entire survey; and
4. to determine the most effective type and size of sampling unit.

In studies of human population, it is often essential to carry out a pilot survey to test a questionnaire. Individuals tend to read ambiguities into quite straightforward-looking questions and to have excessive overreactions to certain types of questions which they consider to be prying into their private affairs. This apart, levels of nonresponse in the pilot survey, difficulties of interpretation of answers to questions, and the feasibility of the intended field procedure can all be determined effectively by a pilot study. Where investigators or interviewers are being used, the pilot study affords an opportunity both to initiate these people into the methods of carrying out the survey and to set up a consistent procedure for dealing with each of the various queries that inevitably come back from interviewers.

Second, population variances contribute significantly to sampling errors, while correct choice of the sampling technique can reduce the sampling errors considerably. Clearly, an outline knowledge of the population variances will contribute toward correct stratification and minimization of sampling error. In addition, where the survey form is set up with categories for people to identify (e.g., income groupings) and the survey forms are to be coded for computer or other systematic analysis, a pilot study which gives indications of population variances will help to ensure that sufficient categories and coding spaces are provided.

The next output of a pilot survey is a guide to the costs and times which will need to be budgeted for the entire survey. The duration of interviews, the time taken for interviewers to travel from one interview to the next, and the response time for self-administered surveys are important aspects of a survey which can rarely be determined accurately beforehand.

Finally, the pilot survey should help to identify the best way of aggregating or disaggregating the population units into sampling units. It is necessary, for this purpose, to carry out the pilot survey on the greatest aggregate of population units which might be considered for sampling units, and to collect information in a manner that will permit complete disaggregation.

These are the basic uses of pilot surveys. More important is the issue of how to set up such a survey. The first point is that the pilot survey should be carried out on individuals or households (if these are the subjects of the survey) who will *not* be included again in the major survey. Most people object very strongly to being asked to respond to a similar survey twice in a very short time, and the public image of the responsible institution or person and the general response level to the survey will both suffer considerably as a result.

The next problem is how to sample for the pilot survey. The sampling technique should, as far as practicable, be identical to the sampling technique to be used for the entire survey. On the basis of the output that was detailed, two opposing things are required for the pilot-study sample: it has to be representative of the total population, and it has to be concentrated so that some reasonable estimates can be made of the population variances. At the same time, the sample size has to be very small. Two basic ways of achieving this are: if the main sample is to be stratified, to concentrate the pilot study in one or two strata only where the biggest variability is expected; alternatively, a form of multistage sampling should allow concentration of the pilot-study sample into the areas of greatest importance. The analyst must guard against a highly concentrated sample in a completely unrepresentative section of the population. Further guidelines on sampling for pilot studies are presented in Yates.[1]

The final point is that the use of a pilot study requires clear economic, practical, and procedural justification. A small-scale survey can rarely be improved in efficiency economically by the use of a pilot study. It is often better to proceed with a nonoptimal survey design since this will still produce sufficiently good results which could only be improved by inordinately large further expenses.

Definition of Survey Population

Usually, the survey purpose will define the survey population broadly in terms of geographical extent. The major concern in this specific item is the decision as to whether all units of the population of the geographical area should be included as the survey population or whether obviously minority or marginal sections of the population can be omitted from it. For instance, should a household survey attempt to include itinerants of no fixed abode? Or should they be omitted from the definition of the population? Similarly, in a survey of employees of a firm, should onc attempt to include in the survey population those who are sick or on vacation at the time of the survey?

Clearly, most of such marginal categories will add considerable complications to the survey technique if they are to be included. That does not mean that they should be excluded just for the sake of simplicity, but rather that their inclusion or exclusion should be carefully weighed against costs and survey intentions. If these marginal categories of the population really are important to the survey purpose and to the completeness of the analysis, they should be included. At the same time, the errors due to their omission should also be assessed. If these errors are smaller than or comparable to the general sampling errors, even claims of completeness or compatibility with other data sources are of no consequence.

If these marginal groups are to be included, considerable gains can be made by isolating them in separate strata and using whatever sampling technique is most appropriate. This obviates any serious impact of their inclusion on the rest of the sampling process. Frequently, the sampling frame will predefine the population. For instance, a list of households prepared by the local-property-taxing administration will not include any means of identifying the itinerant population. In these cases, if the marginal groups are to be included, the frame will have to be extended by some means to include them.

Definition of Information to Be Collected

This phase of the survey design is one of the most important and most neglected. Far too frequently, surveys are undertaken with the philosophy: "let's ask all we can, and then see what we can get out at the end." As found so often before, there are conflicts in the desirable aims of the amount of information to be collected, and a working compromise must be reached. So one should aim both at economy in the amount of information sought, commensurate with the survey purposes, and at comprehensive coverage of relevant population characteristics.

As long as the concern is with samples, two information requirements must be defined: sufficient information must be collected to define adequately the survey population, and sufficient information must be obtained to answer questions which the survey is designed to answer. To fulfill both these requirements, the following approach can be set up.

First, determine the population measurements that are necessary to define the population used for the survey (e.g., distributions of income and car ownership, of employment, etc.). These measurements should be determined as the minimum necessary for the degree of definition of the population that is essential for the purpose of the study.

After this, all the details of the information required for the specific problem that is being studied should be listed comprehensively. On the basis of this listing, any related problems that could also be tackled should be considered in terms of any additional data requirements, which can then be added to the basic list. At this stage, it is important also to identify the expected analytical procedures to ensure that the data collected are complete and defined in terms appropriate for the planned analysis. Again, this reinforces the need to develop appropriate hypotheses before beginning any of the data-collection and analysis activity. The list should then be assessed in terms of economy and practicability of collecting the information and the necessity of including each item. With this assessment, it should be possible to make final decisions as to what should be included. The final selection of information should be such that the information to be collected forms a complete entity.

A number of cautions are advisable when the survey is of human populations. First, it is necessary to try to ensure that the information to be collected appears to be relevant to the survey purpose. This is necessary for two reasons. First, the subjects of the survey will react adversely to being asked to provide information they feel is irrelevant to the survey purpose. Also, where interviewers are being used, it should be clear to the interviewer that all the information is necessary and why. A good basic rule is, if the interviewers cannot be convinced of the relevance of any of the information, one should seriously question its inclusion.

Second, questions on income, marital status, and similar personal matters frequently are necessary items, but also evoke severely adverse reactions. However, if these questions are placed at the conclusion of a questionnaire or interview, a considerably better response is usually obtained. In the first place, if the respondent refuses to answer these questions, the refusal will not occur until all the rest of the questionnaire or interview has been answered. Thus some valid, though somewhat incomplete, data can be obtained from the respondent. Second, by the time the respondent has answered all the previous questions, he or she is somewhat conditioned to answering, and is far less likely to react adversely than if asked personal questions first.

A further caution is to be sure that the questions being asked will provide the required answers. For instance, if it is intended to find out peoples' preference for a particular make and style of auto and their reasons for it, answers to these questions alone will not be indicative of real preference and reason. They will tend to be much more a reflection of current advertising and sales techniques. To really determine the preference and reason, probes into past experience in terms of other makes and styles of autos previously owned will be a much surer base on which to build statements about preferences.

One final problem is that of the time period for which respondents are asked to provide information. There are no rigid rules on this and arguments can be made for several different approaches. The time period should be regarded in part as a function of the process that is the subject of the survey. The more significant or rare the process, the longer is the time period that may be used. The reader has almost certainly, for example, responded to a survey that asks for details of every hospital stay of the respondent in his or her life. A survey may also successfully ask for details of every car owned by each respondent over the last 10 or 20 years. In the United States, there may be problems in asking such a question over a respondent's entire life, since new car purchases may have been made every two or three years, and older respondents may no longer recall several past purchases. On the other hand, a survey seeking to know what respondents ate for breakfast is likely to encounter serious recall problems if it attempts to go back more than a few days. (Some might even argue that such a survey would be in trouble if it tried to go back further than the day of the survey.) The authors therefore recommend that a general rule should be to match the time period with the frequency of occurrence of the phenomenon and its relative significance in a person's life.

This still leaves a lot of room for discussion, however. In travel surveys, for example, it is often argued that questions should be asked only about the most recent 24-hour period, since details of many trips are forgotten quite rapidly and even entire trips may be forgotten within a day or two. Such an approach, however, runs the risk of suppressing variations in the behavior of individuals and of being subject to errors of telescoping. The latter is likely to occur with a particularly memorable trip (usually a very good or a very bad one), which may have stuck in a person's mind to the extent that it is reported for the immediate past 24 hours, when in fact it occurred long before that. Both of these problems become increasingly serious when the study is concerned with individuals rather than groups.

To obtain detailed information on occurrences over a period of time will necessitate either extremely careful survey design and extensive tight control of interviewing or require repeated surveys over a period of time. Either alternative is often highly disadvantageous and likely to be very expensive.

Another alternative is the use of survey diaries in which respondents are asked to keep a record over a period of time of certain behavior or phenomena. This approach suffers from three potential problems. First, the survey may cause changes in a person's behavior patterns by making him or her focus attention on issues and processes normally handled much more superficially. Second, experience shows that many respondents tire of keeping the diary after a relatively short time, so diaries for longer than one or two weeks in most cases will prove unreliable. Third, the survey diary is an expensive process and requires a high degree of cooperation, leading to severe problems of sample control and representativeness. As previously concluded, the ultimate decision on this question must rest with the survey designer in each particular case, since no hard and fast rules can be made.

Choice of Methods of Collecting Information

Frequently, both the type of population to be surveyed and the type of information required will dictate the method of collection of information. Where a choice exists, a set of rules on types of methods can be used which will normally be applicable. This set of rules could be phrased: observations are preferable to questions; questions on facts and past occurrences are preferable to questions on opinions and future actions.

The basic methods of collection of information about human populations are observation, interview, and self-administered questionnaire. On the basis of the rule on questions, the order in which these three methods were given also represents the order of preference for using them. Little is said here about objective observation, since this is relatively simple.

Where interviewers are to be used, it is essential that they are thoroughly trained in the task and are fully aware of the survey purposes. Careful phrasing of the survey form is necessary to avoid any misinterpretation on their part. It is frequently desirable for the interviewer to be permitted a large amount of flexibility in asking questions, particularly where the desired information is likely to be divulged reluctantly or where opinions are being sought. Nevertheless, it is important also that interviewers obtain information on the identical attributes in all cases.

One aspect of the layout of survey forms was discussed when dealing with asking for personal information. The observations made on this are equally applicable to both interview and self-administered surveys. A further point here concerns the income question. A better response will always be obtained by asking people, whether by interview or questionnaire, to identify themselves as being in a particular income group rather than to state their actual income.

The sequence of questions in both an interview and a questionnaire is important. When asking people about journeys they have made, they should be asked questions which refer to the chronological sequence in which the journey was made. In all cases, questions should be ordered so they effectively pursue a particular train of thought or association of ideas until all the information required on that aspect has been requested.

Substantial gains can be made by using interviewers, as opposed to self-administered surveys, in gaining the confidence of the respondent and in obtaining the specific answers required. When a questionnaire is employed, considerable attention must be paid to gaining the confidence of the respondents, making them feel that their contribution is important, and generally inflating their feeling of self worth. It is extremely important to preface the questionnaire with a brief, nontechnical summary of the purpose for which the survey is being carried out. This summary should take pains to emphasize the importance of each individual's response. Whenever possible, respondents should be allowed complete anonymity. In all cases, complete confidentiality of the survey responses (not the findings) should be promised and given. This does not mean that the data cannot be publicized if done in such a form that no connection can be made between the data and the actual respondents.

It is frequently advantageous to have the introductory summary of survey purposes written as a letter from an important noncontroversial and respected public personage, whose signature appears beneath it. In surveys on places of employment, an even better method of obtaining good returns is to use either the employer's signature or that of an influential union or other person. The use of an established letterhead or the creation of a special one for the survey can contribute greatly to a high-response rate since such a letterhead projects the professional and important nature of the survey.

Each respondent on a self-administered survey should feel his or her contribution is of equal importance to that of anyone else. Thus, even though one may only want information from car owners and may use an initial set of questions to determine these, it is politic to include a section for non-car-owners to anwer which will make them feel of equal importance to the car owners.

In self-administered surveys, simplicity of layout and directness of questions are essential to obtain accurate responses. Long, complex questions and involved conditional questions tend to annoy respondents and seriously to deter a return. Each respondent should also be given the space and opportunity on the questionnaire to express his or her feelings about both the survey and its purpose, and on certain specific aspects of the questions. Space for such observations will tend to encourage response and if guidance on the observations that might be made is given, some useful material may result. Apart from anything else, these observations will often indicate faults in the design of the questionnaire or the survey and will be beneficial in planning future similar surveys.

Choice of Sampling Frame

Sampling frames available may largely define the population and the sampling method. As a result, detailed planning of the survey must usually await an identification of the available sampling frames. If no frames exist, the entire survey becomes a nonsample survey designed both to collect the information for which the survey was originally intended and to set up a sampling frame.

Frames suffer from a number of defects which can be listed as (1) inaccuracy, (2) incompleteness, (3) duplication, (4) inadequacy, and (5) being out of date.

A frame is *inaccurate* if it lists information which is not strictly correct. For instance, the list may describe a particular house as comprising three apartments, where in fact there are two. A list of individuals may contain names of people now dead or, like a telephone directory, enter wrong names against certain addresses.

A frame is *incomplete* when units or individuals are missing altogether. Thus a list of employees which does not include temporary workers is incomplete if the list is to be used as a frame for sampling all employees.

A frame is subject to *duplication* when any units appear more than once. For instance, a list of faculty at a university might include people in a specific academic department and also in a separate list of members in an interdisciplinary research center.

A frame is *inadequate* if it does not include parts of the population which it is intended to survey. The above example under incompleteness is also an example of inadequacy.

Although a frame may be adequate, complete, accurate, and free of duplication when it was constructed, it may be *out of date*. If it is out of date, any or all of the previous faults may exist; being out of date is a common cause of all these defects.

Inaccuracy and inadequacy of the frame are each likely to be apparent before the survey is undertaken or will appear during the conduct of the survey and can be corrected. However, incompleteness and duplication are far less apparent and will frequently lead to noncoverage or overcoverage. This is particularly true since incompleteness tends to apply to specific minority categories of the population. Of the two, duplication is more likely to be found in the course of the survey, while incompleteness may not be detected at all.

Each of these defects is also relative to the survey purpose. A list of households giving the name of the occupants may be inaccurate because some of the names are recorded incorrectly. But if one is only interested in a survey of households, such an inaccuracy is irrelevant. On the other hand, a list of credit card holders which denotes some women incorrectly as married or unmarried would be inaccurate for a survey of all credit card holders. But for a survey of married women holding credit cards, it would be incomplete.

For sampling various populations, a number of different frames are available, all of which are subject to some degree of inaccuracy. In the United States, most towns and cities over 10,000 population have city directories, which are recompiled every one to two years on average. These directories suffer from all of the four basic defects because they tend to be out of date. However, if the analyst is interested principally in a frame for sampling households, the errors are not overly great. New construction and subdivision of buildings generally occur slowly, and the lists are not likely to be very incomplete or inaccurate because of these occurrences. However, they tend to suffer from omissions and indefinite boundaries, so care must be exercised in using them.

Lists compiled by local utilities are usually more up to date and less likely to suffer from omissions. However, they will be deficient in ghettos and slums sometimes, where supply of utilities is not 100%. Also, they are of use primarily in determining street addresses, since policies of charging for utilities vary in rented accommodations.

Another useful address frame is the local area map or town plan. However, the process of renewing, revising, and updating such maps and plans is an extended one, often taking four to five years. Consequently, great care is needed with such sampling frames to repair omissions and to update them.

Lists of individuals cannot usually serve as lists of households, unless the individuals are grouped by household. The most common lists of individuals are city directories or voting lists. Whether or not the latter would be made available to a survey designer is uncertain and would probably depend on the survey purpose and sponsor. In both cases, the lists are likely to be out of date by an average of one year. Apart from this, the existence of various lists will depend on the survey purpose and locality.

Another frame is the population census. Again, its availability for use by a particular survey designer is likely to depend on the survey purpose and sponsor. Also, census details suffer severely from time lapse. Censuses are usually carried out every decade, and the sampling frame of the census is unlikely to be available to any outside organization for at least a year after its compilation.

Yates[2] gives a more detailed discussion of the use of these various frames, although several of the listings he mentions are apparently not paralleled in the United States.

Choice of Sampling Unit and Sampling Method

To a very large extent, the result of the previous decisions will determine the sampling unit and sampling method. Assuming there is a choice of sampling unit, it is advisable to aim toward using the most disaggregate of the possible units. It is always possible at a later stage in the analysis to aggregate units. But it is not normally possible to disaggregate any further than the original sampling units, unless the sample size was grossly oversized and even then will depend on what data were collected and how.

To determine the sampling method, some previous knowledge is needed of the population to be surveyed, or the results of a pilot survey are required. The decision on the sampling method will depend upon a number of factors. The primary factor on which the sampling method will depend will be the accuracy required of the final survey results against the budget constraint. As a generalization, at this point, it can be stated that the budget restraint will require the use of a sampling method which will demand the smallest sample size for the required level of accuracy.

Here also, the problem of repeated surveys is considered. At some point in the survey-planning process, a decision must be made on the question of repetition of the survey. Whatever the population to be surveyed, one question has to be resolved: are the repetitions to involve the same sample, or will resampling take place for every repeat of the survey? In a survey of road traffic, one would expect to resurvey each time since it would be an inordinately difficult task to select the same sample on several occasions.

In surveys of human population, there is clearly a different problem. If individuals are surveyed by place of residence or work, or households are surveyed by location, both alternatives are open to the investigator. In a democratic society, despite an ever-increasing number of surveys of human population and an improving acceptance of them, many people object to being subjected to surveys. Fortunately, most people also have relatively short memories about such things as surveys. However, in situations where surveys are to be repeated more frequently than every two years or so, resampling of the total population is likely to give progressively lower responses. This results from the fact that on each resampling some people are sure to be included who were selected in the previous sample. These people will now have a very high intolerance toward surveys, particularly one asking identical questions. In addition, in this highly mobile society a substantial portion of the original population will have moved out of the survey area.

Under these circumstances, the best solution is to set up a panel. This type of repeated survey procedure is particularly useful for measuring changes in the population. If this is the purpose of the panel, a further advantage accrues from its use: a strictly representative sample is not necessary. The sample only needs to be representative as far as the changes to be measured are concerned. Thus, if a panel is to be set up to measure changes in the levels of car ownership of families, one would construct the panel in the strata such that households whose car-ownership levels are close to what might be considered saturation would have minimal representation on the panel (as a check against the assumption of saturation). Another advantage of the panel is that it can be set up in such a way that those who are selected are made aware at the outset that they will be asked repeatedly to respond to a survey on specific matters. At the same time, it becomes necessary only to ask for details of changes since the last survey was carried out.

A special case of repeated surveying is the before-and-after survey. In this type of survey, one will usually require a truly representative sample of the population. It will also be desirable, if not essential, to survey the same sample in each of the two surveys. On this assumption, one would sample initially as though the survey was a once only type of survey. It would then be necessary to carry out extensive publicity to gain acceptance by the selected sample of both the initial survey and the follow-up survey.

Both for resampling in certain time periods and for the use of survey panels, a serious problem exists because this society has a very high residential mobility rate. This means that the sample composition and size would probably have changes in successive surveys.

Nonresponse

The problem of nonresponse is basically that of being unable to obtain a completed answer from a member of the chosen sample. Since adherence to specific sampling methods and the calculations that go with them demand a strict sampling selection, failure to contact members of the selected sample pose special problems. Since nonresponse is a problem associated with surveys of human populations only, attention is focused on two specific types of survey: self-administered and interview.

In self-administered surveys, the biggest problem is usually nonresponse. In most reports dealing with past experience on this type of survey, average nonresponse is frequently quoted as being of the order of 70%, in cases where the questions are concerned with people's behavior and their socioeconomic characteristics. Careful survey design and correct choice of survey method can often reduce this nonresponse level to 30%.[3] However, when 30% of a sample fails to respond—particularly when the failure is a reflection of a lack of interest in or opposition to the survey objectives, or objection to being surveyed—a considerable bias will be introduced.

There are two possible methods for dealing with nonresponse, depending to some extent on the nature of the sample and the degree of identification of members of the sample. Where the sample comprises all employees of a particular firm or in a particular building, but where the sample is not identified both in the sampling frame and on returned questionnaires, a blanket approach is necessary. This approach can comprise simply a letter sent to all respondents and nonrespondents, thanking respondents for their help and reemphasizing the necessity to the survey aims of response from those who have not yet returned questionnaires. This approach has led to the capture of an additional 50% of the first response rate in certain instances,[4] but it is unlikely to obtain anything from those who are violently opposed to survey inquiries. Alternatively, the original sample can be used to serve as a population for a new survey. In this instance, a sample is chosen from the original sample. The new subsample should contain both respondents and nonrespondents. Interviewers are used on the subsample to do two things: establish who are respondents and who are nonrespondents; and then to elicit from both respondents and nonrespondents details of their reactions to the questionnaire. In the case of nonrespondents, skilled interviewers should obtain answers on most of the questionnaire queries. These answers can be used to compare nonrespondents with respondents and determine what bias exists in the original return of the questionnaires.

Where respondents and nonrespondents can be identified in a self-administered survey, a follow-up interview survey of nonrespondents is the best method. Each nonrespondent is encouraged to give details on as many of the survey questions as possible and to express his or her views on the survey and its purpose. This type of follow-up requires diplomatic interviewers, with instructions not to antagonize the interviewee.

In household interviewing, nonresponse takes one of two forms: noncontact or open refusal to cooperate. In the case of noncontact, the interviewer should try to establish contact by callbacks on several occasions, preferably at different times of the day and early evening. Callbacks are an expensive way of using trained interviewers, so it is usually necessary to limit them to a certain number. A considerably more economic way to solve problems of noncontact is the use of prepaid appointment cards. With this method, the interviewer makes one callback. If unable to make a contact, the interviewer leaves a card that explains the survey and informs the person that he or she is a member of the selected sample, and requests an appointment for a specified time and day. The reverse of the card has a prepaid address format.

When this tactic fails, as in the case of refusals, little can be done to obtain responses from these people. Now the question of makeup sampling arises. That is, a decision must be made whether to accept the level of noncontact and use a smaller sample than originally intended or to continue the sampling process until the desired sample size is achieved. In either case, an indeterminate bias will exist in the results.

Two observations can be made here. To avoid additional sampling, it is not an uncommon practice to select a sample that is large enough to allow for a certain percentage of refusals. This percentage can usually be obtained on the basis of previous experience with a particular type of survey. However, this does not help to reduce the bias that refusals introduce. The second observation is that if a large refusal rate occurs with an interview survey, the survey design, its purpose, or the interviewers are basically at fault.

Field Control and Testing

This section deals with some of the problems and procedures that are likely to be encountered in the practical execution of a survey. An individual's involvement or position in the organization of the survey will determine the level of concern that he or she might have with any of these items: (1) administration of the survey, (2) design of forms, (3) interviewers, (4) field control, and (5) coding and basic tabulation. Although five areas of concern have been identified, these are not isolated items. In each case, there will be considerable overlap from one topic to the next.

Survey Administration

Regardless of the size of the survey, some degree of administration is essential. In both small-scale and large-scale surveys, the administration will select and train interviewers (if they are being used), design the forms, supervise fieldwork, and supervise checking, coding, and analysis of survey responses. A task frequently encountered is that of answering irate or worried householders and, sometimes, fending off the press. When the survey is a household one, it is not unusual for all manner of rumors to start about the reasons for the survey. In the before-study on the Victoria Line in London, much of the initial survey work was in an area of urban renewal. In other words, the local councils were compulsorily purchasing slum property for demolition and replacing it with modern high-rise blocks. Considerable problems were encountered both with people still living in slum areas and with those who had moved into high-rise blocks. Those who were still in slums feared eviction and worried that interviewers were prying and preparing their eviction. In the new high-rise blocks, the interviewers were again mistaken for local council officials, who were checking on the condition of their apartments.

People often have to be assured that the interviewers are bona fide. For this assurance, the administration must have a complete list of the names of the interviewers and the addresses they will visit each day. Undesirable elements of society sometimes take advantage of the survey in progress to gain illegal entry to property, masquerading as interviewers. It is therefore advisable, and possibly essential, to inform the police that a household survey is to be carried out, the number of interviewers concerned, and the day-to-day locations of the households. Local councils should be informed as well as any other organization that people may contact to check the bona fides of the survey. Finally, considerable advantage will accrue if adequate advance publicity in all the mass media is given to the survey and its purpose.

Apart from these public-relations tasks of the administration, it must undertake the entire organization of the survey. On large-scale surveys, it is often advisable to seek the services of an organization that specializes in trained interviewers and supportive clerical staff for surveys. Such services can make considerable reductions on the time and expense necessary in setting up the survey.

Design of Forms

Whether the survey is an interview or a self-administered survey, careful design of the survey form is essential. The major aims of form and presentation should be toward the greatest simplicity, clarity, and ease of use.

In interview surveys, the survey forms are generally used by the interviewer only. The survey form must be convenient to use, both in size and layout. The interviewer should be able to fill in answers with a minimum of writing, since voluminous answers generate errors and misinterpretations and the interviewees tend to become uncooperative if they see the interviewer making copious notes about their answers. Detailed instructions on the forms should not be necessary, but interviewers should be given exhaustive written instructions on the conduct of interviews and completion of forms. Extensive briefing sessions should be held before the interviewers are sent out, in which every possible complication, ambiguity, and contingency can be dealt with. In the briefing session, interviewers should interview each other to find out both how it feels to be interviewed and to gain some experience in interviewing. This process often helps to uncover problems and issues relating to the survey. Forms and interviewing instructions should be flexible enough initially to compensate for possible omissions or errors.

If the survey is one of observation rather than interview, some further points are required. If the observations are to be carried out in the open, the forms must be so designed that weather will not hamper their use. Concomitantly, it is necessary to use a pen or pencil that is impervious to all weather conditions and that will not run or smudge. Again, the amount that the observer has to write should be minimized.

In designing the forms for self-administered surveys, the content, layout, and general presentation of the form are even more important. Whereas a number of these points have been covered previously, a few more can be added here. The amount of effort the respondent has to expend to fill out the form should be minimized. It will be necessary to include detailed instructions on the form, but these must be as simple, short, and concise as possible. In any survey requesting details about people or households, confidentiality of the details must be clearly guaranteed but not overstressed. It is also necessary to strike a note of careful diplomacy in setting up instructions, so that one treats each respondent as less than moderately intelligent (in terms of his or her ability to follow ideas and instructions) but does not let the respondent know of such treatment. Instructions, apart from general ones, should be attached to the questions to which they refer. General instructions should appear at the start of the survey form. The practice used by such people as tax authorities of giving detailed instructions on the back of the form or on a separate sheet is not advisable. This is far more likely to result in the form being filled out without the respondent following instructions, than if the instructions are contained within the form.

The production of forms is important in interview/observation and self-administered surveys. A brief summary of the requirements for interview/observation forms follows: (1) forms should be easy to handle; (2) they should be as compact as possible; (3) they should be backed firmly enough to write on or provided with clipboards; (4) they should be produced on paper which will allow legible recording; (5) they should be laid out clearly and logically; (6) two sides of one sheet are preferable to two sheets; (7) the typing or printing of the forms should be as legible as possible.

The general presentation of the questionnaire-survey form will have a considerable effect on the likelihood of response. The requirements for these forms are: (1) forms should be quite small; postcard forms frequently obtain the best responses; (2) they should be clear, logical, and uncluttered; small survey forms should not be obtained by squeezing vast quantities of print into every square millimeter of the form; (3) they should be printed or duplicated on good-quality paper or card and should have a clean professional look; (4) two sides of one sheet are preferable to two sheets.

A final point on designing survey forms is that whatever the method of analysis to be used, the forms should be set up to allow rapid easy coding into a format suitable for the analysis. If punched-card techniques are to be used, it is useful to provide coding boxes, with card column numbers above them, in a separate block beside the questions. Where respondents are to tick relevant boxes which will be coded as numbers or letters on cards, the printing, below each box, of its code is a considerable asset in quick accurate coding. It also allows for more rapid and correct checking of coded forms.

Even though the use of coding boxes on survey forms is advocated and used by a number of organizations and survey designers, it is a matter of substantial controversy for others. Marketing analysts express particularly vehement opposition to the use of coding boxes on self-administered questionnaires.[5]

Finally, a few miscellaneous items should be discussed which can contribute to the success of the survey. If financially possible, the survey forms should be printed rather than duplicated to contribute to the professional appearance of the survey effort.

To improve the legibility of the survey forms and thereby maintain the respondents' goodwill, attention should be paid to using different type faces, colors, and arrows to indicate hierarchical ordering and sequencing of questions. These considerations are particularly important when conditional questions are involved which force the respondent to skip certain segments of the questionnaire.

Interviewers

The selection and training of field investigators varies considerably with the type and scope of work to be carried out. Where the survey work is an occasional or unique occurrence, it is preferable to obtain field investigators from a firm which specializes in selecting and training people for such jobs. If the survey work is on a continuing basis, it is worthwhile to set up a permanent trained staff of investigators.

The organization of a body of investigators should be such that there is one investigator who is trained and conversant with the type of survey being carried out in charge of each of a number of small groups of investigators. These trained investigators can then be used as supervisors and given a somewhat lighter load of interviews or observations. A team set up in this way has a number of advantages: (1) the supervisors are experienced in the investigation and can usually help in anticipating problems before they arise; (2) the supervisors can be used as a control on less experienced investigators; (3) the supervisors can be used for callbacks to hostile respondents; and (4) the supervisors can act as the basic communication channel between the survey administrators and the field investigators.

Interviewing usually requires working in the evening as well as in the day and can be both physically and mentally exhausting. The quota for each interviewer to complete in a day should therefore not be very great and should be tailored to each investigator's capabilities. It is also advisable to give investigators about one day a week out of the field, preferably working on the coding or analysis of completed forms. This will help to give them a better understanding of the task at hand and of their own contribution. It will also clarify the importance of clear and unambiguous form completion. They should not normally be assigned to work on their own completed forms.

Testing of the interviewers should also be carried out, and this can be done very easily by supervisors. A pilot study is necessary for such testing, in which the supervisors are the control group and which indicates the suitability of the form. Both supervisors and new investigators are given identical forms and the results of interviews are compared. A detailed analysis and comparison of the results should serve to indicate the capabilities of new investigators and any dangers of bias from an investigator. This approach can be extended to testing alternative designs of the survey form, using supervisors as control and as a means of assessing the form alone, and using new investigators to determine their capabilities and differences in reactions to different formats.

Field Control

Field control is only necessary for surveys involving interviews or observations. By setting up teams of investigators under a supervisor, field control is reasonably simple. In-the-field checks can be carried out by the supervisors to check on such items as accuracy of measurement or reporting, reliability, and adherence to the selected sample. Supervisors should carry out such checks on all their teams at frequent intervals. In addition, checks should be made by the administrative staff on both the supervisors and a sample of the work of each team. In all cases, it is not usually practicable or desirable to check every return, so just a sample of the work of each investigator is required.

Checks should be run most frequently at the start of the survey work and then reduced, unless contraindicated. It is often desirable to set the survey up in such a way that only a small amount of the work is carried out in the first day or two, to allow for intensive checking. This will also allow remedial action to be taken before any errors or biases have been perpetrated too far into the survey data.

In interview surveys, points which need particular attention are higher than average nonresponses or refusals to one investigator, much higher or much lower rates of incomplete forms, interviewing of people or households not in the selected sample, and inaccurate or ambiguous form-filling on any question. If an interviewer does not respond quickly to requests to change the manner of operation in respect to such errors, he or she should be dispensed with.

Coding and Basic Tabulation

Before any of the above stages are commenced, one important task has to be carried out as part of the survey design—detailing the basic tabulations that will be produced. It is a serious error of judgment and basically a bad design to set up a survey without a clear plan of the tabulations that will be required. Two basic purposes are served by doing this. In the first place, it is a check on the design of the survey and its various questions. If the desired tabulations could not be produced from the planned survey, it is obvious that some redesigning is necessary. Second, the decision on the required tabulations gives considerable assistance in designing the layout of the form and the coding procedures.

In the discussion on the design of forms, several ideas were mentioned of facilitating rapid and accurate coding by placing codes by the boxes to be checked, and by the use of boxes to represent card columns of computer cards. Of course, this is only one of a number of ways of coding survey forms. Some other available alternatives are:

1. Coding and analyzing on the forms themselves: this is usually only possible when the survey is small and simple.

2. Designing the survey form as a Cope-Chat card: this is a card on which holes can be punched around the edges. Analysis is done by running rods through the cards to sort them and then counting those in each sorted category.

3. Designing the survey forms as computer cards on which coding can be done by direct punching or mark sensing: both this technique and the previous one suffer from the disadvantages of a tendency to damage cards in the survey and of being useful more often in relatively small surveys (both in terms of sample and number of questions).

Coding and punching are major sources of error in survey work. Consequently, it is necessary to use rigid checking devices to ensure accuracy. A useful procedure is as follows.

In the initial coding work, have every coded form checked by a second coder (preferably a supervisor). When sufficient accuracy is being maintained, check a random sample of about 10% of the survey forms. If the work of any coder appears to be inaccurate in this sample, maintain a rigorous check on his or her work. When coded forms are used to give data for punched cards, the cards should be checked by a second operator using a *verifier,* a machine which instead of punching cards, compares what is already punched with the symbol typed on the keyboard. If the two are not the same, a red light comes on and the verifier will not move the offending card to the next column. Alternatively, the computer cards may be stored as a file on the computer and various checks programmed. The files are then corrected by using a modern text editor on the data file. This process is, however, less thorough than the use of a verifier. Although this procedure sounds lengthy, it is usually far less expensive than using poorer checking, with the subsequent necessity of correcting cards or taped data and altering or revising previous analysis and tabulation.

This completes the discussion of the conduct of surveys and initial analysis of data. The material of this and the preceding chapters should be adequate to provide the reader with sufficient knowledge and background to appreciate and generate the data typically required for the statistical analysis procedures discussed in the following chapters. Excellent extensive treatments of the subject of chapters 2–6 is presented in Yates[6] and Kish.[7]

Notes

1. Frank S. Yates, *Sampling Methods for Censuses and Surveys,* 3d ed., London: Charles Griffin and Co., 1971, pp. 99–101.

2. Yates, *Sampling Methods.*

3. Werner Brög and Karl-Heinz Neumann, "The Interviewee as a Human Being," Paper presented to XLIV E.S.O.M.A.R. Seminar on *Ways and New Ways of Data Collection,* Jony-en-Josas, France, November 1977.

4. Brög and Neumann, "The Interviewee."

5. See, for example, Survey Research Center, *Experiments in Interviewing Techniques,* Research Report, U.S.H.E.W. Public Health Service, Health Resources Administration, November 1977; *Journal of Marketing Research* (entire issue), May 1978; and *Journal of Marketing Research,* October 1977.

6. Yates, *Sampling Methods.*

7. Leslie Kish, *Survey Sampling,* New York: John Wiley & Sons, 1965.

7

Bivariate Regression

Some General Remarks on Statistical Model Building

The preceding chapters dealt with the collection of data and processing them to some numerical form, amenable to analysis. Frequently in engineering and applied science, a part of the analysis is the construction of mathematical relationships that can be used to explain or predict certain phenomena. A number of methods are available for constructing such relationships, for example, the use of conceptual or flow-diagram models, mathematical and statistical models, and simulation models.[1] The intention here is to provide some of the background knowledge necessary for the development of statistical models. The use of the term statistical implies the use of sample data to produce some form of mathematical relationship.

The first thing that should be emphasized is that statistical modeling techniques should not be regarded as a means of pulling some relationship out of thin air. This is a frequent and serious misuse of statistical techniques. Rather, the use to which statistical modeling techniques should be put is that of attempting to determine the parameters of a theorized relationship.

In other words, a hypothesis (or hypotheses) must be advanced, expressing a relationship that might be expected between certain phenomena (variables). The statistical techniques described in the remaining chapters of the book represent one means by which certain types of hypotheses may be tested. It is extremely important that the reader understands this idea of hypothesis testing, since it underlies and explains much of the procedure of statistical model building and testing. This point is reiterated in specific contexts of certain statistical methods in the succeeding chapters.

Before detailing the various statistical techniques, it is appropriate to review the reasons for statistical modeling and the general aims of model building. The word "model" is used in the purely statistical sense as invoked in many areas of engineering and applied science. A model may be defined as an abstraction of reality. Specifically, this means that a model is intended to be a simplification of reality, not a replica of reality. In this sense, one may distinguish between a model and a physical law. A physical law is reality, without approximation. Thus Newton's law of gravitation[2] is an exact phenomenological statement. It *is* the reality.

$$F = \frac{GM_1M_2}{d^2} \tag{7.1}$$

The relationship of equation 7.1 (Newton's law of gravitation) is not a model within this definition. It is an accurate, precise, and complete statement of a relationship that always holds. To illustrate more clearly just what a model is, one might consider a relationship between the yield of wheat per acre (W), the rate of fertilizer application (F), the number of years since wheat was last grown on that land (Y), and the number of inches of rain since seeding (R). A simple linear relationship might be

$$W = a_0 + a_1 F + a_2 Y + a_3 R \qquad (7.2)$$

Indeed, this equation may represent the hypothesis to be tested. The questions to be answered by the analyst are: can nonzero values be found for a_0, a_1, a_2, and a_3, and how well does the right side of equation 7.2 predict the observed values of the left side? There is no known law that relates yields of wheat to these three other variables precisely and completely. However, it seems reasonable to expect that some relationship might exist between them. Herein lies the essence of a model. There may be many other variables that will affect the relationship, for example, the quality of the seed, the number of hours of sunshine, the amount of rain and sun at particular growth periods, and the type of soil. Equation 7.2 is, however, an abstraction of reality. It is incomplete, not very precise, and subject to error. But it provides insights into the relationship and may represent a useful relationship for predicting yields of wheat from limited available data.

It should be clear from this illustration that a number of demands must be made of the process for testing the hypothesis represented by equation 7.2. First, it is necessary to be able to test the null hypothesis that the relationship adds nothing to our knowledge and understanding. This is equivalent to saying that values of W in equation 7.2 could be predicted by guesswork (i.e., a random process) as accurately as by use of the model. If this hypothesis can be rejected, it is appropriate to determine whether all the variables in the model are necessary for reasonable prediction of W, whether some variables add no new information, and whether other variables are necessary to obtain good predictions. As will be seen, these concerns are raised for each statistical model-building process treated in the remainder of the book.

Sufficient has not yet been said about the basic properties of a model. It has been emphasized that a model is an abstraction of reality. It is also true that a model should be as accurate as possible, while yet retaining its simplicity. It may be used as a predictive or forecasting tool,[3] or it may be used to gain understanding of a process. For each of these uses, the probably conflicting goals of accuracy and simplicity are desired. For the purpose of subsequent discussions, simplicity may be interpreted as implying parsimony of variables and the use of the least complex of func-

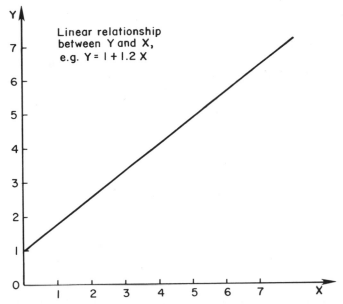

Figure 7–1. A Simple, Bivariate Linear Relationship

tional forms, consistent with the hypothesis to be tested. In this latter respect, a linear relationship (such as that shown in equation 7.2 and figure 7–1) is the simplest. A monotonic, continuous nonlinear function is next (as shown in figure 7–2) and various types of discontinuous nonlinear functions and continuous nonmonotonic functions are next in their degree of complexity.

While retaining these properties, a model must also be useful for its purpose; that is, if predictive, it must be capable of predicting; if explanatory, it must provide explanation and understanding of the phenomenon being modeled. The model must also be economical to use. It should not need the use of data that are extremely difficult and costly to obtain, and it should be cheaper by far to use than real-world experimentation. A model must also be valid. Many interpretations can be given to validity, although principally a dichotomy of meaning may be made between descriptive and predictive models. If a model is intended to be descriptive only, validity may imply logic in the model structure and transferability to other geographic locations or other modeling situations (depending upon the discipline of study). Validity in a predictive model must also imply causality. In other words, the direction of "affect" must be correctly specified in the model. There are a number of other properties that a model should possess, but these are among the most important and serve

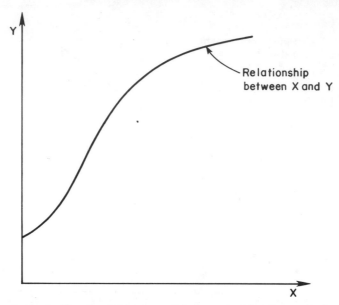

Figure 7–2. A Simple, Bivariate, Monotonic Nonlinear Relationship

adequately to set the stage for the description of the techniques themselves.

This discussion of models should also have pointed up another property possessed by any statistical model—error. Clearly, the preceding statements about accuracy, simplicity, and abstraction of reality all imply the existence of error. Furthermore, models in engineering and applied sciences are generally built with the use of sample data. As discussed in chapter 4, sample data possess sampling errors. In fact, there are three primary sources of error in statistical modeling.[4] These are specification error, measurement error, and calibration error. Specification error is generally unknowable and derives from the inclusion of unnecessary variables or the exclusion of necessary variables. While exhaustive testing may eliminate the former, it is never possible to determine the extent of the latter. Measurement error derives from the sampling error and from impreciseness and inaccuracy of the actual measurements made to generate the data. The measurement error that arises from sampling can be computed, as described in chapter 4, but that arising from impreciseness of actual measurement often cannot be determined, particularly when humans are the subjects of the data collection and model building. Calibration error occurs in the modeling process as a result of the use of data containing measurement errors and the building of a model with specification error.

Measurement error is a property of the data and cannot be ameliorated substantially in the modeling process, with the exception of error propagation (i.e., the means by which a model magnifies or diminishes errors in different variables). One of the goals of statistical model building is to minimize error. Specification error is minimized by careful selection of variables in the model through prior statistical and visual examination of the data, together with a carefully reasoned hypothesis of structure and variable content.

Calibration error is minimized in most statistical techniques as the major principle of model building. The degree to which it can be minimized, however, depends upon a wide range of properties of the data and the model-building technique.

To complete this discussion of errors, it is necessary to consider some specialized terminology. In statistical model building, there are generally two types of variables: dependent and independent. (As discussed later, this taxonomy of variables is not always adequate, particularly in the case of more complex and specialized model-building techniques.) A dependent variable is the variable to be explained or predicted by the model. It therefore depends upon the other variables. Dependency implies a strict one-way causality in almost all cases. A dependent variable is caused to change by changes in the independent variables. The reverse causality, however, must not occur. Changes in the dependent variable cannot cause changes in the independent variables.

The example of equation 7.2 may be used to illustrate this: W, the yield of wheat, is the dependent variable. Its value is changed by the other variables—fertilizer application, number of years since wheat was last grown, and inches of rainfall. These latter three variables are independent variables. If the rate of fertilizer application is changed, the yield of wheat may be expected to change as a result. The yield of wheat, however, will not change the rate of application of fertilizer (in the year) or the amount of rainfall. The unidirectional causality is upheld by this model specification and the meanings of dependent and independent variables should be clear. It is important to note, however, that independence, as applied to variables, does not necessarily imply that the independent variables do not cause changes in each other or change together. This may be a required property in some instances and a preferred property in all instances, but it is not part of the definition of independence. Thus the rate of fertilizer application may depend to some extent on rainfall amounts or the period since wheat was last grown on that land, without violating the independence property of these variables.

To calibrate a model, observations are needed on a sample of values of both dependent and independent variables for the units of concern. The resulting calibrated model would then be used to explain or predict the dependent variable. These sample values will have errors associated with them that will affect the estimation of model parameters in calibration. Various mathematical operations on the independent variables may serve to propagate (magnify) individual errors to a much larger error in the dependent variable, while other operations may reduce errors or at least leave them unchanged. A number of rules can be put forward to reduce error propagation in models.[5] These may be summarized as follows: the preferred mathematical operation is addition, followed by multiplication and division, while the least desirable operations are subtraction and raising variables to powers.

Elementary Bivariate Regression

The simplest form of relationship that can be hypothesized is a linear one, as typified by equation 7.2. Furthermore, the simplest linear relationship is one involving one dependent and one independent variable. The statistical procedure for developing a linear relationship from a set of observations is known as linear regression. The simple case of one dependent and one independent variable is bivariate linear regression, while the use of multiple independent variables is multivariate linear regression. The procedures for estimating linear relationships can be described most economically and understandably by considering the simple bivariate case.

A relationship is hypothesized of the form

$$Y_i = \alpha_0 + \alpha_1 X_i \qquad (7.3)$$

where $Y_i =$ ith value of the dependent variable

           $X_i =$ associated ith value of the independent variable

and $\alpha_0,\ \alpha_1 =$ true values of the linear regression parameters

The parameter values, α_0 and α_1, are those sought by the estimation procedure. The true values are those that can be obtained by estimating the model from the entire population of values of Y and X. In other words, the connotation "true" refers to population values as opposed to sample values, but does not refer to the existence or nonexistence of a linear relationship of the form of equation 7.3. Thus the true value of α_1 may be zero, indicating that there is no relationship between X and Y and this situation is perfectly consistent with the terminology of true values.

In general, only sample data will be available of X and Y. This means that equation 7.3 is the equation sought, but the equation that can be estimated is one based on sample estimates, such as

$$Y_i = a_0 + a_1 X_i + \epsilon_i \qquad (7.4)$$

where Y_i, X_i are as defined before,

a_0, a_1 = sample estimates of α_0 and α_1

   ϵ_i = the error associated with the ith observation

The addition of the error term, ϵ_i, is significant in several respects. First, it is a clear indication of the acceptance of sample error in the calibration process. This error term, however, does not account for measurement or specification errors. Second, it is significant that the error is represented as being additive in the model. This indicates that the error is seen as being effectively independent of the independent variable, X_i, since it does not interact with it. In fact, the calibration procedure of linear regression expressly assumes that the independent variables are known without error for all observations and that the error resides only in the dependent variable, Y_i. Third, the value ϵ_i is assumed to exist, but cannot be measured until the values of a_0 and a_1 are obtained. The existence of ϵ_i is, however, the mathematical basis for obtaining estimates of a_0 and a_1.

Scatter Diagrams

Before attempting to build a linear relationship, such as that presented in equation 7.4, it is necessary to determine whether a linear relationship is appropriate to describe the data. The various measures, discussed in later sections of the chapter, do not indicate whether a linear relationship is the best one to describe a phenomenon. They indicate only how good a fit is obtained by the linear function.

The first step in the analytical process must always be to construct a scatter diagram. Having selected a variable to be the dependent variable, a plot should be constructed of each independent variable against the dependent variable. These plots will generally show fairly clearly how appropriate a linear relationship is, or whether any relationship may be expected.

Some typical, hypothetical scatter diagrams are shown in figures 7–3 through 7–6. Figure 7–3 shows evidence of a positive linear relationship between Y and X. The relationship is positive because increases in X give rise to increases in Y. Figure 7–4 shows the reverse relationship, but still a linear one. Figure 7–5 shows a more-or-less horizontal band of values, indicating no relationship between Y and X. Figure 7–6 indicates a probable nonlinear relationship between Y and X, for which a linear regression would be inappropriate. Having constructed a scatter diagram and finding prima facie evidence of a linear relationship, the analyst may then proceed

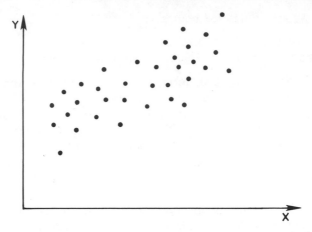

Figure 7–3. Scatter Diagram of a Positive Linear Relationship

Figure 7–4. Scatter Diagram of a Negative Linear Relationship

to construct a regression relationship. If either of the plots of figures 7–5 or 7–6 were obtained, then no linear-regression procedure should be applied to the data.

Calibration for Linear Regression

Given equation 7.4, the problem is now to determine a solution method for a_0 and a_1. Clearly, any solution should seek to minimize the error of calibration. (Measurement and specification errors must be considered as

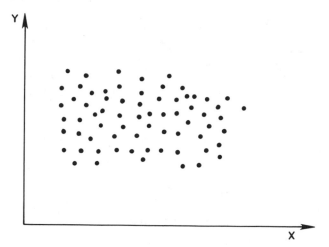

Figure 7–5. Scatter Diagram Showing No Relationship between Y and X

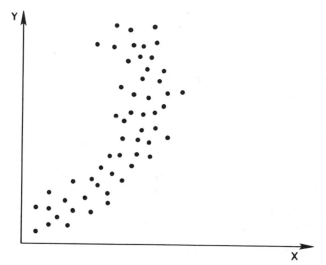

Figure 7–6. Nonlinear Relationship between Y and X

given for an assumed model and data set.) By definition, the sum of the error terms for the population is zero,

$$\sum_{N} \epsilon_i = 0 \qquad (7.5)$$

where N = population size.

This is so because the true model is defined as the one that would be obtained for the population (equation 7.3) in which ϵ_i is absent. Hence, both the sum and average of the error terms over the population must be zero. In turn, this implies that in any sample there must be a range of values of ϵ_i that are distributed across both negative and positive values.

These properties lead to the conclusion that minimizing the sum of the errors for the sample data is not an effective procedure. The absolute sum of the errors (if the sample is indeed drawn correctly) must approach zero as sample size is increased. In fact, in any large sample (where large may be defined as being in excess of 100 observations), the deviation of the sum of the error values from zero will be very small. It may also be noted that a strict minimization of the sum of the error values would lead to a search for values of a_0 and a_1 that would produce extremely large negative values of ϵ_i, since minus infinity is the smallest number known. Such a procedure would lead to estimates of a_0 and a_1 that would be as far from the true values as possible. Indeed, the calibration would generate a value of plus infinity for a_0 and zero for a_1, thus generating all values of ϵ_i as minus infinity.

To avoid trivial or absurd calibration results, it would appear that the best procedure would be to minimize the square of the error terms. By squaring the terms, all values to be summed become positive. Hence, the absolute minimum value becomes zero, occurring only when all values of ϵ_i are also zero. Thus minimizing the sample sum of squared error terms leads to a calibration that is consistent with the true population model. The least-squares approach, as this procedure is termed, also has another important property that is discussed in chapter 13—that of providing maximum-likelihood estimates of the regression coefficients that are unbiased for large samples.

Suppose there are a set of n observations of the values of Y_i and X_i. The hypothesized relationship between X_i and Y_i is that of equation 7.4. This equation may be rewritten to express ϵ_i in terms of the observed values of Y_i and X_i and the unknowns, a_0 and a_1,

$$\epsilon_i = Y_i - a_0 - a_1 X_i \qquad (7.6)$$

The sum of the squares of the deviations, S, from the true line is

$$S = \sum_n \epsilon_i^2 = \sum_n (Y_i - a_0 - a_1 X_i)^2 \qquad (7.7)$$

where n = sample size.

The least-squares method then states that the linear-regression line is that which minimizes the sum of the squares of the deviations, that is, the true line is that for which $\sum_n \epsilon_i^2$ is a minimum. Obtaining estimated values of a_0 and a_1 is a mathematical operation of no great complexity.

To find the minimum value of S, partial derivatives of S are taken with respect to each coefficient in the equation, as shown in equations 7.8[a] and 7.9.

$$\frac{\partial S}{\partial a_0} = -2\Sigma(Y_i - a_0 - a_1 X_i) \tag{7.8}$$

$$\frac{\partial S}{\partial a_1} = -2\Sigma X_i(Y_i - a_0 - a_1 X_i) \tag{7.9}$$

Minimization of S occurs when the partial derivatives of S are zero and the second derivatives are positive. Hence, a_0 and a_1 may be found by setting equations 7.8 and 7.9 to zero.

$$\Sigma(Y_i - a_0 - a_1 X_i) = 0 \tag{7.10}$$

$$\Sigma X_i(Y_i - a_0 - a_1 X_i) = 0 \tag{7.11}$$

With the necessary mathematical manipulations, the solutions for a_0 and a_1 may be obtained.

$$a_1 = \frac{\Sigma X_i Y_i - n\bar{Y}\bar{X}}{\Sigma X_i^2 - n\bar{X}^2} \tag{7.12}$$

$$a_0 = \bar{Y} - a_1\bar{X} \tag{7.13}$$

Proof. The proof of this result may be shown in a few steps. The summations in equations 7.10 and 7.11 are first taken through the individual terms, after first multiplying out the terms in parentheses in equation 7.11 by X_i.

$$\Sigma Y_i - na_0 - a_1\Sigma X_i = 0 \tag{7.14}$$

$$\Sigma X_i Y_i - a_0\Sigma X_i - a_1\Sigma X_i^2 = 0 \tag{7.15}$$

Equations 7.14 and 7.15 are termed the normal equations of the regression. The summation of a variable is n times its mean value. Hence equations 7.14 and 7.15 can be rewritten

$$n\bar{Y} - na_0 - a_1 n\bar{X} = 0 \tag{7.16}$$

$$\Sigma X_i Y_i - a_0 n\bar{X} - a_1\Sigma X_i^2 = 0 \tag{7.17}$$

Equation 7.16 may be divided throughout by n and rearranged to express a_0 in terms of a_1, $\bar{X}$, and $\bar{Y}$.

$$a_0 = \bar{Y} - a_1\bar{X} \tag{7.18}$$

---

[a] For simplicity, the limit of summation, n, has been dropped from the notation. It should be assumed, hereafter, to be n, unless otherwise specified.

Substituting this result in equation 7.17 and collecting terms in a_1 on the left side yields

$$a_1(\Sigma X_i^2 - n\bar{X}^2) = \Sigma X_i Y_i - n\bar{X}\bar{Y} \tag{7.19}$$

Hence results of equations 7.12 and 7.13 are proved.

To ensure that this solution minimizes the error, it is necessary to look at the second partial derivatives; these are

$$\frac{\partial^2 S}{\partial a_0^2} = 2 \tag{7.20}$$

$$\frac{\partial^2 S}{\partial a_1^2} = 2\Sigma X_i^2 \tag{7.21}$$

Since ΣX_i^2 must be positive for all real values of X_i, equations 7.20 and 7.21 are always positive. Hence the solutions obtained in equations 7.12 and 7.13 are minimum square-error solutions for all real values of X_i and Y_i.

Geometry of Linear Regression

It is often helpful to visualize the linear-regression procedure as a geometric problem. The hypothesized relationship of equation 7.4 is shown in figure 7–7. From this, it can be seen that a_0 is the intercept on the Y-axis and a_1 is the slope of the line. By virtue of equation 7.13, the point $(\bar{X}, \bar{Y})$ must be on the line, as shown in figure 7–7.

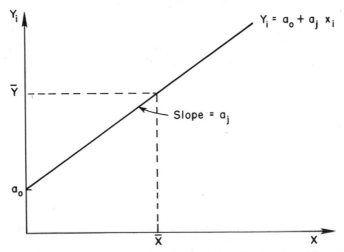

Figure 7–7. Geometric Representation of the Regression Hypothesis

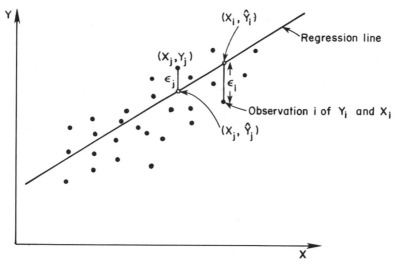

Figure 7–8. Scatter Diagram of n Data Points

The linear-regression problem is shown in figure 7–8 as the problem of finding the best straight line that fits the scatter of data points, representing the n observations of X_i and Y_i. The assumption that the errors, ϵ_i, are in the Y_i values only and that the X_is are known without error indicates that the ϵ_is are the distances in the Y-dimension between the observed point (X_i, Y_i) and the vertical projection of that point on the regression line $(X_i, \hat{Y}_i)$, as shown in figure 7–8. Thus, the least-squares solution is the one that minimizes the sum of the squares of these vertical displacements.

One reason for assuming the errors to reside in the Y_i values and not in the X_i values relates to the use of the resulting model for prediction of Y_i values. Since the errors are not known, it is necessary to assume that any new values of X_i, for which values of Y_i are to be predicted, are known exactly. If the reverse assumption were made, that all of the error were in the X_is, it would be necessary to know that error to predict Y_i values. This is shown in figure 7–9. The value $\tilde{Y}_i$ would be obtained if the error were assumed to be zero. The best estimate of $\hat{Y}_i$, however, requires a knowledge of the error η_i. As shown by the comparative positions of $\hat{Y}_i$ and $\hat{Y}_i'$, this knowledge of the error is critical. Clearly, the geometry of this shows that the independent variable *must* be assumed to be known without error.[6]

This discussion leads to further insights into the definition or meaning of dependent and independent variables. Clearly, the independent variables are those to be used for prediction, while the dependent variable is the one to be predicted. Hence, the direction of causality is imposed on the regression model.

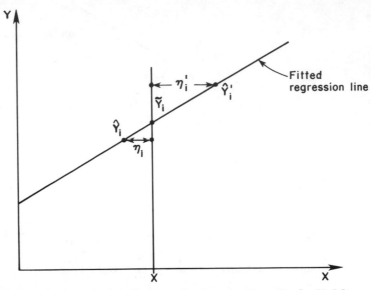

Figure 7–9. Prediction if Error Is Assumed to Be in X_i Measures

Some Properties of the Regression Model

It is useful to examine some alternative ways of expressing the solution for a_1, equation 7.12. The variance of a variable w, for which there are n sample observations, is

$$\widehat{var}(w) = \frac{1}{n-1} \Sigma(w - \bar{w})^2 \qquad (7.22)$$

This equation is an unbiased estimate when n is small. For large values of n, the variance may be approximated.

$$\widehat{var}(w) = \frac{1}{n} \Sigma(w - \bar{w})^2 \qquad (7.23)$$
$$n \to \infty$$

Consider the expansion of the right side of equation 7.23.

$$\frac{1}{n} \Sigma(w - \bar{w})^2 = \frac{1}{n} \Sigma(w^2 - 2w\bar{w} + \bar{w}^2) \qquad (7.24)$$

Summing each term on the right side of equation 7.24 provides

$$\frac{1}{n} \Sigma(w^2 - 2w\bar{w} + \bar{w}^2) = \frac{1}{n} \Sigma w^2 - \frac{2}{n} \Sigma w\bar{w} + \frac{1}{n} \Sigma \bar{w}^2 \qquad (7.25)$$

Because $\bar{w}$ is a constant over all observations, equation 7.25 can be simplified to equation 7.26, also observing that Σw is $n\bar{w}$.

$$\frac{1}{n} \Sigma (w - \overline{w})^2 = \frac{1}{n} \Sigma w^2 - \frac{2}{n} \overline{w} \cdot n\overline{w} + \frac{1}{n} \cdot n\overline{w}^2 \qquad (7.26)$$

Simplifying equation 7.26, by gathering terms in $\overline{w}^2$, the variance of w may be expressed

$$\frac{1}{n} \Sigma (w - \overline{w}^2) = \frac{1}{n} \Sigma w^2 - \overline{w}^2 \qquad (7.27)$$

The covariance of two variables, v and w, may be written

$$\text{cov}(v,w) = \frac{1}{n} \Sigma (w - \overline{w})(v - \overline{v}) \qquad (7.28)$$

By a similar process of manipulation, equation 7.28 can be rewritten

$$\text{cov}(v,w) = \frac{1}{n} \Sigma wv - \overline{wv} \qquad (7.29)$$

Thus it can be seen that the solution for a_1 may be written

$$a_1 = \frac{\Sigma (X_i - \overline{X})(Y_i - \overline{Y})}{\Sigma (X_i - \overline{X})^2} \qquad (7.30)$$

$$a_1 = \frac{\text{cov}(X,Y)}{\text{var}(X)} \qquad (7.31)$$

Sum of Residuals

The first property of interest concerns the sum of the residuals or error terms for the regression model. The residuals for the calibration data are the differences between the regression estimate, $\hat{Y}_i$, and the observed value of the dependent variable, Y_i. This difference, as shown previously, is the error term, ϵ_i.

$$\epsilon_i = Y_i - \hat{Y}_i \qquad (7.32)$$

The sum of the residuals is

$$\sum_i \epsilon_i = \sum_i (Y_i - \hat{Y}_i) \qquad (7.33)$$

To determine this value, it is necessary to substitute for $\hat{Y}_i$, using the regression equation

$$\hat{Y}_i = \overline{Y} - a_1\overline{X} + a_1 X_i \qquad (7.34)$$

Substituting equation 7.34 into equation 7.33 yields

$$\sum_i \epsilon_i = \sum_i (Y_i - \overline{Y} + a_1\overline{X} - a_1 X_i) \qquad (7.35)$$

Summing the terms on the right side of equation 7.35 yields

$$\sum_i \epsilon_i = \sum_i Y_i - n\overline{Y} + a_1 n\overline{X} - a_1 \sum_i X_i \qquad (7.36)$$

However, $\sum_i Y_i$ is $n\bar{Y}$ and $\sum_i X_i$ is $n\bar{X}$. Hence, it is clear that the right side of equation 7.36 is zero.

$$\sum_i \epsilon_i = 0 \tag{7.37}$$

This property is important primarily because it indicates that a potential test of a regression model is not a test. Specifically, resubstituting the values, X_i, used to calibrate the model for a test of predictive power is a very weak test. Clearly, each individual prediction, $\hat{Y}_i$, can be examined against the observed value, Y_i. However, summation of the $\hat{Y}_i$ values will yield, exactly, the summation of the Y_i values. This will occur regardless of how good or how poor the model fit may be. Thus summed predictions from the calibration data provide no test of goodness-of-fit.

Analysis of Variance of the Regression Model

A second property, of interest in assessing the goodness of fit of the linear-regression model, is derived from a univariate analysis of variance (ANOVA).[7] In standard analysis of variance, the subject of analysis is the sum of squares between and within groups of the sample data. The standard use of ANOVA is thus to analyze and perform statistical tests on categorizations of data. In applying ANOVA to regression, the within-group sum of squares is replaced by the sum of squares of the regression and the between-group sum of squares by the residual sum of squares. To demonstrate the validity of this transformation, it is necessary to show that the regression and residual sums of squares add up to the total sum of squares. The total sum of squares is defined as

$$\text{SST} = \sum_i (Y_i - \bar{Y})^2 \tag{7.38}$$

In words: the total sum of squares is the sum of the squares of the observed values, Y_i, for the sample about the sample mean. Likewise, the regression sum of squares, equation 7.39, is the sum of the squares of the regression estimates, $\hat{Y}_i$, for the sample about the sample mean.

$$\text{SSR} = \sum_i (\hat{Y}_i - \bar{Y})^2 \tag{7.39}$$

Finally, the residual sum of squares, or *error* sum of squares, is equal to the sum of the squares of the regression estimates, $\hat{Y}_i$, about the observed values, Y_i, for the sample.

$$\text{SSE} = \sum_i (\hat{Y}_i - Y_i)^2 \tag{7.40}$$

Note that the error sum of squares is the sum of the squared error terms, minimized in the regression procedure

$$\text{SSE} = \sum_i \epsilon_i^2 \tag{7.41}$$

Now it is necessary to prove the identity

$$\sum_i (Y_i - \bar{Y})^2 = \sum_i (\hat{Y}_i - \bar{Y})^2 + \Sigma(\hat{Y}_i - Y_i)^2 \qquad (7.42)$$

To prove this identity, it is necessary to expand the summation terms on the right side of equation 7.42. First, the regression sum of squares is expanded.

$$\Sigma(\hat{Y}_i - \bar{Y})^2 = \sum_i \hat{Y}_i^2 - 2\sum_i \hat{Y}_i \bar{Y} + n\bar{Y}^2 \qquad (7.43)$$

Since $\bar{Y}$ is a constant and $\sum_i \hat{Y}_i$ is equal to $n\bar{Y}$, equation 7.43 simplifies to

$$\Sigma(\hat{Y}_i - \bar{Y})^2 = \sum_i \hat{Y}_i^2 - n\bar{Y}^2 \qquad (7.44)$$

Next, the error sum of squares can be expanded.

$$\sum_i (\hat{Y}_i - Y_i)^2 = \sum_i \hat{Y}_i^2 - 2\Sigma \hat{Y}_i Y_i + \Sigma Y_i^2 \qquad (7.45)$$

To proceed further, it is necessary to substitute for $\hat{Y}_i$ in terms of the observations, X_i, and the regression parameters, a_0 and a_1.

$$\hat{Y}_i = a_0 + a_1 X_i \qquad (7.46)$$

Thus, the terms $\Sigma \hat{Y}_i^2$ and $\Sigma \hat{Y}_i Y_i$ can be rewritten

$$\sum_i \hat{Y}_i^2 = \sum_i [a_0 + a_1 X_i]^2 \qquad (7.47)$$

$$\Sigma \hat{Y}_i Y_i = \sum_i Y_i [a_0 + a_1 X_i] \qquad (7.48)$$

Using equation 7.13 for a_0, equation 7.47 becomes

$$\Sigma \hat{Y}_i^2 = a_1^2 [\sum_i X_i^2 - n\bar{X}^2] + n\bar{Y}^2 \qquad (7.49)$$

Substituting equation 7.30 for a_1 yields

$$\Sigma \hat{Y}_i^2 = \frac{\sum_i [(X_i - \bar{X})(Y_i - \bar{Y})]^2}{\sum_i (X_i - \bar{X})^2} + n\bar{Y}^2 \qquad (7.50)$$

By a similar process, equation 7.48 becomes

$$\Sigma \hat{Y}_i Y_i = \frac{[\sum_i (X_i - \bar{X})(Y_i - \bar{Y})]^2}{\Sigma(X_i - \bar{X})^2} + n\bar{Y}^2 \qquad (7.51)$$

In other words, the terms $\Sigma \hat{Y}_i^2$ and $\Sigma \hat{Y}_i Y_i$ are identical. Substituting equations 7.44 and 7.45 for the right side of equation 7.42 yields

$$\Sigma(\hat{Y}_i - Y_i)^2 + \Sigma(\hat{Y}_i - \bar{Y})^2 = 2\sum_i \hat{Y}_i^2 - 2\Sigma \hat{Y}_i Y_i + \Sigma Y_i^2 - n\bar{Y}^2 \qquad (7.52)$$

Noting the results of equations 7.50 and 7.51, equation 7.52 reduces to

$$\Sigma(\hat{Y}_i - Y_i)^2 + \Sigma(\hat{Y}_i - \bar{Y})^2 = \Sigma Y_i^2 - n\bar{Y}^2 \tag{7.53}$$

However, equation 7.54 is also true.

$$\Sigma(Y_i - \bar{Y})^2 = \Sigma Y_i^2 - n\bar{Y}^2 \tag{7.54}$$

Hence we have proved that equation 7.42 always holds, for any set of values of Y_i and therefore that ANOVA can be applied to analyzing the results of a regression estimation. Because of the importance of the relationship just proved, it is restated here both mathematically and in words.

$$SST = SSR + SSE \tag{7.55}$$

$$\begin{array}{c} \text{total sum} \\ \text{of squares} \end{array} = \begin{array}{c} \text{regression} \\ \text{sum of squares} \end{array} + \begin{array}{c} \text{error sum} \\ \text{of squares} \end{array} \tag{7.56}$$

The appropriate ANOVA is a type 1 problem,[8] in which the error sum of squares is equivalent to the within-group sum of squares and the regression sum of squares is equivalent to the between-group sum of squares. This is shown in table 7–1. The null hypothesis is that all values $\hat{Y}_i$ are equal to the mean, $\bar{Y}$, for all i. In other words, this assumes that the true equation for the regression is equation 7.57, but where a_1 is zero and a_0 is therefore equal to $\bar{Y}$ (equation 7.13).

$$\hat{Y}_i = a_0 + a_1 X_i \tag{7.57}$$

Thus the null hypothesis is equivalent to hypothesizing that there is no relationship between the independent and the dependent variable. Under such conditions, the estimates of the error sum of squares and the total sum of squares will be approximately the same, allowing for sampling errors. As in standard ANOVA, dividing the regression mean square by the error mean square provides a quotient that is distributed like F with 1 and $(n-2)$ degrees of freedom. If the value of the quotient F is greater

Table 7–1
ANOVA for a Bivariate Regression

Source of Variation	Sum of Squares	Degrees of Freedom	Mean Square
Regression	$\Sigma(\hat{Y}_i - \bar{Y})^2 = SSR$	1	SSR
Error	$\Sigma(\hat{Y}_i - Y_i)^2 = SSE$	$n-2$	$SSE/(n-2)$
Total	$\Sigma(Y_i - \bar{Y})^2 = SST$	$n-1$	$SST/(n-1)$

than the table value of F at significance level α, the null hypothesis may be rejected with $(100-\alpha)\%$ confidence. Thus writing MSR for the regression mean square and MSE for the error mean square, the test is carried out by calculating

$$Q = \frac{\text{MSR}}{\text{MSE}} \tag{7.58}$$

Suppose a significance level of 5% is chosen, then the null hypothesis can be rejected with 95% confidence if equation 7.59 holds true;

$$Q \geqslant F_{0.05(1,n-2)} \tag{7.59}$$

where $F_{0.05(1,n-2)}$ = table value of F at 0.05 significance level with 1 and $n-2$ degrees of freedom. This F test provides an overall statistical test of the regression estimation, by testing the hypothesis that there is no relationship between the dependent and independent variables.

Correlation

The ANOVA of the linear regression suggests another possible measure of goodness of fit. The basis of the linear-regression procedure is to minimize the error sum of squares. Clearly, the absolute minimum error sum of squares is zero. From equation 7.55 and the ANOVA table, table 7–1, such a situation requires that the regression sum of squares equals the total sum of squares. In other words, the ratio of the regression sum of squares to the total sum of squares will approach unity as the error sum of squares approaches zero.

Similarly, the maximum value that the error sum of squares can take is to equal the total sum of squares. Under these conditions, the regression sum of squares must approach zero and the ratio of the regression and total sums of squares will approach zero. When the ratio approaches one, it may be concluded that the regression line passes through almost all of the observed points, that is, the model fits the data perfectly. When the ratio approaches zero, the regression line must be (in a bivariate regression) horizontal (that is, parallel to the X-axis) through the value $\bar{Y}$. This implies no fit between model and data. Hence this ratio may be interpreted as a measure of goodness-of-fit for the regression equation. The ratio is called the coefficient of determination and is denoted R^2; mathematically, it is

$$R^2 = \frac{\text{SSR}}{\text{SST}} = \frac{\Sigma(\hat{Y}_i - \bar{Y})^2}{\Sigma(Y_i - \bar{Y})^2} \tag{7.60}$$

Values of R^2 must lie between 0 and 1. The closer they are to unity, the better is the fit, while the closer they are to zero, the poorer is the fit. It is clear that the coefficient of determination also signifies the fraction of the total sum of squares that is captured by the linear-regression model. Therefore, it is not uncommon to see R^2 expressed as a "percentage of variance explained" by the regression. The use of the word "explained" is not quite correct here. More correctly, it should be the proportion captured by the model. This distinction, while seeming a semantic nicety, is in fact rather important. Use of the word "explained" tends to imply a measure of predictive ability, because of the way in which people loosely interchange explanation and prediction. The measure, however, relates only to the calibration data and should not be extended to any prediction.

The square root of the coefficient of determination is called the correlation coefficient and is denoted R. Computation of R, however, is more complex than just taking the square root of equation 7.60. In the case of the correlation coefficient for a bivariate regression, or between a pair of variables, the sign of the correlation coefficient provides information in addition to its value. The sign cannot be determined, however, by simply taking the square root of the coefficient of determination. To compute the correlation coefficient, the numerator of equation 7.60 may be rewritten in terms of the observed values of X_i and Y_i, using equations 7.44 and 7.50.

$$\text{SSR} = \frac{[\sum_i (X_i - \bar{X})(Y_i - \bar{Y})]^2}{\sum_i (X_i - \bar{X})^2 \sum_i (Y_i - \bar{Y})^2} + n\bar{Y}^2 - n\bar{Y}^2 \qquad (7.61)$$

Taking the square root of R^2, after substituting equation 7.60 gives

$$R = \frac{\sum_i (X_i - \bar{X})(Y_i - \bar{Y})}{\sqrt{[\sum_i (X_i - \bar{X})^2 \sum_i Y_i - \bar{Y})^2]}} \qquad (7.62)$$

For large samples, equation 7.62 states that the correlation coefficient is equal to the covariance of X and Y divided by the square root of the product of the variances of X and Y (that is, the product of the standard deviations of X and Y).

$$R = \frac{\text{cov}(X,Y)}{\sigma_x \sigma_y} \qquad (7.63)$$

Since a standard deviation is always considered to be the positive root of the variance, the variable that determines the sign of R is clearly the covariance. As noted previously, the covariance also determines the sign of the coefficient, a_1. Thus the sign of R indicates the direction of the relationship between X and Y, while its magnitude indicates the strength of that relationship. If the covariance, and hence both the correlation

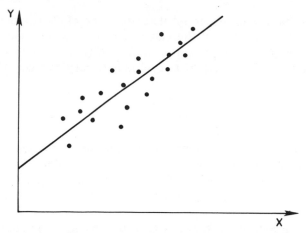

Figure 7–10. Positive Covariance between Y and X

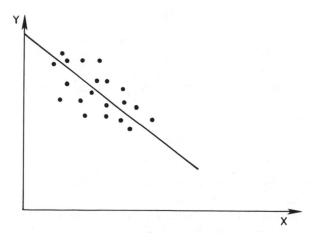

Figure 7–11. Negative Covariance between Y and X

coefficient and the regression coefficient, is negative, this implies that as X increases, Y decreases. Similarly, if the sign of these three quantities is positive, it indicates that increases in X cause increases in Y. The differences between the positive and negative signs of the covariance are illustrated in figures 7–10 and 7–11.

It is worth noting that equation 7.63 also provides an alternative method for computing the coefficient of determination for a bivariate regression. The coefficient of determination is

$$R^2 = \frac{[\mathrm{cov}(X,Y)]^2}{\sigma_x^2 \sigma_y^2} \tag{7.64}$$

It has already been observed that the value of R^2 lies between zero and one. It also follows that the value of R lies between minus one and plus one. As with R^2, the closer the value of R to one (minus or plus), the better is the fit, while the closer the value is to zero, the worse is the fit.

Statistical Measures of Regression

In the preceding section, a number of properties of the regression model have been explored, from which a concept of goodness-of-fit has been developed—the correlation coefficient. This measure, however, is insufficient to assess the model. A simple, if extreme, example suffices to demonstrate this.

Suppose a data set comprises two observations, as shown in table 7–2. Applying linear regression to this provides the model

$$Y_i = -6.16 + 0.7619X_i \qquad (7.65)$$

Computing, from this data set, the variances, covariances, and the correlation coefficient, the values of table 7–3 are obtained. It is observed that, not surprisingly, the correlation coefficient is 1.00—perfect fit. Suppose now that the data of table 7–2 were a random subsample from a larger data set of four observations, shown in table 7–4. The linear-regression equation is

$$Y_i = -1.505 + 0.44X_i \qquad (7.66)$$

The regression statistics from this model are shown in table 7–5. As can be seen, the correlation coefficient is lower, at 0.95, and the variances and covariances have all changed in value. Furthermore, the regression equation of equation 7.66 is markedly different from that of equation 7.65. Clearly, this example could be extended by adding further and further data points and each such addition would show further changes in the model and further reductions in the value of the correlation coefficient. Thus the question must be raised as to how reliable the estimates of a

Table 7–2
Two-observation Data Set

Observations	
X	*Y*
13.6	4.2
15.7	5.8

Table 7–3
Regression Statistics for Table 7–2

Measure	Value
cov (X, Y)	0.84
$V(X)$	1.102
$V(Y)$	0.64
R	1.00

Table 7–4
Augmented Data Set of Table 7–2

Observations	
X	Y
13.6	4.2
15.7	5.8
17.3	5.9
10.6	3.2

Table 7–5
Regression Statistics for Table 7–4

Measure	Value
cov(X, Y)	2.69
$V(X)$	6.29
$V(Y)$	1.28
R	0.95

regression equation are, since they are obviously a function of the number of data points used to construct the estimates. Two basic measures of reliability are desired. First, the correlation coefficient clearly varies with sample size and a measure of reliability is needed for it. Second, the coefficients in the linear-regression model are also a function of sample size and need additional information.

Reliability of the Correlation Coefficient

To assess the statistical reliability of the entire linear-regression relationship, it is necessary to determine the probability that the correlation coefficient could have been obtained from random uncorrelated data. The previous example showed that if only two data points are available, a correlation of 1.00 would be obtained even if those points were totally unrelated to each other. As with all statistical tests, two types of error are possible: type I error and type II error.[9] Type I error is the error of rejecting a truth, while type II error is the error of accepting a falsehood. In this case, type I error would be the error of rejecting a linear-regression equation on the grounds that it could have been generated from random uncorrelated data, while type II error would be that of accepting the linear-regression equation as showing a real relationship when in fact the data are random and unrelated. The standard procedure of statistical testing is to minimize type I error, subject to a given level of type II error. To do this, it is necessary to consider the properties of the correlation coefficient.

Suppose a number of independent random samples were chosen from the same population, for each of which a linear-regression relationship was estimated. The above examples indicate that the value of the correlation coefficient will be different for each sample. In other words, there is a distribution of values of the correlation coefficient for any given sample size. If the distribution of this correlation coefficient is known, the probability of a particular value arising by chance from uncorrelated data can be determined.

In general, it is not practical to draw a sufficiently large number of samples to determine the distribution of the correlation coefficient. However, referring to the ANOVA (table 7–1), the ratio of any two independent mean squares is distributed like F with the degrees of freedom of the respective sums of squares, from which the mean squares are calculated. The test for the correlation coefficient can be derived from considering the analysis of variance, for which an F statistic has already been calculated (equation 7.58). To see the relationship between this F test and the correlation coefficient, some rearrangement of equation 7.60 is necessary.

$$R^2 = \frac{\text{SSR}}{\text{SST}} \qquad (7.67)$$

From the identity (equation 7.55), the coefficient of determination can also be written

$$R^2 = 1 - \frac{\text{SSE}}{\text{SST}} \qquad (7.68)$$

Rearranging both these equations to express, respectively, SSR and SSE, results in

$$\text{SSR} = R^2(\text{SST}) \qquad (7.69)$$

$$\text{SSE} = (1-R^2)\text{SST} \qquad (7.70)$$

The F statistic is

$$F = \frac{\text{SSR}/\nu_{\text{SSR}}}{\text{SSE}/\nu_{\text{SSE}}} \qquad (7.71)$$

Thence, the F statistic may be rewritten in terms of R^2.

$$F = \frac{R^2/\nu_{\text{SSR}}}{(1-R^2)/\nu_{\text{SSE}}} \qquad (7.72)$$

Thus the F test for the total regression is effectively a test of the correlation coefficient. It must be stressed that the reporting of the correlation coefficient, R, or the coefficient of determination, R^2, is meaningless without the F statistic. Given the demonstrated dependence of R and R^2 on sample size, this requirement for the F statistic should be clear. It should also be noted that no assumptions have yet been required about the distribution of any elements of the linear-regression relationship or the data from which it is derived.

Significance of the Regression Coefficient

In the examples presented earlier in this section, not only the correlation coefficient changed with the sample, but also the regression coefficient, a_1, and the constant a_0. From the sample of two points, the regression coefficient was 0.762, while for four observations it was 0.44. As discussed for the correlation coefficient, the basic concept for carrying out the significance test is that of having a knowledge of the distribution. In this case, there is not an ANOVA that can circumvent the need for knowing a distribution and the only way to define the distribution is to draw a large number of independent samples. This is an expensive process and is not guaranteed to define the distribution precisely enough to generate the required statistical measures. The basic regression equation is

$$Y_i = a_0 + a_1 X_i + \epsilon_i \qquad (7.73)$$

The regression coefficient, a_1, is

$$a_1 = \frac{\Sigma(X_i - \bar{X})(Y_i - \bar{Y})}{\Sigma(X_i - \bar{X})^2} \qquad (7.74)$$

The variance of the regression coefficient is

$$V(a_1) = \frac{V(Y_i)}{\sum_i (X_i - \bar{X})^2} \qquad (7.75)$$

The proof of this is given later in the chapter. The important point here is that the distribution that must be defined is that of the ratio of Y_i

and X_i. The solution process for linear regression assumes, however, that the X_is are known without error. Hence, the distribution that must be determined is that of Y_i. Consider the error term, ϵ_i, in equation 7.73. As demonstrated earlier in the chapter, ϵ_i represents the measurement error in Y_i. It was also shown that the sum of the ϵ_i values is zero for this regression, implying that the mean value of ϵ_i is also zero. Any given error term, ϵ_j, associated with a value Y_j, can be considered to have been drawn from a distribution of such values for Y_j. The standard assumption for regression is that this distribution is a normal distribution, with a mean of zero for all values of j. In addition, it is assumed that the distributions all have the same variance.

Figure 7–12 shows a graphical representation of this assumption about the error term. The measured Y_j values are assumed to be drawn from identical normal distributions, as shown in the figure, where the distributions are of the measurement error in each measured Y_j. The specific values of ϵ_j are considered to be independent of each other.

$$\text{cov}(\epsilon_i, \epsilon_j) = 0 \text{ for all } i \text{ and } j \qquad i \neq j \qquad (7.76)$$

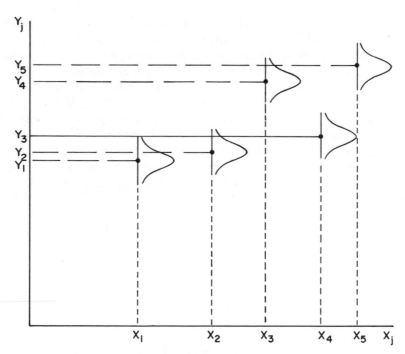

Figure 7–12. Graphical Representation of the Distributional Assumption about the Error Term

The distribution assumption can be stated as equation 7.77, where the equal variance is σ^2.

$$\epsilon_j \sim N(0,\sigma^2) \tag{7.77}$$

Given these properties, the expected value of Y_j for a specific value of X_j is

$$E(Y_j) = a_0 + a_1 X_j \tag{7.78}$$

The variance of Y_j is thus given by the variance of ϵ_j, since X_j is assumed to be known without error. Thus the variance of Y_j is

$$V(Y_j) = \sigma^2 \quad \text{for all } j \tag{7.79}$$

It should also be noted that given the independence assumption for ϵ_i and ϵ_j, shown in equation 7.76, the values of Y_i and Y_j are also independent and the covariance of any two values, Y_i and Y_j, is zero.

Returning now to equation 7.75, the variance of a_1 is

$$V(a_1) = \frac{\sigma^2}{\sum_i (X_i - \bar{X})^2} \tag{7.80}$$

It also follows that a_1 will be normally distributed if ϵ_i is normally distributed. However, σ cannot be determined accurately from a sample. Therefore, the estimated value, s, obtained from the sample is used to calculate the estimated standard error of a_1.

$$\text{est. s.e. } (a_1) = \frac{s}{\sqrt{\sum(X_i - \bar{X})^2}} \tag{7.81}$$

Hence statistical tests of the regression coefficient may be made by using the t distribution as a sample approximation of the normal distribution. First, confidence limits may be set on the estimate of the regression coefficient. If the ϵ_i are all from the same normal distribution, with mean zero and s.d. σ, it can be shown that the $100(1-\alpha)\%$ confidence limits for a_1 are

$$a_1 \pm \frac{t_{(n-2,1-\alpha/2)}s}{\sqrt{[\sum(X_i - \bar{X})^2]}} \tag{7.82}$$

where $t_{(n-2,1-\alpha/2)}$ is the $100(1-\alpha/2)$ percentage point of a t distribution with $(n-2)$ degrees of freedom (the number of degrees of freedom associated with s^2).

Alternatively, hypotheses may be tested of the difference between various values of a_1. The general hypothesis would be:

$$H_1 : \alpha_1 = a_1$$

$$H_0 : \alpha_1 \neq a_1$$

In words, hypothesis one is that the true value of the regression coefficient, α_1, is equal to that found from the sample data, a_1. The null hypothesis is that α_1 is not equal to a_1. The test of the hypothesis is made by computing a value, t'.

$$t' = \frac{(a_1 - \alpha_1)}{\text{est. s.e. } (a_1)} \tag{7.83}$$

If the calculated value of t' is greater than or equal to the table value of t at the $100(1-\alpha/2)$ percentage level and with $(n-2)$ degrees of freedom, hypothesis one can be rejected with $100(1-\alpha)\%$ confidence.

A special case of this test arises when α_1 is set as zero. In this instance, the test is effectively a determination of whether or not the variables Y and X are related to each other. If the true value of a_1 is zero, there is no relation between Y and X. Thus the value of equation 7.84 would be calculated.

$$t'' = \frac{a_1}{\text{est. s.e. } (a_1)} \tag{7.84}$$

If the value t'' is greater than the appropriate table value of t, the hypothesis that the true value of a_1 is zero can be rejected with $100(1-\alpha)\%$ confidence.

Two interesting points may be made about these tests. First, if the confidence limits are determined, the test of equation 7.84 is superfluous for the $100(1-\alpha)\%$ confidence level. If the confidence limits include zero, that is, the lower bound is negative and the upper bound is positive, the hypothesis that the regression coefficient is not significantly different from zero at $100(1-\alpha)\%$ confidence cannot be rejected.

The second point relates to the value of the standard error of a_1. Reviewing equation 7.81, it is clear that the larger is the value of $\Sigma(X_i - \bar{X})^2$, the smaller will be the value of the standard error. It follows, therefore, that the standard error would be minimized if the data were located in two clusters, one for large positive values of X_i and one for large negative clusters. This follows, since two such clusters would give a mean value of $X, \bar{X}$, close to zero and thus each value of $(X_i - \bar{X})^2$ becomes very large. If the collection of data can be designed in such a way, it is clear that a dumbbell-shaped distribution of data points is far preferable to a central cluster of points.

Proof of Equation 7.75. It is appropriate now to return to a proof of equation 7.75. To do this, it is necessary first to note a property of the variance of a complex function. Consider the function defined by

$$F(Z) = \beta_1 Z_1 + \beta_2 Z_2 + \beta_3 Z_3 + \ldots + \beta_m Z_m \tag{7.85}$$

Assuming that each β is invariant and that each Z_k has a variance $V(Z_k)$ and that the covariances are all zero, the variance of the function of equation 7.85 is

$$V[F(Z)] = \beta_1^2 V(Z_1) + \beta_2^2 V(Z_2) + \beta_3^2 V(Z_3) + \ldots + \beta_m^2 V(Z_m) \quad (7.86)$$

Recapping the expression for the regression equation, given earlier as equation 7.74, a_1 is

$$a_1 = \frac{\Sigma(X_i - \bar{X})(Y_i - \bar{Y})}{\Sigma(X_i - \bar{X})^2} \quad (7.87)$$

This may be rewritten as equation 7.88, by partitioning the numerator.

$$a_1 = \frac{\Sigma Y_i(X_i - \bar{X})}{\Sigma(X_i - \bar{X})^2} - \frac{\Sigma \bar{Y}(X_i - \bar{X})}{\Sigma(X_i - \bar{X})^2} \quad (7.88)$$

In the second element of equation 7.88, the $\bar{Y}$ can be taken outside the summation, since it is a constant for all i, thus yielding

$$\frac{\Sigma \bar{Y}(X_i - \bar{X})}{\Sigma(X_i - \bar{X})^2} = \frac{\bar{Y}\Sigma(X_i - \bar{X})}{\Sigma(X_i - \bar{X})^2} \quad (7.89)$$

However, $\Sigma(X_i - \bar{X})$ is equal to zero. Therefore equation 7.89 is zero and equation 7.88 reduces to

$$a_1 = \frac{\Sigma(X_i - \bar{X})Y_i}{\Sigma(X_i - \bar{X})^2} \quad (7.90)$$

This equation may be rewritten in the form of equation 7.85 as

$$a_1 = \frac{(X_1 - \bar{X})}{\Sigma(X_i - \bar{X})^2}Y_1 + \frac{(X_2 - \bar{X})}{\Sigma(X_i - \bar{X})^2}Y_2 + \frac{(X_3 - \bar{X})}{\Sigma(X_i - \bar{X})^2}Y_3 + \ldots$$

$$+ \frac{(X_n - \bar{X})}{\Sigma(X_i - \bar{X})^2}Y_n \quad (7.91)$$

Since the terms $(X_k - \bar{X})/\Sigma(X_i - \bar{X})^2$ are invariant, this equation is completely analogous to equation 7.85. Hence the variance of a_1 may be written as equation 7.92, bearing in mind the earlier assumption of independence of the Y_is.

$$V(a_1) = \frac{(X_1 - \bar{X})^2}{[\Sigma(X_i - \bar{X})^2]^2} V(Y_1) + \frac{(X_2 - \bar{X})^2}{[\Sigma(X_i - \bar{X})^2]^2} V(Y_2)$$

$$+ \frac{(X_3 - \bar{X})^2}{[\Sigma(X_i - \bar{X})^2]^2} V(Y_3) + \ldots + \frac{(X_n - \bar{X})^2}{[\Sigma(X_i - \bar{X})^2]^2} V(Y_n) \quad (7.92)$$

However, the variances of the values Y_k are all identical and equal to

σ^2. Therefore equation 7.92 can be rewritten

$$V(a_1) = \frac{[(X_1-\bar{X})^2 + (X_2-\bar{X})^2 + (X_3-\bar{X})^2 + \ldots + (X_n-\bar{X})^2]\sigma^2}{[\Sigma(X_i-\bar{X})^2]^2} \quad (7.93)$$

The part of the numerator in square brackets, however, is equal to $\Sigma(X_i-\bar{X})^2$. Hence equation 7.93 can be rewritten as equation 7.94, which is identical to equation 7.80.

$$V(a_1) = \frac{\sigma^2}{\Sigma(X_i-\bar{X})^2} \quad (7.94)$$

Significance of the Constant

In addition to determining the significance and precision of the regression coefficient(s), it is also of value to determine the significance and precision of the constant, a_0. This is done by an analogous process to that used for determining the significance of the regression coefficient. As before, it is necessary to determine the standard error of a_0. The constant may be expressed

$$a_0 = \bar{Y} - a_1\bar{X} \quad (7.95)$$

The variance of a_0 is then given by equation 7.96 as will now be proved.

$$V(a_0) = \frac{\sigma^2\Sigma X_i^2}{n\Sigma(X_i-\bar{X})^2} \quad (7.96)$$

Substituting equation 7.91 in equation 7.95 for the regression coefficient, a_1, results in

$$a_0 = \bar{Y} - \frac{\bar{X}[(X_1-\bar{X})Y_1 + (X_2-\bar{X})Y_2 + \ldots + (X_n-\bar{X})Y_n]}{\Sigma(X_i-\bar{X})^2} \quad (7.97)$$

By an analogous process, the variance of a_0 may be written

$$V(a_0) = \sigma^2\left\{\left[\frac{-\bar{X}(X_1-\bar{X})}{\Sigma(X_i-\bar{X})^2}\right]^2 + \left[\frac{-\bar{X}(X_2-\bar{X})}{\Sigma(X_i-\bar{X})^2}\right]^2 + \ldots\right.$$

$$\left. + \left[\frac{-\bar{X}(X_n-\bar{X})}{\Sigma(X_i-\bar{X})^2}\right]^2\right\} + \frac{\sigma^2}{n} \quad (7.98)$$

Note that the variance of $\bar{Y}$ is σ^2/n. Gathering the terms gives

$$V(a_0) = \frac{\sigma^2\Sigma\{\bar{X}^2(X_i-\bar{X})^2\}}{[\Sigma(X_i-\bar{X})^2]^2} + \frac{\sigma^2}{n} \quad (7.99)$$

This equation may be rewritten

$$V(a_0) = \frac{\sigma^2\{n\Sigma(X_i-\bar{X})^2\bar{X}^2 + [\Sigma(X_i-\bar{X})^2]^2\}}{n[\Sigma(X_i-\bar{X})^2]^2} \qquad (7.100)$$

The numerator can be rewritten by noting that $\Sigma(X_i-\bar{X})^2$ is a common factor of both parts of the numerator. This term is also common to the denominator, thus leading to a simplification

$$V(a_0) = \frac{\sigma^2\{n\bar{X}^2 + \Sigma(X_i-\bar{X})^2\}}{n\Sigma(X_i-\bar{X})^2} \qquad (7.101)$$

Expanding the second term of this numerator yields equation 7.102, which is identical to equation 7.96.

$$V(a_0) = \frac{\sigma^2\Sigma X_i^2}{n\Sigma(X_i-\bar{X})^2} \qquad (7.102)$$

Again, given that only an estimate, s, of σ can be determined from a sample, the estimated standard error can be computed.

$$\text{est. s.e. of } a_0 = s\sqrt{[\Sigma(X_i)^2/n\Sigma(X_i-\bar{X})^2]} \qquad (7.103)$$

The same significance test, using the t distribution, can be performed, as shown for the regression coefficient, a_1. Likewise, confidence limits can be established for a_0 as for a_1, using the t distribution. The implications of the test, however, are somewhat different for a regression coefficient. Testing against the hypothesis that the true value of a_0 is zero is a test to determine whether or not the regression line passes through the origin. Finding that a_0 is not significantly different from zero is not therefore a test of the regression relationship.

Estimation Error of the Regression Relationship

The final measurement of use in assessing the regression relationship is that of the standard error of estimate. Given that the values of X are assumed to be known without error and that standard errors have been computed for the regression coefficient and the constant, a standard error can be determined for an estimate, $\hat{Y}_k$, from the regression line.

It must be noted, first, that the implications of the preceding sections are that the estimated values of a_1 and $\bar{Y}$ are each drawn from a family of such values, described by normal distributions. To estimate the error in a regression esimate, $\hat{Y}_k$, involves therefore a two-step process. In the first step, one can esti ate the error in $\hat{Y}_k$, given specific values of a_1 and $\bar{Y}$. Subsequently, one can estimate the error in $\hat{Y}_k$ that arises from the family of a_1 and $\bar{Y}$ values, on which the estimate of $\hat{Y}_k$ is based. In other words, in the first step, one assumes that a_1 and $\bar{Y}$ are true values with an error distribution, while the second step adjusts for the fact that a_1 and $\bar{Y}$ are not true values.

For the first step, the error is calculated in the following way; the estimate of $\hat{Y}_k$ is

$$\hat{Y}_k = \bar{Y} + a_1(X_k - \bar{X}) \qquad (7.104)$$

It may be noted that direct use of the standard error of the constant is unnecessary, since the constant can be expressed as a function of $\bar{Y}$ (subject to error), a_1 (subject to error), and $\bar{X}$ (known without error). The variance of $\hat{Y}_k$ is

$$V(\hat{Y}_k) = V(\bar{Y}) + (X_k - \bar{X})^2 V(a_1) \qquad (7.105)$$

As noted earlier, the variance of $\bar{Y}$ is equal to σ^2/n. The variance of a_1 is given by equation 7.95. Substituting both of these in equation 7.105 yields

$$V(\hat{Y}_k) = \frac{\sigma^2}{n} + \frac{(X_k - \bar{X})^2 \sigma^2}{\Sigma(X_i - \bar{X})^2} \qquad (7.106)$$

The estimated standard error of the value $\hat{Y}_k$ is

$$\text{est. s.e. } (\hat{Y}_k) = s \sqrt{\left[\frac{1}{n} + (X_k - \bar{X})^2/\Sigma(X_i - \bar{X})^2\right]} \qquad (7.107)$$

It should be noted that this value must be calculated for each estimate, $\hat{Y}_k$, using the value X_k that produced it. It is also clear that the value of the standard error will be smallest at the mean, since X_k will then equal $\bar{X}$, the second term of equation 7.107 will be zero, and the standard error of estimate (minimum) will be

$$\text{min est. s.e. } (\hat{Y}_k) = s\sqrt{[1/n]} \qquad (7.108)$$

This equation also shows that the estimated standard error will decline with increasing sample size, n. The form of the confidence limits on the estimate, $\hat{Y}_k$, is shown in figure 7–13. It can be seen that the standard error increases rapidly as X_k becomes more remote from $\bar{X}$.

The variance and standard error just calculated represent the first step in the error calculation for $\hat{Y}_k$. It is now necessary to undertake the adjustment for the range of possible a_1 and $\bar{Y}$ values that could be obtained from independent samples. Since a_1 and $\bar{Y}$ are both based on the observed set of values of Y, from which arises the errors in estimating a_1 and $\bar{Y}$, the correction process involves the addition of the variance of the observed values of Y to the variance of the estimated $\hat{Y}_k$. Thus the variance of the

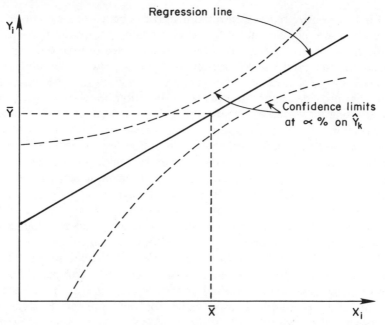

Figure 7–13. Shape of the Confidence Limits on Estimates of Y from the Regression Line

prediction of an individual observation is

$$V(\hat{Y}_k) + \sigma^2 = \sigma^2\left\{1 + \frac{1}{n} + \frac{(X_k-\bar{X})^2}{\Sigma(X_i-\bar{X})^2}\right\} \tag{7.109}$$

The estimated standard error is

$$\text{est. s.e. } (\hat{Y}_k) = s\left[1 + \frac{1}{n} + \frac{(X_k-\bar{X})^2}{\Sigma(X_i-\bar{X})^2}\right]^{1/2} \tag{7.110}$$

Summary

This section has provided a number of measures of goodness of fit for linear-regression equations. These measures were shown to be needed because of the variation in the regression estimates resulting from changes in sample size or the choice of different samples of the same size. It is worthwhile to summarize briefly the various measures developed and the contribution they make to assessing the reliability of the regression relationship.

First, the entire regression relationship can be assessed by computing the correlation coefficient and the coefficient of determination. The coefficient of determination provides a direct measurement of the proportion of the observed sample variance of the dependent variable that is captured or explained by the regression equation. The correlation coefficient provides a measure of the association between the dependent and independent variables, where the sign of the coefficient also indicates the direction of the association. However, since both the correlation coefficient and the coefficient of determination are determined from the sample variance, both measures are dependent upon the sample size and will also vary across independent samples of the same size drawn from a single population. Hence, these two measures also require some statistical-significance measure that will permit a judgment to be made as to whether the computed values could have occurred by chance from uncorrelated data or represent a nonrandom relationship.

Second, and in response to the need for a statistical measure for the correlation, it was shown that an F statistic could be computed from the ratio of the regression and residual or error mean squares. This F statistic may be compared with a table value for the appropriate degrees of freedom and the selected probability, α, of type I error. Since the degrees of freedom are a function of the sample size and (as described in chapter 8) the number of independent variables, the F statistic provides an absolute measure of goodness of fit.

Third, a statistical test was developed, based on the t distribution, for the regression coefficient. Three principal uses were described for this test. In the first place, the regression coefficient may be tested for a significant difference from zero. If the hypothesis that the coefficient is equal to zero cannot be rejected, the regression has not established a relationship between the variable X and the variable Y. Also, the test may be used to compare a given regression coefficient with that obtained from a different sample or with a prior hypothesized value. Thus, for example, samples may be drawn on two separate occasions and the resulting regression coefficients compared statistically to determine whether the coefficient value has changed between the two occasions. The third use is to establish confidence limits on the regression coefficient, thus providing a direct measure of the reliability of that coefficient.

Fourth, a similar statistical test, also based on the t distribution, was derived for the constant of the regression. While the statistical test is similar, its implications are different. The test against zero is a test to determine whether or not the regression line passes through the origin. For example, suppose Y is the number of home-based trips produced by a given unit of area and X is the population of the area. The lack of population in an area should imply that no home-based trips are produced.

A nonzero constant in the linear relation implies, however, that some base number of trips will be produced from a vacant area. Clearly, this is counterintuitive and the result of finding a nonzero constant throws some doubt on the reliability of the relationship produced. Apart from this, the test against some other given value and the establishment of confidence limits are of identical use to those for the regression coefficient.

Finally, a measure was developed of the estimated standard error of an estimate from the regression. This measure was developed both for the mean estimate and for an individual estimate. It may be noted, from this measure, that the standard error increases rapidly as the estimate is made further from the mean of the sample. It must be noted that this standard error of estimate is only valid for the sample observations of X and not for values outside the observed range.

A final note is in order on these measures. In the case of a bivariate regression only, either the F test or the t test for the regression coefficient is redundant. First, it is clear that a failure of the t test for a difference between the regression coefficient and zero implies no nonrandom relationship between Y and X. This is the identical test being carried out with the F test. Second, it may be noted that there is a relationship between the table values of F and t, as shown,

$$F_{(1,n-2)\alpha} = t^2_{(n-2)\alpha/2} \tag{7.111}$$

In other words, a table value of F with 1 and $(n-2)$ degrees of freedom at the $\alpha\%$ level is equal to the square of the table value of t with $(n-2)$ degrees of freedom for a one-tailed test at $\alpha/2\%$, or a two-tailed test at $\alpha\%$. Third, the relationship clearly holds between the calculated F statistic and the square of the t statistic for the regression coefficient. From equation 7.84, the square of the computed t is

$$t^2 = \frac{a_1^2 \Sigma(X_i - \bar{X})^2}{\sigma^2} \tag{7.112}$$

Similarly, the calculated F value for the regression, given in equation 7.71, can be written

$$F = \frac{(n-2)\Sigma(\hat{Y}_i - \bar{Y})^2}{\Sigma(\hat{Y}_i - Y_i)^2} \tag{7.113}$$

This equation can be shown to reduce to equation 7.114, making the appropriate substitutions for $\hat{Y}_i$.

$$F = \frac{a_1^2 \Sigma(X_i - \bar{X})^2}{\sigma^2} \tag{7.114}$$

This is identical to t^2 in equation 7.112, hence confirming the redundancy of one of these tests for a bivariate regression.

Table 7–6
Random Number Data Set for Example 1

Y	X	Y	X
67	20	22	91
66	6	37	95
58	58	64	61
37	82	5	1
10	78	51	81
24	64	81	82
79	27	35	12
72	11	77	63
36	24	63	28
44	40	3	0

Some Examples of the Application of Linear Regression

Two examples may be useful to illustrate the procedures dealt with in this chapter. In the first example, 40 random numbers have been drawn and assigned arbitrarily to variables X and Y, as shown in table 7–6. The first 10 numbers were assigned to Y, the next 10 to X, the next 10 to Y, and the last 10 to X. From the table, $\bar{Y}$ is 46.55 and $\bar{X}$ is 46.20. The resulting regression equation is

$$\hat{Y}_k = 46.05 + 0.011X_k \tag{7.115}$$

The various sums of squares needed to compute the other relevant measures are given in table 7–7. From these, a number of goodness-of-fit measures can be computed, as shown in table 7–8. Not surprisingly, the correlation coefficient (R) and coefficient of determination (R^2) are very small. The F statistic is not significant even for $\alpha = 10\%$. Likewise, the t

Table 7–7
Sum of Squares for Data of Table 7–6

Measure	Value
$\bar{X}$	46.20
$\bar{Y}$	46.55
ΣY_i^2	55179.0
ΣX_i^2	62940.0
$\Sigma X_i Y_i$	43233.0
$\Sigma(X_i-\bar{X})^2$	20251.2
$\Sigma(Y_i-\bar{Y})^2$	11840.95
$\Sigma(X_i-\bar{X})(Y_i-\bar{Y})$	220.8

Table 7–8
Goodness-of-Fit Measures for the Model of Equation 7.112

Statistic	Value
R^2	0.000203
R	+0.014
F	0.00345
t_{a_1}	0.061
t_{a_0}	4.555

statistic for the regression coefficient, a_1, is not significant even at $\alpha = 40\%$. However, the constant term does have a significant value (i.e., significantly different from zero) at better than $\alpha = 0.1\%$.

It may also be noted that the constant is almost equal to the mean value of both X and Y, indicating that equation 7.115 represents approximately a horizontal line through the mean of Y and X, as would be expected. It is interesting to compare the confidence limits at $\alpha = 5\%$ for the constant.

$$95\% \text{ confidence limits} = 46.05 \pm 21.24$$
$$= 24.81 - 67.29 \qquad (7.116)$$

Hence the horizontal line lies, with 95% confidence, somewhere between 24.8 and 67.3.

The second example represents some real data on trips per household and cars per household for a study area divided into 40 zones.[10] The data are presented in table 7–9.

The resulting linear-regression equation between trips per household (Y) and cars per household (X) is

$$Y_i = 0.387 + 6.665X_i \qquad (7.117)$$

The measures needed to compute the goodness-of-fit statistics are shown in table 7–10. The goodness-of-fit statistics are presented in table 7–11. From these, it can be seen that the correlation coefficient and coefficient of determination are much higher than in the previous example. In addition, the F statistic is significant at better than 1% (the table value being 9.0 for 1 and 38 degrees of freedom and at 0.5%). Hence, the model represents a real nonrandom relationship at far better than 99.6% confidence. It also follows that the coefficient, a_1, is highly significant, the t value at 99.9% being about 3.6. One may also note that the F value in table 7–11 is indeed the square of the t score for a_1. Finally, it may be seen that the constant is not significantly different from zero, as shown by the very low t value of 0.24. This suggests that when there are no cars in a household no trips are made. (It should be noted that the figures used are

Table 7–9
Transport Data Set for Example 2

Trips/ Household	Cars/ Household	Trips/ Household	Cars/ Household
8.5	1.2	10.9	1.4
6.9	1.4	8.0	1.1
9.9	1.4	6.3	1.2
9.9	1.6	10.0	1.4
10.8	1.3	7.9	1.3
10.9	1.6	9.3	1.4
10.1	1.6	9.6	1.5
10.0	1.5	12.3	1.7
8.3	1.1	9.8	1.7
8.3	1.1	9.8	1.6
8.2	1.2	10.5	1.5
7.6	1.2	8.7	1.3
8.4	1.3	10.1	1.4
15.8	1.7	11.9	1.7
9.1	1.4	13.4	1.8
12.2	1.3	11.3	1.7
13.0	1.7	8.8	1.6
9.0	1.2	10.6	1.5
11.5	1.4	9.9	1.4
10.6	1.5	10.6	1.6

Table 7–10
Means and Sums of Squares for Data of Table 7–9

Measure	Value
$\bar{X}$	1.4375
$\bar{Y}$	9.9675
ΣX_i^2	84.11
ΣY_i^2	4106.49
$\Sigma X_i Y_i$	582.82
$\Sigma(X_i-\bar{X})^2$	1.454
$\Sigma(Y_i-\bar{Y})^2$	132.448
$\Sigma(X_i-\bar{X})(Y_i-\bar{Y})$	9.6888

Table 7–11
Goodness-of-Fit Statistics for Model of Equation 7.114

Statistic	Value
R^2	0.487
R	0.698
F	36.07
t_{a_1}	6.02
t_{a_0}	0.24

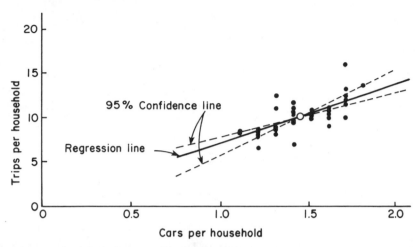

Figure 7–14. Plot of Regression Line and Confidence Limits for Second Example

mean values for 40 zones, so a zero entry would mean no cars in an entire zone.) Confidence limits may be determined for the regression coefficient.

$$\text{confidence limits } (95\%) = 4.416 \text{ to } 8.914 \qquad (7.118)$$

These confidence limits are illustrated in figure 7–14. Similarly, one can compute the estimated standard error of values of $\hat{Y}_k$. Some example values are shown in table 7–12 and 95% confidence limits, based on these, are illustrated in figure 7–15. As can be seen, the high significance of the regression is reflected in the relatively tight and fairly constant confidence bands, which diverge only slightly over the range from 0.5 to 2.2 on cars per household.

In conclusion, these two examples demonstrate the usefulness of the

Table 7–12
Standard Errors of Estimate for Equation 7.114

X_k	$\hat{Y}_k$	*Standard Error*
0.5	3.72	±1.705
0.8	5.72	±1.526
1.1	7.72	±1.403
1.3	9.05	±1.361
1.4375	9.9675	±1.353
1.6	11.05	±1.365
1.7	11.72	±1.384
2.0	13.72	±1.489
2.2	15.05	±1.595

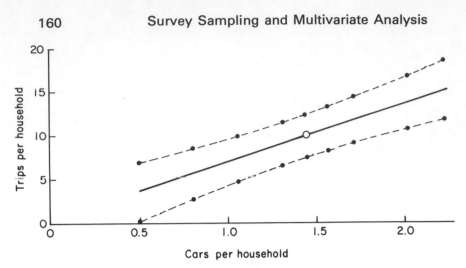

Figure 7–15. Plot of 95% Confidence Limits on Estimates of Trips per Household

goodness-of-fit measures to determine how good a regression relationship is. In the case of the random numbers, it showed that no relationship was achieved, while the data on trips and cars were shown to yield a strong and potentially useful relationship. In the next chapter, these various concepts and relationships are extended to the multivariate case, and some important properties of that case are examined.

Notes

1. A definition of these various models, particularly in the context of transportation planning, is in P.R. Stopher and A.H. Meyburg, *Urban Transportation Modeling and Planning,* Lexington, Mass.: Lexington Books, D.C. Heath and Co., 1975, pp. 27–30.

2. Isaac Newton, *Philosophiae Naturalis Principia Mathematica,* London, 2d ed., 1713, Book III, Prop. VII and Corr. II, p. 414.

3. Stopher and Meyburg have provided a definition of the terms "prediction" and "forecasting," in *Urban Transportation Modeling,* p. 26.

4. William Alonso, "The Quality of Data and Choice and Design of Predictive Models," *Highway Research Board Special Report 97,* National Academy of Sciences, Washington, D.C., 1968, pp. 178–192; Stopher and Meyburg, *Urban Transportation Modeling,* pp. 41–44.

5. Stopher and Meyburg, *Urban Transportation Modeling,* p. 44.

6. It also follows that an assumption of error in both X_i and Y_i will

equally lead to a situation in which prediction requires the knowledge of an observable error in the X_is. In no case is knowledge necessary of the magnitude of the error in the dependent variable to produce estimates of its values from the regression line.

7. Edwin L. Crow, Francis A. Davis, and Margaret W. Maxfield, *Statistics Manual,* New York: Dover Publications, 1960, chapter 5; Sylvain Ehrenfeld and Sebastian B. Littauer, *Introduction to Statistical Method,* New York: McGraw-Hill Book Co., 1964, chapter 9.

8. Crow, Davis, and Maxfield, *Statistics Manual,* pp. 118, 122–123.

9. Ehrenfeld and Littauer, *Statistical Method,* pp. 244–253. A more detailed discussion of hypothesis testing and type I and type II errors is in chapter 13 of this book.

10. Peter R. Stopher, *Sample Project Manual, U.T.P. Package,* Version of 1970–71, unpublished manual.

8

Multivariate Regression

Introduction

The last chapter provided a detailed treatment of bivariate regression, that is, a linear relationship between only two variables. However, it is relatively unlikely that a useful model can be obtained from a bivariate relationship, because this would generally suggest a much too close relationship of variables to be plausible. For example, in the case of the model of trips per household as a function of cars per household, only 48.7% of the variance in the observed trips per household was captured by the variable cars per household. It is clearly not very satisfactory to be able to account for less than half of the variance in the model. Thus it appears extremely probable that real modeling efforts (as opposed to the illustrative examples discussed in the last chapter) will require a more complex relationship to be established.

Additional complexity can involve two basic alternatives: a more complex relationship between the variables, that is, a nonlinear relationship; or the use of more than one independent variable. The second alternative is both simpler and the subject of this chapter. The extension of bivariate linear regression to multivariate linear regression is principally a matter of simple mathematical generalizations of the relationships developed in chapter 7.

Before undertaking these generalizations, it is useful to consider the concepts and visualization of the multivariate linear relationship. Bivariate regression involved fitting a straight line to data in a two-dimensional space. Multivariate regression involves the fitting of an $(n-1)$-dimensional surface to data in n-dimensional space. It should be noted that the surface is a plane, since the relationship is linear. Illustrations of higher dimensionality problems cannot, of course, be provided. However, the extension is conceptually relatively simple.

The three-dimensional problem may be examined in another light. The relationship postulated is

$$Y_i = \alpha_0 + \alpha_1 X_{1i} + \alpha_2 X_{2i} + \epsilon_i \tag{8.1}$$

If one of the two independent variables, X_1 or X_2, were to be held constant, the relationship could be written

$$Y_i = \alpha_0' + \alpha_1 X_{1i} + \epsilon_i \tag{8.2}$$

(X_{2i} = fixed value).

Alternatively, equation 8.3 could be used;

$$Y_i = \alpha_0'' + \alpha_2 X_{2i} + \epsilon_i \qquad (8.3)$$

(X_{1i} = fixed value).

In these two equations, α_0' and α_0'' represent adjusted constant terms that take account of the effect of the variable whose value is fixed. Thus in equation 8.2, assuming that the fixed value of X_{2i} is some arbitrary value γ, then α_0' is

$$\alpha_0' = \alpha_0 + \alpha_2 \gamma \qquad (8.4)$$

Similarly, if the fixed value of X_{1i} in equation 8.3 is θ, then α_0'' is

$$\alpha_0'' = \alpha_0 + \alpha_1 \theta \qquad (8.5)$$

The relationships of equations 8.2 and 8.3 could be represented then as a family of straight lines in a two-dimensional space. Note that the lines are parallel, since at each value of X_2 that variable has no further effect on the value of Y_i as X_{1i} changes in value. In engineering terms, the lines represent a series of projections of the surface onto the two-dimensional surface formed by the Y and X_1 axes.

Conceptually, the multivariate regression represents a postulate that some phenomenon, Y, is related to several other variables, all of which may be considered to cause, individually, changes in the value of Y. For example, one may postulate that the likelihood of precipitation on a particular day is a function of the atmospheric pressure, the pressure gradient, the direction of the wind, the relative humidity, the location of the point of concern, and the upper-air movement pattern. Each of these elements may be considered to be an independent variable, while the likelihood of precipitation is the dependent variable. No one of the independent variables will by itself provide a good estimate of the likelihood of precipitation, but taken together the variables may provide a reasonably good estimator of the likelihood. It must be noted, however, that complete explanation of the phenomenon is still not likely to be achieved. As discussed in chapter 7, such complete explanation would be of the nature of a physical law. In most applications in the social sciences and engineering, such laws are not likely to be found or even sought. Rather, the goal is to find a good model of a phenomenon.

It is important at this point to reconsider the concepts of dependence and independence. As stated in chapter 7, the basic concept of dependence and independence is one of causality. Thus changes in an independent variable cause changes in the dependent variable, while changes in the dependent variable do not cause changes in the independent variable. Thus in the weather-forecasting example, humidity may not be an appropriate independent variable, since precipitation will cause changes in humidity (although it may not be strictly possible to say that the likelihood of precipitation will cause changes in humidity). In the context of multivariate regression, independence does *not* mean that the independent variables are unrelated to each other. Hence it is not a statement that the variables are independent of each other. However, if any two or more independent variables are strongly related to each other, problems will arise. First, one may consider that if there are two or more highly related independent variables, the use of them in a model creates a redundancy. Conceptually, such a redundancy is undesirable and may contribute to some problem in understanding how the phenomenon being modeled behaves in reality. Statistically (discussed later in the chapter), the presence of such redundancy will lead to incorrect estimates of the coefficients of all such interrelated variables. Hence it may be stated that high intercorrelations are conceptually and statistically undesirable, but the principle of independence by itself does not exclude such intercorrelations. Such variables should therefore be checked for and, if found, excluded from the proposed linear relationship.

Multivariate Linear Regression

Given the basic concepts of multivariate linear regression, attention can now be given to the estimation of the coefficients of the regression equation. In general, a model of the form of equation 8.6 may be postulated as representing the multivariate linear-regression equation.

$$Y_i = \alpha_0 + \alpha_1 X_{1i} + \alpha_2 X_{2i} + \ldots + \alpha_m X_{mi} + \epsilon_i \qquad (8.6)$$

To calibrate such a model, the procedure is to minimize the sum of the squared error terms, ϵ_i, to find the values of the coefficients and the constant. Rearranging equation 8.6 to isolate ϵ_i on one side produces

$$\epsilon_i = Y_i - \alpha_0 - \alpha_1 X_{1i} - \alpha_2 X_{2i} - \ldots - \alpha_m X_{mi} \qquad (8.7)$$

The process is then completely analogous to that described for bivariate linear regression. Equation 8.7 is squared and summed over all observations, i, and partial differentials taken with respect to each of the unknown values, α_k. The result of this procedure is a set of normal equations of the following form:

$$\Sigma(Y_i - a_0 - a_1 X_{1i} - a_2 X_{2i} - \cdots - a_m X_{mi}) = 0$$

$$\Sigma X_{1i}(Y_i - a_0 - a_1 X_{1i} - a_2 X_{2i} - \cdots - a_m X_{mi}) = 0$$

$$\cdots \quad \cdots \quad \cdots \quad \cdots \quad \cdots \quad \cdots \quad \cdots \quad \cdots \quad \cdots \quad \cdots \quad \cdots \quad \cdots \qquad (8.8)$$

$$\cdots \quad \cdots \quad \cdots \quad \cdots \quad \cdots \quad \cdots \quad \cdots \quad \cdots \quad \cdots \quad \cdots \quad \cdots$$

$$\Sigma X_{mi}(Y_i - a_0 - a_1 X_{1i} - a_2 X_{2i} - \cdots - a_m X_{mi}) = 0$$

The set has $(m+1)$ equations, all of which are linear in the unknown values, a_0 through a_m. Thus a unique solution may be found by solving this set of linear equations as was done before. As an example, consider the case in which m equals 2. The normal equations are

$$\Sigma(Y_i-a_0-a_1X_{1i}-a_2X_{2i}) = 0 \qquad (8.9)$$

$$\Sigma X_{1i}(Y_i-a_0-a_1X_{1i}-a_2X_{2i}) = 0 \qquad (8.10)$$

$$\Sigma X_{2i}(Y_i-a_0-a_1X_{1i}-a_2X_{2i}) = 0 \qquad (8.11)$$

As before, the first normal equation provides an identity in terms of the means.

$$\bar{Y} - a_0 - a_1\bar{X}_1 - a_2\bar{X}_2 = 0 \qquad (8.12)$$

Rearranging this equation to yield a definition of the constant, a_0, produces

$$a_0 = \bar{Y} - a_1\bar{X}_1 - a_2\bar{X}_2 \qquad (8.13)$$

The remainder of the solution becomes algebraically tedious and is best solved by matrix algebra. Defining the vector of observations of Y_i as $\mathbf{Y}$, the matrix of observations of X_{1i} and X_{2i} as $\mathbf{X}$, the vector of coefficients as $\mathbf{a}$, and the vector of error terms, ϵ_i, as $\boldsymbol{\epsilon}$, the solution for the coefficients is[1]

$$\mathbf{a} = (\mathbf{X'X})^{-1}\mathbf{X'Y} \qquad (8.14)$$

This is the general solution of the multivariate regression.

Properties of the Regression

In chapter 7, a number of properties of the bivariate linear-regression model were described and discussed. Without exception, the multivariate linear-regression model holds these properties.

First, the sum of the residuals is zero, regardless of the number of independent variables used. This follows from the first of the normal equations. The sum of the residuals is given by equation 8.15, which is a rearrangement of equation 8.6, summed over all observations, i.

$$\Sigma\epsilon_i = \Sigma(Y_i-\alpha_0-\alpha_1X_{1i}-\alpha_2X_{2i}- \cdots -\alpha_mX_{mi}) \qquad (8.15)$$

However, the first normal equation is

$$\Sigma(Y_i-a_0-a_1X_{1i}-a_2X_{2i}- \cdots -a_mX_{mi}) = 0 \qquad (8.16)$$

Hence it follows that the sum of the errors (residuals) is zero.

$$\Sigma\epsilon_i = 0 \qquad (8.17)$$

The implications of this property are again the same as those discussed for bivariate linear regression.

Second, the analysis-of-variance procedure may be applied again, as for the bivariate case, where the same relationship exists between the sums of squares.

$$\begin{array}{c}\text{sum of squares} \\ \text{about the mean}\end{array} = \begin{array}{c}\text{sum of squares} \\ \text{about the regression}\end{array} + \begin{array}{c}\text{sum of squares} \\ \text{due to the regression}\end{array} \quad (8.18)$$

Mathematically, these sums of squares are defined as before.

$$\Sigma(Y_i - \bar{Y})^2 = \Sigma(Y_i - \hat{Y}_i)^2 + \Sigma(\hat{Y}_i - \bar{Y})^2 \qquad (8.19)$$

However, the computations become more involved since the estimates of Y, $\hat{Y}_i$ are now more complex.

$$\hat{Y}_i = a_0 + a_1 X_{1i} + a_2 X_{2i} + \cdots + a_m X_{mi} \qquad (8.20)$$

The expressions for the numbers of degrees of freedom of the sums of squares must now be modified, or generalized, to the multivariate case. It was shown that the sum of squares about the mean has $(n-1)$ degrees of freedom, the residual sum of squares has $(n-2)$ degrees of freedom, and the regression sum of squares has 1 degree of freedom. From the arguments used to obtain these and equation 8.20, it can be seen that the sum of squares about the mean still has $(n-1)$ degrees of freedom. However, the estimates of Y_i, the $\hat{Y}_i$, now require the evaluation of m regression coefficients, so the residual sum of squares has $(n-m-1)$ degrees of freedom and the regression sum of squares has m degrees of freedom. Thus, in the general case, the mean squares may be written.

$$\text{mean square about mean (total)} = \frac{\text{sum of squares about mean}}{n-1} \quad (8.21)$$

$$\text{regression mean square} = \frac{\text{regression sum of squares}}{m} \quad (8.22)$$

$$\text{residual mean square} = \frac{\text{residual sum of squares}}{(n-m-1)} \quad (8.23)$$

Given these relationships, the analysis-of-variance table can be constructed in its general form for a multivariate-linear regression, as shown in table 8–1. It also follows that an F test can be constructed from ANOVA, as for the bivariate regression. The F value for the regression is again computed as the ratio of the regression mean square to the error (or residual) mean square, and has m and $(n-m-1)$ degrees of freedom. Thus the test for overall significance of the regression is

$$Q = \frac{\text{SSR}/m}{\text{SSE}/(n-m-1)} \geq F_{\alpha(m,n-m-1)} \qquad (8.24)$$

Table 8–1
ANOVA for Multivariate Regression

Source of Variation	Sum of Squares	Degrees of Freedom	Mean Square
Regression	$\Sigma(\hat{Y}_i-\bar{Y})^2 = \mathrm{SSR}$	m	SSR/m
Error (residual)	$\Sigma(Y_i-\hat{Y})^2 = \mathrm{SSE}$	$n-m-1$	$\mathrm{SSE}/(n-m-1)$
Total	$\Sigma(Y_i-\bar{Y})^2 = \mathrm{SST}$	$n-1$	$\mathrm{SST}/(n-1)$

If equation 8.24 is true, that is, Q is greater than or equal to the table value of F at a selected value of α (e.g., 5%), the null hypothesis that there is no relationship between the dependent variable, Y, and the set of independent variables, $X_1, X_2, \ldots X_m$, may be rejected with $(100-\alpha)\%$ confidence. It should be noted that this test does not test the contribution of any individual independent variable, X_k, but examines the entire relationship with all m independent variables.

To test the significance of each individual variable in the regression equation, t tests can again be conducted on each coefficient, as described for the bivariate regression, and the computation is exactly the same. The general form of the computation of a t score to determine if a coefficient is significantly different from zero is

$$t_k = \frac{a_k\sqrt{\Sigma(X_{ki}-\bar{X}_k)^2}}{s} \tag{8.25}$$

where a_k = coefficient of the kth variable
  s = standard error of an observation of Y
  X_{ki} = ith observation of the kth variable
and $\bar{X}_k$ = mean of the kth variable

The constant may be tested by the same procedure as before. The degrees of freedom of the t tests are all $(n-2)$.

Correlation in Multivariate Regression

For the entire multivariate regression equation, a correlation coefficient, comprising the square root of the ratio of the regression sum of squares to the total sum of squares of the dependent variable, can again be determined. This correlation coefficient is denoted as R or $R_{y|1,2,\ldots,m}$. It represents the correlation between the dependent variable and all the independent variables. R or $R_{y|1,2,\ldots,m}$ = multiple correlation coefficient between Y and m independent variables. Similarly, R^2 is defined as the multiple coefficient of determination and provides an estimate of the amount of the observed variance in the dependent variable that is captured by the set of independent variables used.

A second type of correlation coefficient also exists in multivariate regression. This is the simple correlation coefficient, denoted r_{yj} or r_{ij}. It represents the correlation between any pair of variables, ignoring all other variables. These simple correlations are symmetric.

$$r_{yj} = r_{jy} \tag{8.26}$$

$$r_{ij} = r_{ji} \tag{8.27}$$

A matrix of these simple correlations for all variables is usually a standard part of the output of a computer-program package for multivariate regression. Its use is discussed later. These simple correlation coefficients are computed as the ratio of the sums of squares which would be used for bivariate regression between the two variables.

Finally, a third correlation coefficient can be defined in multivariate regression: the partial correlation coefficient. This is the correlation coefficient between the dependent variable and any one independent variable after taking account of the effect of all the other independent variables in the equation. It is denoted $r_{yj \mid 1,2,...,m \neq j}$, and is computed

$$r_{yj \mid 1,2,...,m \neq j} = \sqrt{\frac{SSR_{X_1,X_2,...,X_m} - SSR_{X_1,X_2,...,X_{m \neq j}}}{SST}} \tag{8.28}$$

Thus the partial correlation is the correlation between an independent variable and the residual variance of the dependent variable (after the deduction of the variance "explained" by the other independent variables).

In summary, the multiple correlation coefficient provides an overall goodness-of-fit measure for the entire multivariate-regression equation. The simple correlation coefficient provides both a check on the basic usefulness of each independent variable (the simple correlation with the dependent variable) and provides a diagnostic of potential multicollinearity problems among the independent variables. The partial correlation coefficient provides an indication of the worthwhileness of adding any one independent variable in the presence of the others. Both the simple and partial correlation coefficients assist in the selection of independent variables.

Calibration Methods

In bivariate regression, there is a single simple process of calibration. The values of the constant and the single coefficient are computed directly from the equations presented in chapter 7, together with the relevant statistical measures. The process is simple and offers no serious problems. This is not the case for multivariate regression. While prior hypotheses and scatter diagrams provide a basis for structuring the multivariate regression, the existence of multicollinearities among the independent variables may not be revealed until the multivariate-regression model is calibrated. In addition, redundancies of some of the independent variables do not show up in scatter diagrams and are frequently not apparent when the prior hypotheses are constructed. The detection of problems in the independent variables is rather complex, however. Even in a simple case of three independent variables, it would be necessary to test all possble combinations of variables, taken one, two, or three at a time. With three variables, there are seven possible models that would need to be tested. Clearly, if there is a reasonable number of independent variables hypothesized to cause changes in a dependent variable, there could be a very large number of models to be tested (6 variables generates 63 models to be tested, 10 variables generates 922 models, etc.).

What is needed, for multivariate regression, is a systematic procedure to search for the best model from a hypothesized set of independent variables. The procedure must be able to reduce substantially the number of models that must be tested to select the best model. Two basic procedures have been developed for this and one hybrid method. These are called, respectively, Backward Elimination, Forward Selection, and Stepwise Estimation. Each method hinges on the concept and calculation of the partial correlation coefficients.

Backward Elimination

The Backward Elimination procedure is carried out as follows. First, all the independent variables considered to be relevant are entered into an equation. These variables are examined and eliminated one by one from the regression, subject to the equation continuing to fulfill three specified criteria. These criteria are generally couched in terms of F values and coefficient t scores. This procedure is carried out using a partial F test.

A partial F test is analogous to the computation of a partial correlation. The partial correlation coefficient is calculated by taking the square root of the ratio of the difference in regression sums of squares with and without the independent variable in question to the total sum of squares for the dependent variable (equation 8.28). The partial F value is

$$F_k = \frac{\text{MSR}_{(k)} - \text{MSR}_{(k-1)}}{\text{MSE}} \tag{8.29}$$

where F_k = partial F value for the kth variable
   $MSR_{(k)}$ = regression mean square with k variables in the equation,
   $MSR_{(k-1)}$ = regression mean square without the kth variable in the equation
and   MSE   = residual mean square with k variables in the equation

The number of degrees of freedom of the partial F value will, in this case, be computed as the difference in the numbers of degrees of freedom of the regression sum of squares with k variables and the regression sum of squares with $(k-1)$ variables, and the residual mean square will have $(n-k-1)$ degrees of freedom. In equation 8.29, the number of degrees of freedom for F_k will be 1 and $(n-k-1)$. In such a case, the value of F computed from this ratio of mean squares would be compared with that in the table for a particular significance level, say the 95% significance level. If the F value found from the ratio of the mean squares were greater than that given in the table, one would conclude that the addition of the variable k had added significantly to the "explanatory" power of the equation. If the partial F value so calculated was less than the value in the table, it would be concluded that the addition of this variable had not improved significantly the "explanatory" power of the equation.

Returning now to the Backward Elimination procedure, the process continues by computing the partial Fs of all the variables as though they were the last to enter the equation. From these, the variable with the lowest partial F below some predetermined value, F_0, is discarded from the equation. The regression equation is then recomputed without the variable and partial F values are again calculated for each variable still in the equation. Again, the variable with the lowest partial F below F_0 is discarded, and this process is continued until all the partial Fs are greater than the required value.

An example is useful to demonstrate the method. Suppose six variables denoted $X_1, X_2, \ldots, X_6$ are hypothesized as causing changes in a dependent variable Y. First, a model is constructed with all six independent variables.

$$Y_i = a_0 + a_1 X_{1i} + a_2 X_{2i} + a_3 X_{3i} + a_4 X_{4i} + a_5 X_{5i} + a_6 X_{6i} \quad (8.30)$$

Next partial F scores are computed for each of the six variables, assuming the other five are already in the equation. These values are examined and the smallest one is checked against the relevant table value, say, for a 95% confidence level. If it fails the test, that variable is excluded. In this example, suppose X_4 has the smallest partial F score and is not significant at 95% (the chosen confidence level). This variable is discarded and the model of equation 8.31 is constructed.

$$Y_i = b_0 + b_1 X_{1i} + b_2 X_{2i} + b_3 X_{3i} + b_5 X_{5i} + b_6 X_{6i} \quad (8.31)$$

The process continues until the variable with the lowest partial F score is significant at the 95% level. Thus the model of equation 8.32 might result, in which all of the remaining variables make a significant contribution (at 95%) to the explanatory power of the model.

$$Y_i = d_0 + d_2 X_{2i} + d_3 X_{3i} + d_6 X_{6i} \qquad (8.32)$$

Forward Selection

The Forward Selection method is the exact reverse of the Backward Elimination procedure. Instead of starting with all of the variables in the model, the Forward Selection method starts with none of the variables in the model. Instead, partial Fs are calculated for each variable not yet in the equation, and the one with the highest value is added, until the highest partial F value of any of the variables not yet in the equation is less than some predetermined significance level.

Using the same example, simple correlations and F scores are computed between each independent variable and the dependent variable. Suppose X_3 has the highest correlation. The first model step is

$$Y_i = a_0 + a_3 X_{3i} \qquad (8.33)$$

At the next step, the partial F scores are computed for the other five variables with X_3 in the equation. Suppose X_2 now has the highest partial F score and it is also significant at the desired level, the model now becomes

$$Y_i = b_0 + b_2 X_{2i} + b_3 X_{3i} \qquad (8.34)$$

The process continues until no partial F scores are significant at the desired level. In the example, the final model should be

$$Y_i = d_0 + d_2 X_{2i} + d_3 X_{3i} + d_6 X_{6i} \qquad (8.35)$$

This will not always be the case, however, if there are significant intercorrelations among the independent variables. Where there are, say, two intercorrelated variables, one of the two would be added by the Forward Selection method, while it may be the same variable that is deleted by the Backward Elimination method. Thus the two methods will not necessarily result in the same model.

Stepwise Regression

Stepwise regression is a procedure intended to combine the procedures of Forward Selection and Backward Elimination in such a way as to generate a unique model. Each step of the Stepwise Regression procedure is alternately Forward Selection or Backward Elimination. Thus the Stepwise procedure can be described as follows:

Step 1: Compute simple correlations and F scores and select the independent variable with the highest significant score. Build a model with the selected variable.

Step 2: Compute the partial F score for the variable in the model and reject if it is not significant.[a]

Step 3: Compute partial F scores for all of the variables not yet in the equation. If the highest one is above the required significance value, a new model is built with this variable and the one entered at step 1.

Step 4: Compute partial F scores for the variables in the equation. If either is below the required value, drop it from the equation and reestimate the model without that value.

Step 5: Repeat step 3.

Step 6: Repeat step 4.

The process continues until no more variables can be added and none that are included can be dropped. To ensure that the procedure does not fall into an endless cycle, the F value for including a variable is set slightly higher than the F value for deleting a variable. If the two values are set equal, it would be possible for a variable having that exact F score to be added and deleted repeatedly at successive steps.

The stepwise procedure is clearly a more sophisticated process tailored to the usual situation in which independent variables are significantly correlated with each other. It is important to note that all three methods—Forward Selection, Backward Elimination, and Stepwise Regression—will lead to the same model *if and only if* all the independent variables are completely uncorrelated with each other. Under the normal situation, in which the independent variables have significant but not large correlations with each other, the procedures will lead to different results. In particular, consider the situation in which three of the independent variables, X_r, X_s, and X_t, exhibit the following relationships: all three variables have a high correlation with the dependent variable, Y, with X_r having the highest correlation. X_s and X_t have relatively low correlations with each other but X_s is highly correlated with X_r (after taking account of X_s). In Forward Selection, X_r will be added first and, since only residual

---

[a]This step is redundant but usually takes place because of computer programming simplicity.

variance is examined after that, probably neither X_s nor X_t will be added. Backward Elimination will probably result in the same model, with X_s and X_t having lower partial F scores than X_r and therefore being eliminated earlier. However, in Stepwise Regression it is probable that while X_r will be added first, it may be dropped subsequently when either X_s or X_t is added and the other is considered for addition. If the combined effect of X_s and X_t is greater in explaining variance in Y, they will appear in the stepwise model in place of X_r, resulting in a better model statistically than either of the other two procedures. Since this type of intercorrelation situation is generally fairly common in real-world data, the Stepwise Regression procedure will usually lead to better explanatory models.

A word of caution is, however, necessary in the use of these procedures for regression analysis, particularly Stepwise Regression. The techniques are very much open to misuse. As discussed at the beginning of chapter 7, statistical analysis is basically a tool for testing hypotheses. Thus it is essential that hypotheses are formulated first and then subjected to the relevant statistical analysis to test for validity. There is, however, a great temptation to use these regression procedures as a "fishing rod" with which hopefully to catch a "model." Unfortunately, statistical procedures of this type are completely undiscriminating and are liable when used in this fashion to produce completely fallacious results based upon purely coincidental relationships in a set of sample data. Therefore, it is essential to be very circumspect in using and interpreting regression analysis, particularly where computer programs are available for carrying out Forward Selection, Backward Elimination, or Stepwise Regression procedures.

An Example

In this example, data have been obtained on housing expenditure for a number of households, and an attempt is made to relate this expenditure to certain descriptors of the households. These descriptors include the disposable income of the household, the floor area, the size of the household, and the location and socioeconomic status of the household. The data are shown in table 8–2. A stepwise regression is attempted between household expenditure and the three variables—disposable income, floor area, and household size. The correlation matrix is shown in table 8–3.

From an examination of table 8–3, it can be seen that disposable income and floor area correlate quite highly with housing expenditure and with each other. Household size does not correlate strongly with housing expenditure but does correlate with floor area. The data may not therefore be expected to yield as strong a model as might be desired.

Table 8–2
Data for Examples on Multivariate Regression

House-hold	Housing Expenditure	Disposable Income	Floor Area	Household Size	Location	Socioeco-nomic Status
1	2090	9700	60	4	2	4
2	630	7000	32	3	4	1
3	870	5210	82	5	4	4
4	1110	6840	42	2	4	4
5	1470	7840	85	5	4	2
6	1700	5060	60	4	4	4
7	1330	6260	66	5	5	4
8	1560	5750	40	4	5	4
9	1400	5550	42	5	5	4
10	890	8680	42	3	5	4
11	1640	6670	68	4	3	3
12	1040	4870	54	4	1	2
13	750	4530	47	4	4	4
14	1770	9110	61	3	4	4
15	950	5940	45	2	2	3
16	2870	9110	86	5	2	1
17	2230	13340	75	2	2	1
18	490	5130	20	2	2	4
19	1960	8990	64	4	2	1
20	760	4320	34	3	5	4
21	2090	7230	79	5	4	1
22	1470	6050	30	4	5	4
23	660	5210	24	3	3	4
24	1290	7990	44	4	2	4
25	2090	7020	58	2	1	4

Table 8–3
Correlation Matrix for First Example

Variable	Housing Expenditure	Disposable Income	Floor Area	Household Size
Housing expenditure	1.000	.6397	.6762	.2717
Disposable income		1.000	.4579	−.1355
Floor area			1.000	.5051
Household size				1.000

Proceeding, however, with the Stepwise Regression, the following results are obtained, with housing expenditure as the dependent variable.

On step 1, floor area enters to form

housing exp. = 264.5 + 21.27 (floor area)   (8.36)

The t statistics for the constant and coefficient are 0.96 and 4.40, respectively; the coefficient of determination (R^2) is 0.457; and the F score for the model is 19.38, with 1 and 23 degrees of freedom.

On step 2, disposable income enters with a partial correlation of 0.504. (Household size has a partial correlation of -0.109 at this step.) The model now becomes

$$\text{housing exp.} = -255.8 + 15.25 \text{ (floor area)}$$
$$+ 0.121 \text{ (disposable income)} \qquad (8.37)$$

The t statistics are 0.83, 3.18, and 2.74, respectively; the R^2 is 0.595; and the F score is 16.17, with 2 and 22 degrees of freedom.

On step 3, household size enters with a partial correlation of 0.17. The resulting model is

$$\text{housing exp.} = -511.6 + 11.98 \text{ (floor area)}$$
$$+ 0.141 \text{ (disposable income)}$$
$$+ 81.17 \text{ (household size)} \qquad (8.38)$$

The t statistics are 1.15, 1.89, 2.77, and 0.80, respectively; the R^2 is 0.607; and the F score is 10.82, with 3 and 21 degrees of freedom.

From this, it can be seen that the model at step 2, equation 8.37, is the best, as may have been anticipated from table 8–3. The intercorrelation between household size and floor area clearly results in an inferior model in step 3, and the t statistic of household size is not significant. In equation 8.37, both coefficients are significant at better than 99% confidence, although the constant is not.

It should be noted that the regression model must be interpreted only within the range of values of the dependent and independent variables occurring in the data of table 8–2. Clearly, a zero value of disposable income and floor area (neither of which is likely to be a real value) would generate a negative housing expenditure, which does not make sense. This, however, is well outside the range of the data. One should also note that, although the constant is a large value, it is not significantly different from zero, even at only 50% significance. Another example of stepwise regression is provided later in the chapter, after the discussion on dummy variables.

Polynomial Regression

The discussion of regression so far has concentrated on strictly linear relationships, using a linear additive model in first-order powers of the variables. However, linear regression is not restricted to such relationships. The first potential polynomial form of regression is one in which the dependent variable is a function of various powers of independent variables.

$$Y_i = a_0 + a_1 X_{1i}^{n_1} + a_2 X_{2i}^{n_2} + a_3 X_{3i}^{n_3} + \dots + a_m X_{mi}^{n_m} \qquad (8.39)$$

It is not necessary that all of the Xs be distinct. For example, $X_j^{n_j}$ and $X_k^{n_k}$ could be X_3^3 and X_3^5. It is, of course, necessary that the same variable be raised to different powers if it is entered more than once. It is also not necessary for the powers to be integer. However, it is not possible to solve simultaneously for the coefficients, as, and the powers, ns. The powers must be hypothesized and the coefficients may then be found by linear-regression procedures. The simplest way to see how this form of regression works is by algebraic substitution.

Let $X_{1i}^{n_1} = Z_{1i}$

$X_{2i}^{n_2} = Z_{2i}$

$X_{3i}^{n_3} = Z_{3i}$

.   .
.   .
.   .

$X_{mi}^{n_m} = Z_{mi}$

Equation 8.39 can then be rewritten

$$Y_i = a_0 + a_1 Z_{1i} + a_2 Z_{2i} + a_3 Z_{3i} + \dots + a_m Z_{mi} \qquad (8.40)$$

It is now clear that this represents a standard linear-regression problem. It is also clear that this principle can be applied to any situation in which an algebraic transform can be accomplished. Thus a variety of functional forms of independent variables can be used, such as $\log X_{ji}$, $e^{X_{ji}}$, $X_{ji} X_{ki}$, $1/X_{ji}$, $\sin X_{ji}$, etc. In each case, a simple algebraic substitution may be made to reduce the problem to a simple linear-regression problem. One should note, however, that all these models are linear in parameters. Thus this is not any form of nonlinear regression but is, rather, a more complex form of linear regression.

An alternative situation arises, however, when one wishes to consider a truly nonlinear relationship, such as

$$Y_i = a_0 X_{1i}^{a_1} X_{2i}^{a_2} X_{3i}^{a_3} \dots X_{mi}^{a_m} \qquad (8.41)$$

It is possible, with certain restrictions, to calibrate such a model by linear regression. In the case of equation 8.41, the model can be made linear by taking logarithms of both sides of the equation.

$$\ln Y_i = \ln a_0 + a_1 \ln X_{1i} + \ln X_{2i} + a_3 \ln X_{3i} + \dots + a_m \ln X_{mi} \qquad (8.42)$$

The model is now linear in parameters and a simple substitution, in which $Z_{ki} = \ln X_{ki}$, will clearly permit calibration by linear regression. However, some caution is needed here. First, the true model is

$$\ln Y_i = \ln \alpha_0 + \alpha_1 \ln X_{1i} + \alpha_2 \ln X_{2i} + \alpha_3 \ln X_{3i} + ... + \alpha_m \ln X_{mi} + \epsilon_i \quad (8.43)$$

It may be recalled that the linear-regression procedure requires no assumptions on distributions of parameters for estimation of those parameters. However, the execution of statistical tests of the calibrated model requires the assumption of a normal distribution for the error term, ϵ_i. If this is normally distributed, equation 8.43 can be subjected to statistical tests, but these do not apply to the untransformed model, equation 8.41. This is clear since the error term of equation 8.41 is e^{ϵ_i} which is not normally distributed if ϵ_i is. It also follows that the estimates of standard errors for equation 8.43 will not apply to equation 8.41. Finally, there may be problems from the distribution of data points resulting from the transformation of the dependent and independent variables, where the distribution is extended at the low value end and foreshortened at the high-value end.

Apart from these cautionary remarks, it is possible to estimate by linear regression any relationship that can be transformed to a linear form. Principally, this would be a product form of model, but any other form that can be made linear in parameters would be a candidate.

Dummy Variables

Up to now, the independent and dependent variables have been assumed to take values over a continuous range. It is not uncommon, however, to wish to introduce independent variables which can take only two or more discontinuous, discrete values. The values taken on by the dummy variables may or may not bear some relation to the values of some physical variable.

In conceptual terms, a dummy variable in a linear regression has the same effect on the model produced as that of stratifying the data on the basis of that variable and computing two separate regressions, with one exception. In stratification of the data, each of the coefficients in the two models could be different, and each model has wider confidence limits because it is based on less than all the data. Thus the use of dummy discontinuous variables is equivalent to stratified regressions with all coefficients constrained to be equal except for the constant term, and with the confidence limits of the entire data set.

There are several important properties of discontinuous dummy variables. First, suppose a travel model is to be built using a variable relating to time of day. Suppose also the time of day needs to be split only into peak and offpeak, so that a dummy variable, Z_1, is proposed of the form

$$Z_1 = \begin{cases} 0 & \text{for offpeak times} \\ 1 & \text{for peak times} \end{cases} \tag{8.44}$$

The entire model is of the form

$$Y_i = a_0 + a_1 X_{1i} + a_2 X_{2i} + a_3 X_{3i} + a_4 Z_{1i} \tag{8.45}$$

The coefficients, a_0, a_1, a_2, a_3, and a_4 can be evaluated as before. Two models can then be postulated from equation 8.45.

$$Y_i = a_0 + a_1 X_{1i} + a_2 X_{2i} + a_3 X_{3i} + a_4 \tag{8.46}$$

$$Y_i = a_0 + a_1 X_{1i} + a_2 X_{2i} + a_3 X_{3i} \tag{8.47}$$

Now consider the situation in which the model is calibrated using aggregate data at a zonal level. The data comprise a number of offpeak and peak trips, say n_1 and n_2, respectively. Given these data, two alternatives are possible for the values of Z_1. Values could be assigned of nZ_1 to Z_1, so for peak trips $Z_1 = n_2$, and for offpeak trips $Z_1 = 0$.

Another useful value set would be to assign values to Z_1 such that the sum of squares of Z_1 is unity.[2] This could be achieved by assigning to Z_1 the value shown in equation 8.48 for offpeak trips.

$$Z_1 = \frac{-n_2}{\sqrt{n_1 n_2 (n_1 + n_2)}} \tag{8.48}$$

Similarly, for peak trips the value shown in equation 8.49 would be assigned to Z_1.

$$Z_1 = \frac{n_1}{\sqrt{n_1 n_2 (n_1 + n_2)}} \tag{8.49}$$

Using these values, the mean of Z_1 is

$$\bar{Z}_1 = \frac{1}{(n_1 + n_2)} \left[\frac{-n_2 n_1}{\sqrt{(n_1 n_2)(n_1 + n_2)}} + \frac{n_1 n_2}{\sqrt{n_1 n_2 (n_1 + n_2)}} \right] = 0 \tag{8.50}$$

The variance is

$$\sum_{i=1}^{n_1+n_2} (Z_{1i} - \bar{Z}_1)^2 = \sum_{i=1}^{n_1+n_2} Z_{1i}^2 = \frac{n_1 n_2^2}{n_1 n_2 (n_1 + n_2)} + \frac{n_2 n_1^2}{n_1 n_2 (n_1 + n_2)} \tag{8.51}$$

Gathering terms, equation 8.51 may be simplified

$$\sum_{i=1}^{n_1+n_2} (Z_{1i} - \bar{Z}_1)^2 = \frac{n_2 + n_1}{n_1 + n_2} = 1 \tag{8.52}$$

It is useful to consider the implications of different values assigned to the dummy variable, Z_1. If values of 0 and 1 are used, it is assumed that for $Z_1 = 0$, the constant term is a_0; and that for $Z_1 = 1$, the constant term is (a_0+a_4) (from equation 8.49). Now consider the case if the dummy variable were assigned values of 1 and 2. In this case, the implied constants are, respectively, (a_0+a_4) and (a_0+2a_4). This form implies specifically that there is a linear relationship between the effects of $Z_1 = 1$ and $Z_1 = 2$ on the model. This presupposes a specific change between peak and offpeak trips in equation 8.45, given by the coefficient of Z_1.

Consider next the situation when a variable can take on three discrete values. Suppose another variable is to be added to equation 8.45 relating to income, and three income ranges are to be used, as shown in table 8–4.

The correct way to represent this in the model is to use two dummy variables, Z_2 and Z_3, such that values of 0 or 1 can again be assigned to each dummy variable. Thus the income groups are

$Z_2 = 0, Z_3 = 0$ for income group 1

$Z_2 = 1, Z_3 = 0$ for income group 2

$Z_2 = 0, Z_3 = 1$ for income group 3   (8.53)

Equation 8.46 now becomes

$$Y_i = a_0 + a_1 X_{1i} + a_2 X_{2i} + a_3 X_{3i} + a_4 Z_{1i} + a_5 Z_{2i} + a_6 Z_{3i} \quad (8.54)$$

For aggregated data, values may be assigned in a number of ways. The dummy scores may be multiplied by the number of people for which that score is appropriate. Or values may be assigned that will give a unit sum of squares to each of Z_2 and Z_3. Other alternatives are possible. Consider for a moment an alternative which is rarely correct. Assume that only one dummy variable is used, with values $Z_2 = 1$ for income group 1, $Z_2 = 2$ for income group 2, and $Z_2 = 3$ for income group 3. In this case, equation 8.54 becomes

$$Y_i = a_0 + a_1 X_{1i} + a_2 X_{2i} + a_3 X_{3i} + a_4 Z_{1i} + a_5 Z_{2i} \quad (8.55)$$

This clearly assumes that incomes in group 2 have twice the effect of incomes in group 1, and that incomes in group 3 have three times the

Table 8–4
Example of Income Ranges

Income Group	Income Range
1	Less than $10,000
2	$10,000 to $25,000
3	Over $25,000

effect. The use of Z_2 and Z_3, however, yielded effects of 0, a_5, and a_6 instead of a_5, $2a_5$, and $3a_5$. Thus the use of Z_2 with the three integer values presupposes the effective weights of the income groups in the model. Since it creates a constraint on the value a_5, it is therefore generally an undesirable and frequently incorrect form for the dummy variable.

In general, any number of levels of a variable to be used as a dummy variable can be entered in a regression equation. If there are r levels of the dummy variable, $(r-1)$ dummy variables will be needed.

An Example

The data used for the Stepwise Regression example can be used to demonstrate the composition and use of dummy variables. Referring to table 8–2, the location and socioeconomic status can be reformulated as dummy variables. Location takes the five integer values of 1, 2, 3, 4, and 5, while there are four socioeconomic status values—1, 2, 3, 4. Clearly, these integer values provide no cardinal and probably no ordinal information on the two variables. Hence formulation as dummy variables offers the only useful way to include these characteristics in a model of housing expenditure. To this end, seven dummy variables were created. Variables for location are given in table 8–5, and those for socioeconomic status in table 8–6.

A Stepwise Regression procedure was applied once more to this problem. The results of the first 5 steps are shown in tables 8–7 through 8–11. There are two principal observations: first, if a 95% significance level is used as the cutoff for a coefficient, step 2 still provides the best model and is identical to equation 8.37, which was the best model in the previous example. If one sets a less stringent cutoff, the second observation emerges—the dummy variables begin to enter, replacing the highly intercorrelated variable household size. It may also be noted that some

Table 8–5
Dummy Variables for Location

Location	Dummy Variables			
	L1	L2	L3	L4
1	0	0	0	0
2	1	0	0	0
3	0	1	0	0
4	0	0	1	0
5	0	0	0	1

Table 8–6
Dummy Variables for Socioeconomic Status

Socioeconomic Status	Dummy Variables		
	S1	S2	S3
1	0	0	0
2	1	0	0
3	0	1	0
4	0	0	1

instability occurs in the model in subsequent steps. Note, particularly, the *t* score of disposable income in steps 3, 4, and 5, and the *t* score, in the same steps, for L3. These two variables are somewhat unstable in the presence of additional dummy variables.

By step 5, the additional variables are clearly far from significant and it must be concluded that they add very little to the model. However, it is useful to look at steps 9, 10, and 11, as shown in tables 8–12, 8–13, and 8–14. It can be seen here that S3, which entered at step 6, is now removed at step 11, while household size, which entered at step 10, remains in the model. The effect of removing S3 also has a far-reaching effect on other variables in the equation, so that, from step 9 to step 11, disposable income, L3, and S1 all obtain coefficients that are significant at better than 95% confidence. Also, the *t* scores on L2, L4, and L1 rise substantially with the dropping of S3. All other measures of the model display improvement in step 11, compared with step 9, showing the value of the stepwise procedure. Further improvement might have been obtained by a more stringent *F* level for exclusion. Examining step 11, S2 might be a candidate to drop at a twelfth step and may give rise to a further improvement in the rest of the model. This clearly shows the power of the stepwise procedure over either Forward Selection or Backward Elimination.

Some comments are worthwhile on the conceptual or intuitive appeal of the model of table 8–14. Housing expenditure is seen to increase with floor area and disposable income, as would be expected. In size, the coefficients show similar effects of these two variables, as shown by the beta-coefficient values of 0.32 and 0.51 for floor area and disposable income, respectively. (These are the coefficients standardized for the means and standard deviations of the variables and remove the effects of different units of measurement on coefficient size.) Household size has a positive effect on housing expenditure, as might be expected. Again, the beta coefficient is similar in size to the other two variables, at 0.37.

In table 8–5, it may be observed that, ceteris paribus, housing expen-

Table 8–7
First Stepwise Regression with Dummy Variables

DEPENDENT VARIABLE.. HOUSING   EXPENDITURE

MEAN RESPONSE   1404.40000   STD. DEV.   601.59566

VARIABLE(S) ENTERED ON STEP NUMBER 1..   FLAREA

		ANALYSIS OF VARIANCE	DF	SUM OF SQUARES	MEAN SQUARE	F
MULTIPLE R	.676					
R SQUARE	.457	REGRESSION	1.	3971850.57	3971850.57	19.38
ADJUSTED R SQUARE	.434	RESIDUAL	23.	4714165.43	204963.71	
STD DEVIATION	452.729	COEFF OF VARIABILITY	32.2 PCT			

-----VARIABLES IN THE EQUATION-----

VARIABLE	B	STD ERROR B	BETA	ELASTICITY	T	SIGNIFICANCE
FLAREA	21.27	4.83	.676	.812	4.40	.000
(CONSTANT)	264.5	274.32			.96	.345

-----VARIABLES NOT IN THE EQUATION-----

VARIABLE	PARTIAL CORRELATION	F
DISPINC	.504	7.49
HHSIZE	-.110	.268
L1	.340	2.87
L2	-.028	.166
L3	-.432	5.03
L4	.097	.206
S1	-.342	2.91
S2	-.116	.301
S3	.001	.298

Table 8-8
Step Two of Dummy Variable Regression

DEPENDENT VARIABLE..   HOUSING   EXPENDITURE

VARIABLE(S) ENTERED ON STEP NUMBER 2..   DISPOSABLE INCOME

		ANALYSIS OF VARIANCE	DF	SUM OF SQUARES	MEAN SQUARE	F
MULTIPLE R	.771					
R SQUARE	.595	REGRESSION	2.	5169561.78	2584780.89	16.17
ADJUSTED R SQUARE	.558	RESIDUAL	22.	3516454.22	159838.83	
STD DEVIATION	399.798	COEFF OF VARIABILITY	28.5 PCT			

-----VARIABLES IN THE EQUATION -----

VARIABLE	B	STD ERROR B	BETA	ELASTICITY	T	SIGNIFICANCE
FLAREA	15.25	4.80	.485	.418	3.18	.004
DISPINC	.121	.044	.582	.600	2.74	.012
(CONSTANT)	-255.77	307.91			-.83	.415

-----VARIABLES NOT IN THE EQUATION-----

VARIABLE	PARTIAL	F
HHSIZE	.173	.647
L1	.128	.348
L2	-.044	.399
L3	-.352	2.96
L4	.166	.594
S1	-.268	1.62
S2	-.061	.786
S3	.105	.234

Table 8–9
Step Three of Dummy Variable Regression

DEPENDENT VARIABLE..   HOUSING   EXPENDITURE

VARIABLE(S) ENTERED ON STEP NUMBER 3..   L3

		ANALYSIS OF VARIANCE	DF	SUM OF SQUARES	MEAN SQUARE	F
MULTIPLE R	.803					
R SQUARE	.645	REGRESSION	3.	5605201.53	1868400.51	12.74
ADJUSTED R SQUARE	.595	RESIDUAL	21.	3080814.47	146705.45	
STD DEVIATION	383.021	COEFF OF VARIABILITY	27.3 PCT			

------VARIABLES IN THE EQUATION ------

VARIABLE	B	STD ERROR B	T	SIGNIFICANCE	BETA	ELASTICITY
FLAREA	18.36	4.94	3.72	.001	.584	.700
DISPINC	.10	.044	2.27	.034	.345	.496
L3	-305.93	177.53	-1.72	.100	-.242	-.070
(CONSTANT)	-178.01	298.42	-.60	.557		

------VARIABLES NOT IN THE EQUATION------

VARIABLE	PARTIAL	F
HHSIZE	.136	.374
L1	-.003	.163
L2	-.134	.365
L4	.043	.366
S1	-.292	1.87
S2	-.174	.622
S3	.143	.416

Table 8-10
Step Four of Dummy Variable Regression

DEPENDENT VARIABLE..   HOUSING EXPENDITURE

VARIABLE(S) ENTERED ON STEP NUMBER 4..   S1

		ANALYSIS OF VARIANCE	DF	SUM OF SQUARES	MEAN SQUARE	F
MULTIPLE R	.822					
R SQUARE	.676	REGRESSION	4.	5868737.60	1467184.40	10.42
ADJUSTED R SQUARE	.611	RESIDUAL	20.	2817278.40	140863.92	
STD DEVIATION	375.32	COEFF OF VARIABILITY	26.7 PCT			

-------VARIABLES IN THE EQUATION -------

VARIABLE	B	STD ERROR B	T	SIGNIFICANCE	BETA	ELASTICITY
FLAREA	20.52	5.09	4.03	.001	.653	.783
DISPINC	.087	.045	1.94	.066	.298	.428
L3	-309.66	173.98	-1.78	.090	-.245	-.071
S1	-401.82	293.78	-1.37	.187	-.185	-.023
(CONSTANT)	-164.88	292.57	-.56	.579		

------VARIABLES NOT IN THE EQUATION------

VARIABLE	PARTIAL	F
HHSIZE	.151	.441
L1	-.043	.356
L2	-.163	.519
L4	.006	.567
S2	-.229	1.04
S3	.036	.249

Table 8–11
Step Five of Dummy Variable Regression

DEPENDENT VARIABLE.. HOUSING EXPENDITURE

VARIABLE(S) ENTERED ON STEP NUMBER 5.. S2

		ANALYSIS OF VARIANCE	DF	SUM OF SQUARES	MEAN SQUARE	F
MULTIPLE R	.832					
R SQUARE	.693	REGRESSION	5.	6015907.35	1203181.47	8.56
ADJUSTED R SQUARE	.612	RESIDUAL	19.	2670108.65	140532.03	
SDT DEVIATION	374.88	COEFF OF VARIABILITY	26.7 PCT			

--------VARIABLES IN THE EQUATION--------						
VARIABLE	B	SDT ERROR B	BETA	ELASTICITY	T	SIGNIFICANCE
FLAREA	21.76	5.226	.692	.830	4.16	.001
DISPINC	.076	.046	.261	.375	1.66	.113
L3	−360.67	180.789	−.285	−.082	−1.99	.061
S1	−445.99	296.586	−.205	−.025	−1.50	.149
S2	−299.71	292.869	−.138	−.017	−1.02	.319
(CONSTANT)	−112.71	296.643			−.38	.708

-------VARIABLES NOT IN THE EQUATION-------		
VARIABLE	PARTIAL	F
HHSIZE	.068	.835
L1	−.003	.172
L2	−.099	.176
L4	−.076	.104
S3	−.111	.224

Table 8–12
Step Nine of Dummy Variable Regression

DEPENDENT VARIABLE.. HOUSING   EXPENDITURE

VARIABLE(S) ENTERED ON STEP NUMBER 9..   L1

		ANALYSIS OF VARIANCE	DF	SUM OF SQUARES	MEAN SQUARE	F
MULTIPLE R	.847					
R SQUARE	.717	REGRESSION	9.	6229552.04	692172.45	4.23
ADJUSTED R SQUARE	.548	RESIDUAL	15.	2456463.96	163764.26	
SDT DEVIATION	404.678	COEFF OF VARIABILITY	28.8 PCT			

------VARIABLES IN THE EQUATION------

VARIABLE	B	STD ERROR B	T	SIGNIFICANCE	BETA	ELASTICITY
FLAREA	20.48	5.80	3.53	.003	.651	.782
DISPINC	.079	.058	1.36	.193	.271	.389
L3	-664.40	350.39	-1.90	.077	-.526	-.151
S1	-688.91	408.84	-1.69	.113	-.317	-.039
S2	-309.59	405.18	.76	.457	-.142	-.017
S3	-115.63	266.88	-.43	.671	-.094	-.052
L2	-425.33	432.29	-.98	.341	-.234	-.036
L4	-349.07	375.53	-.93	.367	-.253	-.060
L1	-377.29	412.22	-.92	.375	-.273	-.064
(CONSTANT)	351.48	589.22	.60	.560		

------VARIABLES NOT IN THE EQUATION------

VARIABLE	PARTIAL	F
HHSIZE	.329	1.69

Table 8–13
Step Ten of Dummy Variable Regression

DEPENDENT VARIABLE.. HOUSING EXPENDITURE

VARIABLE(S) ENTERED ON STEP NUMBER 10.. HHSIZE

MULTIPLE R	.865
R SQUARE	.748
ADJUSTED R SQUARE	.568
SDT DEVIATION	395.630

ANALYSIS OF VARIANCE	DF	SUM OF SQUARES	MEAN SQUARE	F
REGRESSION	10.	6494692.82	649469.28	4.15
RESIDUAL	14.	2191323.18	156523.08	
COEFF OF VARIABILITY	28.2 PCT			

------VARIABLES IN THE EQUATION------ ------VARIABLES NOT IN THE EQUATION------

VARIABLE	B	STD ERROR B	BETA	ELASTICITY	T	SIGNIFICANCE	VARIABLE	PARTIAL	F
FLAREA	10.14	9.76	.322	.387	1.04	.317			
DISPINC	.15	.077	.503	.722	1.91	.077			
L3	-878.34	379.95	-.695	-.200	-2.31	.037			
S1	-768.08	404.30	-.354	-.044	-.190	.078			
S2	55.98	485.60	.026	.003	.12	.910			
S3	-17.56	271.58	-.014	-.008	-.06	.949			
L2	-895.87	556.16	-.494	-.076	-1.61	.130			
L4	-794.72	502.03	-.576	-.136	-1.58	.136			
L1	-657.10	456.76	-.476	-.112	-1.44	.172			
HHSIZE	205.10	157.59	.367	.532	1.30	.214			
(CONSTANT)	-94.19	670.14			-.14	.890			

Table 8–14
Step Eleven of Dummy Variable Regression

DEPENDENT VARIABLE.. HOUSING EXPENDITURE

VARIABLE(S) REMOVED ON STEP NUMBER 11.. S3

			DF	SUM OF SQUARES	MEAN SQUARE	F
MULTIPLE R	.865	ANALYSIS OF VARIANCE				
R SQUARE	.748	REGRESSION	9.	6494083.79	721559.87	4.94
ADJUSTED R SQUARE	.596	RESIDUAL	15.	2191977.21	146131.81	
STD DEVIATION	382.271	COEFF OF VARIABILITY	27.2 PCT			

------VARIABLES NOT IN THE EQUATION------

VARIABLE	PARTIAL	F
S3	-.017	.417

------VARIABLES IN THE EQUATION------

VARIABLE	B	STD ERROR B	T	SIGNIFICANCE	BETA	ELASTICITY
FLAREA	10.05	9.35	1.08	.299	.320	.384
DISPINC	.15	.069	2.16	.047	.509	.731
L3	-879.67	366.59	-2.40	.030	-.696	-.200
S1	-756.98	353.71	-2.14	.049	-.348	-.043
S2	73.14	392.84	.19	.855	.033	.004
L2	-903.00	526.71	-1.71	.107	-.498	-.077
L4	-802.20	472.03	-1.70	.110	-.581	-.137
L1	-656.91	441.33	-1.49	.157	-.476	-.112
HHSIZE	207.93	146.29	1.42	.176	.372	.539
(CONSTANT)	-123.47	477.21	-.26	.799		

F-LEVEL OR TOLERANCE LEVEL INSUFFICIENT FOR FURTHER COMPUTATION diture will be greatest in location 1 then followed, in order, by location 2, 3, 5, and 4. Similarly, using table 8–6, housing expenditure will be greatest by people in socioeconomic group 3, then 1, and then 2. Since S3 (socioeconomic group 4) did not enter the equation, this group would be added, effectively, to group 1, where both S1 and S2 are zero. Without knowing exactly what these groups are (which was not reported in the original data) it is not possible to state whether this is reasonable. Examination of the full correlation matrix (not reported here) does not reveal any further insights on this result. One must therefore conclude that further judgment on the reasonableness of these results cannot be made. In general, however, the model seems to have some conceptual and intuitive appeal. It may be noted that the constant is negative and the same cautionary comments made before are in order here.

Weighted Regression

In linear regression, each observation is accorded an equal weight in helping to determine the values of the parameters of the linear relationship. However, situations may arise in which it is desirable to assign different weights to different observations. Such a situation could arise because some observations may have more error associated with them, or because observations are in fact aggregations of observations where different numbers of data points may be used to derive each aggregate observation. In the latter case, the reliability (or variance) of each aggregate observation will be a direct function of the number of data points used to obtain the aggregate value.

In the social sciences, the principal reasons for needing weighted regression are associated with sample data from surveys. First, with the use of survey data, differential response rates may be obtained by some socioeconomic classification, which will lead to different variances being associated with some observations than with others. Second, the survey data may be aggregated before use as the data set for a regression. If each value in the aggregated set is composed of variable numbers of observations, the variances will not be the same throughout the aggregated set.

Similar situations may arise in engineering, for example, where data are collected on different occasions and measurement error may vary on each occasion. Another example would again arise when using aggregated data, where different numbers of observations are used to obtain each aggregate data point. Suppose the simple linear relationship of equation 8.56 is to be fitted on data for which the variance of each observation is different.

$$Y_i = \alpha_0 + \alpha_1 X_{1i} + \epsilon_i \tag{8.56}$$

It is assumed, therefore, that the individual error terms have the properties that $E(\epsilon_i) = 0$ for all i, but that $V(\epsilon_i) = \sigma_i^2$. [The usual assumption, in contrast, is that $E(\epsilon_i) = 0$ and $V(\epsilon_i) = \sigma^2$ for all i.] Assume that each σ_i^2 can be expressed as $V_i \sigma^2$, where σ^2 is the constant variance of all the values, and V_i represents a weighting of each variance such that for all i equation 8.57 is true.

$$\sigma_i^2 = V_i \sigma^2 \tag{8.57}$$

In other words, it is assumed that the variable variance, σ_i^2, can be expressed as a compound of a constant variance, σ^2, and a variable factor, V_i.

Clearly, if the original equation is divided by $\sqrt{V_i}$, then ϵ_i is converted into $\epsilon_i / \sqrt{V_i}$ which has a mean of 0 and variance of σ^2, as originally hypothesized in simple linear regression. Using this stratagem yields

$$Y_i/\sqrt{V_i} = a_0/\sqrt{V_i} + a_1 X_{1i}/\sqrt{V_i} + \epsilon_i/\sqrt{V_i} \tag{8.58}$$

where $E(\epsilon_i/\sqrt{V_i}) = 0$ because $E(\epsilon_i) = 0$

and $\quad V(\epsilon_i/\sqrt{V_i}) = \dfrac{1}{V_i} V(\epsilon_i)$

The variance of the error term may be written

$$V(\epsilon_i/\sqrt{V_i}) = \frac{1}{V_i} \sigma_i^2 \tag{8.59}$$

Since σ_i^2 is equal to $V_i \sigma^2$, equation 8.59 shows that the variance of the new error term is a constant σ^2. It is clear therefore that this stratagem has produced a linear-regression equation (equation 8.58) which obeys the original assumptions of linear-regression analysis. The change that has achieved this is a transformation of the dependent and independent variables of the original model (equation 8.55). The required transformations are

$$y_i = Y_i/\sqrt{V_i} \tag{8.60}$$

$$x_{1i} = X_{1i}/\sqrt{V_i} \tag{8.61}$$

$$x_{0i} = 1/\sqrt{V_i} \tag{8.62}$$

$$\phi_i = \epsilon_i/\sqrt{V_i} \tag{8.63}$$

Substituting these transformed variables into equation 8.58 yields

$$Y_i = a_0 x_{0i} + a_1 x_{1i} + \phi_i \tag{8.64}$$

This equation may now be solved using standard linear-regression procedures. It is useful to follow through the standard procedure for this simple example to see how the estimates of the regression parameters are different from those obtained from an unweighted regression. The values of the two coefficients are

$$a_1 = \frac{\Sigma 1/V_i \ \Sigma (X_{1i} \ Y_{1i}/V_i) - \Sigma \ Y_i/V_i \Sigma X_i/V_i}{\Sigma (X_{1i}/V_i)^2 - (\Sigma X_{1i}/V_i)^2} \qquad (8.65)$$

$$a_0 = \frac{\Sigma \ Y_i/V_i - a_1 \Sigma X_{1i}/V_i}{\Sigma \ Y_i V_i} \qquad (8.66)$$

These may be compared with the simple linear-regression estimates.

$$a_1 = \frac{\Sigma (X_{1i} Y_i) - \Sigma Y_i \Sigma X_i}{\Sigma X_{1i}^2 - (\Sigma X_{1i})^2} \qquad (8.67)$$

$$a_0 = \frac{\Sigma \ Y_i - a_1 \Sigma X_i}{N} \qquad (8.68)$$

The derivation of equations 8.65 and 8.66 can be shown quite easily. The standard procedure is to minimize the sum of squares of the error terms, $\Sigma \phi_i^2$, of equation 8.64. Rearranging the equation and summing over all observations produces

$$\Sigma \phi_i^2 = \Sigma (y_i - a_0 x_{0i} - a_1 x_{1i})^2 \qquad (8.69)$$

Differentiating equation 8.69 with respect to each of the unknowns, a_0 and a_1, yields

$$\frac{\partial \Sigma (\phi_i)^2}{\partial a_0} = -2\Sigma x_{0i}(y_i - a_0 x_{0i} - a_1 x_{1i}) \qquad (8.70)$$

$$\frac{\partial \Sigma (\phi_i)^2}{\partial a_1} = -2\Sigma x_{1i}(y_i - a_0 x_{0i} - a_1 x_{1i}) \qquad (8.71)$$

As usual, to minimize the sum of the squared error terms, the two partial differential equations are set equal to zero and the equations solved for the unknowns, a_0 and a_1.

$$\Sigma y_i x_{0i} - a_0 \Sigma x_{0i}^2 - a_1 \Sigma x_{1i} x_{0i} = 0 \qquad (8.72)$$

$$\Sigma y_i x_{1i} - a_0 \Sigma x_{0i} x_{1i} - a_1 \Sigma x_{1i}^2 = 0 \qquad (8.73)$$

Rearranging and solving these equations produces equations 8.65 and 8.66. It is useful to investigate some specific applications of weighted regression.

Example 1

Suppose the equation to be estimated is

$$\hat{Y}_i = a_1 X_{1i} \qquad (8.74)$$

Suppose further the variance of each observation may be expressed as

$$\sigma_i^2 = \sigma^2/\omega_i \tag{8.75}$$

The transformations, shown in general in equations 8.60 through 8.63, are given by equations 8.76 through 8.78 for this case.

$$y_i = Y_i\sqrt{\omega_i} \tag{8.76}$$

$$x_{0i} = \sqrt{\omega_i} \tag{8.77}$$

$$x_{1i} = X_{1i}\sqrt{\omega_i} \tag{8.78}$$

Since a_0 is assumed to be zero in equation 8.74, only one normal equation is produced and can be solved immediately for a_1.

$$a_1 = \frac{\Sigma y_i x_{1i}}{\Sigma x_{1i}^2} \tag{8.79}$$

Expressing this equation in terms of the original variables and ω_i produces

$$a_1 = \frac{\Sigma \omega_i Y_i X_{1i}}{\Sigma \omega_i X_{1i}^2} \tag{8.80}$$

Hence, the estimated values of Y are

$$\hat{Y}_k = \left[\frac{\Sigma \omega_i Y_i X_{1i}}{\Sigma \omega_i X_{1i}^2}\right] X_{1k} \tag{8.81}$$

If ordinary regression rather than weighted regression had been carried out, the estimating equation would have had the form

$$\hat{Y}_k = \left[\frac{\Sigma Y_i X_{1i}}{\Sigma X_{1i}^2}\right] X_{1k} \tag{8.82}$$

Example 2

Suppose now that in the previous example the variance of each observation is proportional to the size of the measure X_{1i}. In this case, the error variances are

$$\sigma_i^2 = kX_{1i} \tag{8.83}$$

Taking the previous example, it is possible to express ω_i.

$$\omega_i = \sigma^2/kX_{1i} \tag{8.84}$$

Substituting for ω_i in equation 8.81 yields

$$\hat{Y}_k = \left[\frac{\Sigma(\sigma^2/kX_{1i})\ Y_iX_{1i}}{\Sigma(\sigma^2/kX_{1i})\ X_{1i}^2} \right] X_{1k} \tag{8.85}$$

Simplifying equation 8.85 and noting that σ^2 is a constant that can be taken outside the summation

$$\hat{Y}_k = \left[\frac{\Sigma Y_i}{\Sigma X_{1i}} \right] X_{1k} \tag{8.86}$$

In turn, this equation can be simplified further to

$$\hat{Y}_k = \left[\frac{\bar{Y}}{\bar{X}_1} \right] X_{1k} \tag{8.87}$$

It is also useful to examine a common situation in the social sciences where aggregated data are being used. Suppose there is a set of survey observations of two variables u and v; N observations have been obtained on these variables, and they are aggregated to n aggregate units. The aggregated data are represented by Y and X where for the ith unit the values are

$$Y_i = \frac{1}{m_i} \sum_{j=1}^{m_i} u_{ji} \tag{8.88}$$

$$X_i = \frac{1}{m_i} \sum_{j=1}^{m_i} v_{ji} \tag{8.89}$$

In other words, each unit is assigned the mean values of u and v for the observations in the unit.

$$Y_i = \bar{u}_i \tag{8.90}$$

$$X_i = \bar{v}_i \tag{8.91}$$

If the variances of each original observation are equal and are σ^2, the variance of each value of Y_i and X_i is

$$V(Y_i) = \left[\frac{1}{m_i} \right]^2 [v(u_{1i}) + v(u_{2i}) + \dots + v(u_{mi})i] \tag{8.92}$$

This may be simplified to

$$V(Y_i) = \frac{m_i\sigma^2}{m_i^2} = \frac{\sigma^2}{m_i} \tag{8.93}$$

The variance of each aggregate observation is therefore dependent upon m_i, as would be expected, and an unweighted regression would be incorrect. Suppose the hypothesized relationship is

$$\hat{Y}_i = a_0 + a_1 X_i \tag{8.94}$$

If an unweighted regression were performed, ignoring the variations in the variance, the coefficient estimates would be

$$a_1 = \frac{\Sigma X_i Y_i - n\bar{X}\bar{Y}}{\Sigma X_i^2 - n\bar{X}^2} \tag{8.95}$$

$$a_0 = \bar{Y} - a_1\bar{X} \tag{8.96}$$

In terms of the original parameters these equations can be written

$$a_1 = \frac{\Sigma\bar{u}_i\bar{v}_i - (\Sigma\bar{u}_i \, \Sigma\bar{v}_i)/n}{\Sigma\bar{v}_i^2 - (\Sigma\bar{v}_i)^2/n} \tag{8.97}$$

$$a_0 = (\Sigma\bar{u}_i)/n - a_1\Sigma(\bar{v}_i)/n \tag{8.98}$$

These would be the estimates if the variations in the variances of the individual aggregated data points were ignored. The correct estimates of a_1 and a_0 could be obtained in two ways: each data point could be weighted by the number of observations used to compute it or the weighting technique just developed could be applied. Estimating a_0 and a_1 from the weighting technique first, equation 8.93 yields

$$\sigma_i^2 = \sigma^2/m_i \tag{8.99}$$

Substituting this in equations 8.60 through 8.62 yields

$$y_i = Y_i\sqrt{m_i} = \bar{u}_i\sqrt{m_i} \tag{8.100}$$

$$x_{0i} = \sqrt{m_i} \tag{8.101}$$

$$x_{1i} = X_i\sqrt{m_i} = \bar{v}_i\sqrt{m_i} \tag{8.102}$$

Substituting these relationships in equation 8.65 yields

$$a_1 = \frac{\Sigma m_i \, \Sigma m_i X_i Y_i - \Sigma m_i Y_i \, \Sigma m_i X_i}{\Sigma m_i X_i^2 \Sigma m_i - (\Sigma m_i X_i)^2} \tag{8.103}$$

In terms of the u_is and v_is of equations 8.90 and 8.91, equation 8.103 becomes

$$a_1 = \frac{\Sigma m_i \, \Sigma m_i \bar{u}_i\bar{v}_i - \Sigma\bar{u}_i m_i \, \Sigma\bar{v}_i m_i}{\Sigma m_i \bar{v}_i^2 \Sigma m_i - (\Sigma m_i \bar{v}_i)^2} \tag{8.104}$$

Similarly, using $\bar{u}_i$ and $\bar{v}_i$ and equation 8.66 yields

$$a_0 = \frac{\Sigma\bar{u}_i m_i - a_1 \Sigma\bar{v}_i m_i}{\Sigma m_i} \tag{8.105}$$

Note the differences between equations 8.104 and 8.97, and 8.105 and 8.98. It is clear that an unweighted regression is extensively biased. Alternatively, each estimation could be weighted by the number of observations used to estimate it.

$$Y_i = m_i \bar{u}_i \qquad (8.106)$$

$$X_i = m_i \bar{v}_i \qquad (8.107)$$

From these two equations, the elements of equations 8.95 and 8.96 would be

$$X_i Y_i = m_i \bar{u}_i \bar{v}_i \qquad (8.108)$$

$$X_i^2 = m_i \bar{v}_i^2 \qquad (8.109)$$

Similar substitutions can be used to estimate the means, variances, and covariances. Then using equations 8.95 and 8.96, the estimates of a_1 and a_0 are

$$a_1 = \frac{\Sigma m_i \bar{u}_i \bar{v}_i - (\Sigma m_i \bar{u}_i)(\Sigma m_i \bar{v}_i)/\Sigma m_i}{\Sigma m_i \bar{v}_i^2 - (\Sigma m_i \bar{v}_i)^2/\Sigma m_i} \qquad (8.110)$$

$$a_0 = \frac{\Sigma m_i \bar{u}_i}{\Sigma m_i} - a_1 \frac{\Sigma m_i \bar{v}_i}{\Sigma m_i} \qquad (8.111)$$

On comparing equations 8.110 and 8.111 with 8.104 and 8.105, respectively, the formulations for a_0 and a_1 are seen to be identical in each case.

A Practical Example

As an illustration of the effect of weighting or not weighting a regression problem, the data used in the earlier examples were used again. For this example, these data are grouped (aggregated) by location and housing type and the variables of housing expenditure, floor area, disposable income, and household size are averaged for each location and housing type. The numbers of observations in each location are shown in table 8–15. It should be clear that the reliability of the estimates of the average values of each variable will be different for each location. If this fact is ignored and a model built on the average data, without weighting, the results would be as shown in table 8–16. This regression is significant at 98.1%.

Because of the small sample size, the results appear to be misleadingly good. However, the significance of the results is not very striking.

Table 8–15
Number of Observations for Each Location and Type

Group		
Location	Type	No. of Obs.
1	1	1
1	2	1
2	1	1
2	2	5
3	1	1
3	2	2
4	1	7
4	2	1
5	1	4
5	2	2

Table 8–16
Results of Unweighted Regression

Variable Coefficients (t Score)					
Disposable Income	Floor Area	Household Size	Constant	R^2	F (d.f.)
0.146 (2.1)	19.88 (2.8)	−143.4 (1.2)	−43.5 (0.1)	0.787	7.38 (3,6)

Only floor area has a significant coefficient, although disposable income is close to the 95% level. The results of this aggregate model are fairly similar to the results reported in equation 8.38. The coefficient of disposable income is similar (0.146 compared to 0.191), floor area is also similar (19.88 compared to 11.98), but the coefficient of household size is quite different (−143.4 compared to 81.2), as is the constant (−43.5 compared to −511.6).

Carrying out the linear regression by weighting the observations results in the estimates shown in table 8–17, which differ little from those of table 8–16, but compare well with the models based on 25 observations (equations 8.36 through 8.38). The estimates obtained in table 8–17 are more correct than those in table 8–16. Both disposable income and floor area have coefficient values that are statistically no different from those of equation 8.38. Household size and the constant are substantially different (even significantly so, despite the relatively small t scores and large standard errors). However, the results are closer than those of the unweighted regression.

Table 8–17
Results of Weighted Regression

Variable Coefficients (t Score)					
Disposable Income	Floor Area	Household Size	Constant	R^2	F (d.f.)
0.138 (3.13)	11.70 (1.93)	−145.41 (2.01)	599.9 (2.47)	.923	23.97 (3,6)

Analysis of Residuals

This section deals with some diagnostic information that can be obtained when a linear-regression relationship has been fitted. It will be remembered that the residuals are

$$\epsilon_i = Y_i - \hat{Y}_i \qquad (8.112)$$

First, as shown earlier in the chapter, the sum of the residuals for the set of observations is zero. Similarly, it is assumed that the mean of the residuals is zero, and this assumption holds since the sum of residuals is zero.

If a frequency plot of the residuals is examined, this should not contraindicate a normal distribution with zero mean.[3] It may not resemble such distribution closely, but should be acceptable unless there is a concentration of values at some extreme or an indisputable skewness. The best test of this is to use a table of random normal deviates and plot two or three sets of these, using the same numbers of observations as the original number of residuals. This will indicate whether or not the plot is acceptable.

If the data represent a change over time, the residuals can be plotted on a time scale. If the general effect of this plot is not a horizontal band, it must be assumed that the data had some long-term variation over time which has not been taken into account. If consistent variations occur within the horizontal band, it may be concluded that there is a short-term effect which has not been accounted for.

Again, if the residuals are plotted against $\hat{Y}_i$, a horizontal-band effect should be obtained. If not, the model is in some way invalidated. If the residuals lie in a band which opens out with increasing $\hat{Y}_i$, it would appear that some of the observations should be weighted since the variance of the residual is not constant.

Other plots may be used to determine effects in the model which are invalid. It is also possible to use calculations instead of visual plots, but this is not usually recommended.

An analysis of residuals is, however, not foolproof, particularly in multivariate regression. Errors in the derived coefficients caused by correlations between the independent variables or errors of measurement of these variables may lead to residual plots which appear to indicate other factors as the source of error in the model. Such plots should therefore be used with considerable caution and with careful reference to the other errors known to exist in the model. A more detailed discussion of this topic may be found elsewhere.[4]

Notes

1. N.R. Draper and H. Smith, *Applied Regression Analysis,* New York: John Wiley & Sons, pp. 104–107; J. Johnston, *Econometric Methods,* New York: McGraw-Hill Book Co., 1963, p. 109; Arthur S. Goldberger, *Econometric Theory,* New York: John Wiley & Sons, 1964, p. 158.

2. Draper and Smith, *Applied Regression Analysis,* pp. 134–136; Johnston, *Econometric Methods,* pp. 221–228.

3. Draper and Smith, *Applied Regression Analysis,* chapter 3.

4. Ibid.

9

Simultaneous Equations

Concepts

The models so far considered in this text are simple linear models with a single dependent variable and one or more independent variables. In this chapter, a more complex situation is envisaged. First, it is necessary to review the definition of an independent variable. As stated in chapter 7, an independent variable causes changes in the dependent variable but is not itself affected by changes in the dependent variable. Thus one might develop a model of grain yield as a function of rainfall. It is clear that rainfall will affect the grain yield but grain yield will not affect the amount of rainfall. Grain yield is a dependent variable and rainfall an independent variable in this model.

Consider now a situation in which one wishes to develop a model to estimate the number of trips made by a household. The model might be developed as a function of the number of people in the household, the number of cars owned, the number of workers in the household, and the type of dwelling unit. In turn, however, it may be necessary to find an estimating equation for the number of cars owned. This may be postulated to be a function of the household income, the number of licensed drivers, and the number of trips made. It is clear now that at least two variables in these two postulated models violate the conditions of independence. Number of trips made appears as both a dependent and an independent variable, as does car ownership. By the normal rules of linear regression, this situation is inadmissible. It would be incorrect to estimate these two equations as two completely separate and independent models. However, the models may be estimated correctly by estimating them simultaneously, that is, by use of simultaneous equations.

To pursue the notions of simultaneous equations, it is necessary to introduce two new terms for variables to replace those of dependent and independent variables. It should be clear from this example that such terms may become somewhat ambiguous in the case of simultaneous equations, where certain variables may be so-called independent variables in one model and dependent variables in others.

First, an *exogenous* variable is one whose value is determined outside the system of simultaneous equations. In other words, in the above examples, variables such as the number of workers in the household, the number of people in the household, and household income, are all exogenous variables. Second, an *endogenous* variable is one whose value is determined within the system of simultaneous equations. In the examples, trips made and cars owned are endogenous variables. In the simple regression cases considered in chapters 7 and 8, the dependent variables are endogenous variables and the independent variables are exogenous variables. In the more complex case of simultaneous equations, all variables on the left side of the equations are endogenous variables, while those on the right side will be a mixture of exogenous and endogenous variables.

The analysis of simultaneous-equation solutions is somewhat complex, and necessitates the use of matrix algebra to derive general solutions. In the next section, a classic simultaneous-equation problem from economics is used as a basis for developing the basic rationale of simultaneous-equation solutions. In the following section, a matrix treatment of general solution methods is described. The reader who wishes to know only what the method is and where it might be applicable, without necessarily applying it, should find the next section sufficient and the succeeding one, giving the general solutions, may be omitted.

A Simple Example

The model used to illustrate the solution methods for simultaneous equations is the simple Keynesian model of the economy,[1] which is

$$C = \alpha + \beta Y \tag{9.1}$$

$$Y = C + I \tag{9.2}$$

where C = expenditure on consumption
$\quad\quad Y$ = income
and $\quad I$ = investment

The model states that income is equal to the sum of expenditure and savings—an equilibrium model—and that consumption expenditure is proportional to income—standard theoretical consumer behavior. Equations 9.1 and 9.2 can be rewritten as statistical estimation models.

$$C_t = \alpha + \beta Y_t + \epsilon_t \tag{9.3}$$

$$Y_t = C_t + I_t \tag{9.4}$$

The relationship in equation 9.4 is an exact one, and therefore has no error term. Conventionally, it is assumed that I_t is an exogenous variable, controlled by public authorities independently of C_t and Y_t. This then determines Y_t and C_t as endogenous variables. As usual, ϵ_t is specified as being normally distributed with variance σ^2, mean zero, and where ϵ_t and

ϵ_{t+s} $(s\neq0)$ are independent. That is, $E(\epsilon_t) = 0$, $V(\epsilon_t) = \sigma^2$, and $cov(\epsilon_t,\epsilon_{t+s}) = 0$, so long as s is nonzero. In classical least squares it is assumed that the independent variables are independent of the error term. The coefficients, α and β in equation 9.3, could be estimated by classical least squares, if this condition is met. Substituting equation 9.3 in equation 9.4 results in

$$Y_t = \alpha + \beta Y_t + \epsilon_t + I_t \tag{9.5}$$

Gathering the Y_ts on one side yields

$$Y_t = \frac{\alpha}{1-\beta} + \frac{\epsilon_t}{1-\beta} + \frac{I_t}{1-\beta} \tag{9.6}$$

It can be shown that Y_t is not independent of ϵ_t, and therefore violates the classical regression assumptions. First, the expected value of Y_t is

$$E(Y_t) = \frac{\alpha}{1-\beta} + \frac{I_t}{1-\beta} \tag{9.7}$$

The covariance between Y_t and ϵ_t which would normally be assumed to be zero is

$$cov(\epsilon_t, Y_t) = E\left[\epsilon_t \left(\frac{\alpha}{1-\beta} + \frac{\epsilon_t}{1-\beta} + \frac{I_t}{1-\beta} - \frac{\alpha}{1-\beta} - \frac{I_t}{1-\beta}\right)\right] \tag{9.8}$$

This is so since the covariance is

$$cov(\epsilon_t, Y_t) = E\left[(\epsilon_t - E(\epsilon_t))(Y_t - E(Y_t))\right] \tag{9.9}$$

The expected value of ϵ_t (i.e., the mean of ϵ_t) is of course assumed to be zero. By substituting Y_t from equation 9.6 and $E(Y_t)$ from equation 9.7, equation 9.8 results. This can be simplified to

$$cov(\epsilon_t, Y_t) = \frac{1}{1-\beta} E(\epsilon_t^2) \tag{9.10}$$

Because the expected value of ϵ_t^2 is Δ^2, equation 9.10 is clearly nonzero. Therefore the normal assumption of linear regression, that the covariance between ϵ_t and Y_t is zero, is violated for equation 9.3 and estimates of α and β will be biased for any sample size. Thus there is a need to define a solution procedure for obtaining unbiased estimates of the regression coefficients.

The Reduced Form

The first possible process would be to remove the offending variable. Clearly, I_t is not correlated with the error term, since it is an exogenous variable. The reduced form of the model is found by expressing the endogenous variables as functions of the exogenous variables only. First, equation 9.6 can be substituted into equation 9.3, producing

$$Y_t = \frac{\alpha}{1-\beta} + \frac{I_t}{1-\beta} + \frac{\epsilon_t}{1-\beta} \qquad (9.11)$$

This expresses Y_t in terms of I_t only. The same can be done for C_t, by substituting equation 9.4 in equation 9.3, producing

$$C_t = \alpha + \beta(C_t + I_t) + \epsilon_t \qquad (9.12)$$

Gathering terms and expressing equation 9.12 in terms of C_t, yields

$$C_t = \frac{\alpha}{1-\beta} + \frac{\beta I_t}{1-\beta} + \frac{\epsilon_t}{1-\beta} \qquad (9.13)$$

Equations 9.12 and 9.13 are the reduced forms of equations 9.3 and 9.4. Given the normal assumptions on the error term, conventional least-squares regression can be applied to the equations to estimate the three compound coefficients, $\alpha/(1-\beta)$, $\beta/(1-\beta)$, and $1/(1-\beta)$. The estimates of these coefficients are unbiased.

$$\frac{\beta}{1-\beta} = \frac{\Sigma(I_t-\bar{I})(C_t-\bar{C})}{\Sigma(I_t-\bar{I})^2} \qquad (9.14)$$

$$\frac{\alpha}{1-\beta} = \bar{C} - \bar{I}\left\{\frac{\Sigma(C_t-\bar{C})(I_t-\bar{I})}{\Sigma(I_t-\bar{I})^2}\right\} \qquad (9.15)$$

$$\frac{1}{1-\beta} = \frac{\Sigma(I_t-\bar{I})(Y_t-\bar{Y})}{\Sigma(I_t-\bar{I})^2} \qquad (9.16)$$

From equation 9.14, an estimate of β can be obtained, denoted β^*.

$$\beta^* = \frac{\Sigma(I_t-\bar{I})(C_t-\bar{C})}{\Sigma(I_t-\bar{I})^2 + \Sigma(I_t-\bar{I})(C_t-\bar{C})} \qquad (9.17)$$

Similarly, from equations 9.14 and 9.15, an estimate, α^*, can be obtained.

$$\alpha^* = \frac{\bar{C}\Sigma(I_t-\bar{I})^2 - \bar{I}\Sigma(I_t-\bar{I})(C_t-\bar{C})}{\Sigma(I_t-\bar{I})^2 + \Sigma(I_t-\bar{I})(C_t-\bar{C})} \qquad (9.18)$$

Equation 9.17 is consistent with equation 9.16 also. Equation 9.16 can be rearranged to yield

$$\beta^* = \frac{\Sigma(I_t-\bar{I})(Y_t-\bar{Y}) - \Sigma(I_t-\bar{I})^2}{\Sigma(I_t-\bar{I})(Y_t-\bar{Y})} \qquad (9.19)$$

Now to show the consistency between equations 9.16 and 9.17, it is necessary to substitute for the terms in Y_t, terms in I_t and C_t. Returning to the original model, equation 9.20 provides an estimate of Y_t in terms of C_t and I_t (see equation 9.4).

$$Y_t = C_t + I_t \tag{9.20}$$

Adjusting each variable to its respective mean yields

$$(Y_t - \bar{Y}) = (C_t - \bar{C}) + (I_t - \bar{I}) \tag{9.21}$$

This follows also since, given equation 9.20, equation 9.22 is also true.

$$\bar{Y} = \bar{C} + \bar{I} \tag{9.22}$$

Multiplying throughout equation 9.21 by $(I_t - \bar{I})$ and summing over all observations yields

$$\Sigma(I_t - \bar{I})(Y_t - \bar{Y}) = \Sigma(C_t - \bar{C})(I_t - \bar{I}) + \Sigma(I_t - \bar{I})^2 \tag{9.23}$$

Substituting equation 9.23 into equation 9.19 yields

$$\beta^* = \frac{\Sigma(C_t - \bar{C})(I_t - \bar{I})}{\Sigma(C_t - \bar{C})(I_t - \bar{I}) + \Sigma(I_t - \bar{I})^2} \tag{9.24}$$

This equation is identical to equation 9.17, thus confirming that equations 9.14 and 9.16 yield consistent estimates for β^*.

Although estimates of α^* and β^* have now been obtained, it is necessary to manipulate the expressions for them to yield two further important items of information. In the first place, it is desirable to know whether the estimates, α^* and β^*, are biased or unbiased estimates of α and β, and whether, if biased, the estimates are consistent. Second, several alternative solution methods are derived subsequently, and it is desirable to express each solution in the same format for comparison purposes.

It can be shown, first, that the estimates of α^* and β^* are biased but consistent estimators. This is shown in the following procedure. By using equation 9.23, equations 9.17 and 9.18 can be rewritten

$$\beta^* = \frac{\Sigma(I_t - \bar{I})(C_t - \bar{C})}{\Sigma(I_t - \bar{I})(Y_t - \bar{Y})} \tag{9.25}$$

$$\alpha^* = \frac{\bar{C}\Sigma(I_t - \bar{I})^2 - \bar{I}\Sigma(I_t - \bar{I})(C_t - \bar{C})}{\Sigma(I_t - \bar{I})(Y_t - \bar{Y})} \tag{9.26}$$

These two expressions can be simplified by adopting the notation of equations 9.27 through 9.29.

$$\Sigma(I_t - \bar{I})^2 = n\,V(I) \tag{9.27}$$

$$\Sigma(I_t - \bar{I})(C_t - \bar{C}) = n\,cov(I,C) \tag{9.28}$$

$$\Sigma(I_t - \bar{I})(Y_t - \bar{Y}) = n\,cov(I,Y) \tag{9.29}$$

Similar expressions can be used for other sums of squares and cross products. To determine bias and consistency, the least-squares solution of equation 9.13 may be rewritten by substituting equation 9.15 in equation 9.13.

$$C_t = \bar{C} - \frac{\beta}{1-\beta}\bar{I} + \frac{\beta}{1-\beta}I_t + \frac{\epsilon_t}{1-\beta} \tag{9.30}$$

By gathering terms, multiplying throughout by $(I_t - \bar{I})$, and summing over all observations, and using the notation of equations 9.27 through 9.29, equation 9.30 can be transformed.

$$\text{cov}(C,I) = \frac{1}{1-\beta}[\beta V(I) + \text{cov}(\epsilon,I)] \tag{9.31}$$

By a similar process, equation 9.11 can be transformed.

$$\text{cov}(Y,I) = \frac{1}{1-\beta}[V(I) + \text{cov}(\epsilon,I)] \tag{9.32}$$

Rewriting equation 9.25 in the new notation yields

$$\beta^* = \frac{\text{cov}(C,I)}{\text{cov}(Y,I)} \tag{9.33}$$

Substituting equations 9.31 and 9.32 into equation 9.33 produces

$$\beta^* = \frac{\beta V(I) + \text{cov}(\epsilon,I)}{V(I) + \text{cov}(\epsilon,I)} \tag{9.34}$$

However, the $\text{cov}(\epsilon,I)$ approaches 0 as n approaches infinity. Hence, as n approaches infinity, the value of β^* approaches the true value of β.

$$\lim_{n\to\infty} \beta^* = \beta \tag{9.35}$$

Therefore, for a finite sample, β^* is a biased estimate of β, but is consistent since the bias approaches zero as the sample gets larger.

This approach is sometimes known as Indirect Least Squares, but it does not yield unbiased estimates of β and α. It is not always applicable to simultaneous-equations formulations. In general, this method is only applicable when the structural relationships are exactly identified.

The proof of the reduction of equation 9.30 to equation 9.31 and of equation 9.11 to equation 9.32 can be demonstrated in the following steps. Rearranging equation 9.30 produces

$$(C_t - \bar{C}) = \frac{\beta}{1-\beta}(I_t - \bar{I}) + \frac{\epsilon_t}{1-\beta} \tag{9.36}$$

Multiplying equation 9.36 by $(I_t - \bar{I})$ yields

$$(C_t - \bar{C})(I_t - \bar{I}) = \frac{\beta}{1-\beta}(I_t - \bar{I})^2 + \frac{\epsilon_t(I_t - \bar{I})}{1-\beta} \tag{9.37}$$

Finally, summing equation 9.37 over all observations produces

$$\Sigma(C_t-\bar{C})(I_t-\bar{I}) = \frac{\beta}{1-\beta}\Sigma(I_t-\bar{I})^2 + \frac{1}{1-\beta}\Sigma(\epsilon_t)(I_t-\bar{I}) \qquad (9.38)$$

Using the notation of equations 9.27 through 9.29, equation 9.38 can be written as equation 9.31. In a similar manner, equation 9.11 can be written

$$Y_t = \bar{Y} - \frac{1}{1-\beta}\bar{I} + \frac{I_t}{1-\beta} + \frac{\epsilon_t}{1-\beta} \qquad (9.39)$$

Gathering terms results in

$$Y_t - \bar{Y} = \frac{1}{1-\beta}(I_t-\bar{I}) + \frac{\epsilon_t}{1-\beta} \qquad (9.40)$$

Multiplying throughout by (I_t-I) and summing over all observations produces

$$\Sigma(Y_t-\bar{Y})(I_t-\bar{I}) = \frac{1}{1-\beta}\Sigma(I_t-\bar{I})^2 + \frac{1}{1-\beta}\Sigma\epsilon_t(I_t-\bar{I}) \qquad (9.41)$$

Using the simpler notation, equation 9.41 can be written

$$\mathrm{cov}(Y,I) = \frac{1}{1-\beta}[V(I)+\mathrm{cov}(\epsilon,I)] \qquad (9.42)$$

This is identical to equation 9.32.

Two-stage Least Squares

This is a more generally applicable method of solution than the Indirect Least-Squares method. The major difficulty in solving simultaneous equations is caused by the correlation between Y and ϵ when Y is the independent variable in the consumption relation. Two-stage least squares involves a two-stage estimating procedure which attempts to "purge" Y of its dependence on ϵ before estimating α and β in the first relation. This purging is accomplished by proposing a relationship between Y and the only exogenous variable, I.

$$Y_t = \hat{\pi}_1 + \hat{\pi}_2 I_t + u_t \qquad (9.43)$$

From equation 9.43, the ordinary least-squares solution produces estimates for π_1 and π_2.

$$\hat{\pi}_1 = \bar{Y} - \hat{\pi}_2\bar{I} \qquad (9.44)$$

$$\hat{\pi}_2 = \frac{\mathrm{cov}(Y,I)}{V(I)} \qquad (9.45)$$

The estimated equation for Y_t in terms of I_t is

$$\hat{Y}_t = \hat{\pi}_1 + \hat{\pi}_2 I_t \tag{9.46}$$

Substituting the estimated $\hat{Y}_t$s into the consumption relation, equation 9.3, results in

$$C_t = \alpha + \beta(\hat{Y}_t) + (\epsilon_t + \beta u_t) \tag{9.47}$$

Now $\hat{Y}_t$ is an exact function of I_t and is uncorrelated with ϵ_t, while u_t is uncorrelated with I_t. Thus $\hat{Y}_t$ is uncorrelated with the total error term $(\epsilon_t + \beta u_t)$ in equation 9.47. Ordinary least-squares estimation can now be applied directly to equation 9.47. The estimates of α and β (which are denoted $\hat{\alpha}$ and $\hat{\beta}$) are

$$\hat{\beta} = \frac{\text{cov}(\hat{Y}, C)}{V(\hat{Y})} \tag{9.48}$$

$$\hat{\alpha} = \bar{C} - \hat{\beta}\bar{\hat{Y}} \tag{9.49}$$

In this formulation, there is a variance of $\hat{Y}$ and a covariance involving $\hat{Y}$. However, $\hat{Y}$ is not one of the original variables. It is necessary therefore to replace $\hat{Y}$ in equations 9.48 and 9.49. Thus, in equation 9.43, substituting for $\hat{\pi}_1$ results in

$$\hat{Y}_t = \bar{Y} - \hat{\pi}_2 \bar{I} + \hat{\pi}_2 I_t \tag{9.50}$$

This may be rewritten

$$\hat{Y}_t - \bar{Y} = \hat{\pi}_2(I_t - \bar{I}) \tag{9.51}$$

Multiplying throughout by $(C_t - \bar{C})$ and summing over all observations produces

$$\text{cov}(\hat{Y}, C) = \hat{\pi}_2 \text{cov}(I, C) \tag{9.52}$$

Similarly, squaring and summing over all observations produces

$$V(\hat{Y}) = \hat{\pi}_2^2 V(I) \tag{9.53}$$

Substituting these expressions for those containing Ys in equation 9.48 results in

$$\hat{\beta} = \frac{\hat{\pi}_2 \text{cov}(I, C)}{\hat{\pi}_2^2 V(I)} \tag{9.54}$$

However, $\hat{\pi}_2$ is

$$\hat{\pi}_2 = \frac{\text{cov}(Y, I)}{V(I)} \tag{9.55}$$

Using equations 9.52 and 9.53 produces equation 9.56 as the final comparable version of equation 9.54.

$$\hat{\beta} = \frac{\text{cov}(I,C)}{\text{cov}(Y,I)} \tag{9.56}$$

In this case, it can be seen that this is identical to the Indirect Least-Squares estimate, given in equation 9.33. It follows that the value of $\hat{\alpha}$ will be identical to α^*. Again, it needs to be reemphasized that this occurs in this case because of the form of the original equations.

Least-Variance Ratio

A third approach to solving simultaneous equations is one termed the Least-Variance Ratio. Reexamining the original models, equations 9.3 and 9.4, equation 9.5 implies that I does not enter the consumption relation in addition to Y. Hence equation 9.57 may be proposed.

$$C_t = \alpha + \beta Y_t + \gamma I_t + \phi_t \tag{9.57}$$

But equation 9.3 suggests that $\gamma = 0$. If equation 9.57 was calibrated from a finite sample, a nonzero value for γ would be obtained and a smaller residual error will be obtained for equation 9.3. The least-variance ratio theorem states that the estimators of α and β should be chosen so that the ratio of residual variances of equation 9.3 to equation 9.57 is minimized. To proceed with this method, it is necessary to obtain expressions of the residual variances, whose ratio is to be minimized. From equations 9.57 and 9.3, the ratio to be minimized is

$$\tau = \frac{\Sigma \epsilon_t^2}{\Sigma \phi_t^2} \tag{9.58}$$

From equations 9.57 and 9.3, it is not possible to express the numerator and denominator of equation 9.58 directly in terms that will permit one to see under what conditions the ratio is minimized. Hence some manipulations are necessary. First, define C_t^{**} and $\bar{C}^{**}$.

$$C_t^{**} = C_t - (\alpha^{**} + \beta^{**} Y_t) \tag{9.59}$$

$$\bar{C}^{**} = \bar{C} - (\alpha^{**} + \beta^{**} \bar{Y}) \tag{9.60}$$

In these equations, α^{**} and β^{**} are the least-variance estimators of α and β. Rearranging equation 9.60 to give an expression for α^{**} results in

$$\alpha^{**} = \bar{C} - \bar{C}^{**} - \beta^{**} \bar{Y} \tag{9.61}$$

Similarly, β^{**} is given by equation 9.62, by subtracting equation 9.60 from equation 9.59.

$$C_t^{**} - \bar{C}^{**} = C_t - \bar{C} - \beta^{**}(Y_t - \bar{Y}) \tag{9.62}$$

Comparing equation 9.62 with equation 9.3, $(C_t^{**} - \bar{C}^{**})$ is the error term, ϵ_t, of equation 9.3. Thus the numerator of the least-variance ratio is expressed in terms of C_t^{**} and it only remains to develop an expression for the denominator in terms of C_t^{**} to be able to understand the behavior of the ratio, τ, and determine under what conditions the ratio is minimized. The numerator of course is

$$\sum_t \epsilon_t^2 = \sum_t (C_t^{**} - \bar{C}^{**})^2 \tag{9.63}$$

Equation 9.57 is rewritten, first, in terms of α^{**} and β^{**}

$$C_t = \alpha^{**} + \beta^{**} Y_t + \gamma I_t + \phi_t \tag{9.64}$$

Rearranging equation 9.64 yields

$$C_t - (\alpha^{**} + \beta^{**} Y_t) = \gamma I_t + \phi_t \tag{9.65}$$

However, the left side of equation 9.65 is C_t^{**}, as given by equation 9.59. Hence, equation 9.65 can be written

$$C_t^{**} = \gamma I_t + \phi_t \tag{9.66}$$

Since $\bar{\phi}$ is zero, one may also write

$$\bar{C}^{**} = \gamma \bar{I} \tag{9.67}$$

From these simple relations, $\hat{\gamma}$ may be estimated by ordinary least squares; the estimate is

$$\hat{\gamma} = \frac{\text{cov}(I, C^{**})}{V(I)} \tag{9.68}$$

From equation 9.66, the sum of the squares of ϕ_t can be written

$$\sum_t \phi_t^2 = \sum_t (C_t^{**} - \hat{\gamma} I_t)^2 \tag{9.69}$$

For this expression to be useful, it is necessary to substitute for γ by using equation 9.68. This cannot be done usefully within the summation, so it is necessary first to expand the right side of equation 9.69 and get the γ values outside the summations. Expanding the right side of equation 9.69 yields

$$\sum (C_t^{**} - \hat{\gamma} I_t)^2 = \sum (C_t^{**})^2 - 2\hat{\gamma} \sum C_t^{**} I_t + \hat{\gamma}^2 (I_t)^2 \tag{9.70}$$

Substituting for γ from equation 9.68 produces

$$\sum (C_t^{**} - \hat{\gamma} I_t)^2 = \sum (C_t^{**})^2 - \frac{2\text{cov}(I, C^{**})}{V(I)} \sum C_t^{**} I_t + \left[\frac{\text{cov}(I, C^{**})}{V(I)} \right]^2 \sum I_t^2 \tag{9.71}$$

From equation 9.67, one may write

$$\bar{C}^{**} - \hat{\gamma}\bar{I} = 0 \qquad (9.72)$$

Therefore, equation 9.73 also follows.

$$n(\bar{C} - \hat{\gamma}\bar{I})^2 = 0 \qquad (9.73)$$

The left side of equation 9.73 may be expanded to equation 9.74, which still equals zero.

$$n(\bar{C})^2 - 2n\hat{\gamma}\bar{C}\bar{I} + n\hat{\gamma}^2\bar{I}^2 = 0 \qquad (9.74)$$

Subtracting the left side of equation 9.74 from the right side of equation 9.71 will not alter the equation, since this is subtracting zero from equation 9.71. However, this will yield

$$\Sigma(C_t^{**} - \hat{\gamma}I_t)^2 = \Sigma(C_t^{**})^2 - n(\bar{C}^{**})^2 - \frac{2\text{cov}(I,C^{**})}{V(I)}$$

$$[\Sigma C_t^{**}I_t - n\bar{C}^{**}\bar{I}] + \left[\frac{\text{cov}(I,C^{**})}{V(I)}\right]^2 [\Sigma I_t^2 - n\bar{I}^2] \quad (9.75)$$

Simplifying the notation of equation 9.75, by dividing throughout by n, produces

$$\frac{1}{n}\Sigma\phi_t^2 = V(C^{**}) - \frac{2[\text{cov}(I,C^{**})]^2}{V(I)} + \frac{[\text{cov}(I,C^{**})]^2}{V(I)} \qquad (9.76)$$

This may be simplified to

$$\frac{1}{n}\Sigma\phi_t^2 = V(C^{**}) - [\text{cov}(I,C^{**})]^2/V(I) \qquad (9.77)$$

From equation 9.63, one can write

$$\frac{1}{n}\Sigma\epsilon_t^2 = V(C^{**}) \qquad (9.78)$$

Hence, the variance ratio, τ, is

$$\tau = \frac{V(C^{**})}{V(C^{**}) - [\text{cov}(I,C^{**})]^2/V(I)} \qquad (9.79)$$

Since $[\text{cov}(I,C^{**})]^2/V(I)$ must always be nonnegative, it follows that the minimum value of τ occurs when this term is zero and τ is 1. This condition now needs to be translated into estimates of α^{**} and β^{**}.

Multiplying equation 9.62 by $(I_t - \bar{I})$ and summing over all observations produces

$$\Sigma(C_t^{**} - \bar{C})(I_t - \bar{I}) = \Sigma(C_t - \bar{C})(I_t - \bar{I}) - \beta^{**}\Sigma(Y_t - \bar{Y})(I_t - \bar{I}) \quad (9.80)$$

Dividing throughout by n yields

$$\text{cov}(I,C^{**}) = \text{cov}(I,C) - \beta^{**}\text{cov}(Y,I) \qquad (9.81)$$

According to the Least-Variance Ratio theorem, the best estimate of β^{**} is obtained when the left side of equation 9.81 is zero; hence β^{**} is

$$\beta^{**} = \frac{\text{cov}(I,C)}{\text{cov}(Y,I)} \qquad (9.82)$$

Comparing this estimate with equations 9.56 and 9.33, it can be seen that, in this example, the solution for β is identical with the estimates from the other two methods. Hence, it appears that the three methods—Indirect Least Squares (ILS), Two-stage Least Squares, and Least-Variance Ratio (LVR)—are consistent with each other. As will be seen in the next sections, the reason for having all three methods is that in any given system of equations some of these methods cannot be applied or are not valid. Under certain circumstances, none of these methods may be acceptable and a fourth one may be needed, called Three-stage Least Squares. This fourth method cannot be demonstrated on this example because the example system is too simple.

General Solution Methods

At the beginning of the chapter, the terms exogenous and endogenous were defined. For most applications, these terms suffice to distinguish the two major types of variables in simultaneous equations. However, in dynamic models where there may be lagged variables a modification to this terminology may be necessary. The lagged values of endogenous variables are already determined and therefore should not be treated as endogenous from the point of view of the solution methods. For the general case, it is useful to define two new terms. The lagged endogenous variables and exogenous variables are termed *predetermined variables* and the current values of the endogenous variables are *jointly dependent variables*.[2]

To study the solution methods in detail, a general linear model is set up and matrix notation is used to manipulate the solutions. Assume there is a set of G relations, the ith one of which may be written

$$\beta_{i1}y_{1j} + \beta_{i2}y_{2j} + \ldots + \beta_{iG}y_{Gj} + \gamma_{i1}x_{1j} + \ldots$$

$$+ \gamma_{iK}x_{Kj} = u_{ij} \qquad (j=1,2,\ldots,n) \qquad (9.83)$$

where y_{ij} = the jointly dependent variables
$\qquad x_{ij}$ = the predetermined variables
and $\quad u_{ij}$ = the disturbances (or error terms)

In general, some of the βs and γs must be zero in each relationship, since otherwise all the relationships would look alike statistically, and one would be unable to distinguish between them. A constant may, of course, be introduced into the relationships by setting one of the xs at constant unity for all j.

In matrix form, the set of relationships may be written

$$\mathbf{B}\mathbf{Y}_j + \mathbf{\Gamma}\mathbf{X}_j = \mathbf{U}_j \tag{9.84}$$

where $\mathbf{B}$ = a $G \times G$ matrix of coefficients of the jointly dependent variables; $\mathbf{\Gamma}$ = a $G \times K$ matrix of coefficients of the predetermined variables; and $\mathbf{Y}_j$, $\mathbf{X}_j$, and $\mathbf{U}_j$ = column vectors of G, K, and G elements, respectively. These vectors and matrices are shown in equations 9.85 through 9.89.

$$\mathbf{Y}_j = \begin{vmatrix} y_{1j} \\ y_{2j} \\ \vdots \\ y_{Gj} \end{vmatrix} \tag{9.85}$$

$$\mathbf{X}_j = \begin{vmatrix} x_{1j} \\ x_{2j} \\ \vdots \\ x_{Kj} \end{vmatrix} \tag{9.86}$$

$$\mathbf{U}_j = \begin{vmatrix} u_{1j} \\ u_{2j} \\ \vdots \\ u_{Gj} \end{vmatrix} \tag{9.87}$$

$$\mathbf{B} = \begin{vmatrix} \beta_{11} & \beta_{12} & \cdots & \beta_{1G} \\ \beta_{21} & & & \beta_{2G} \\ \vdots & & & \vdots \\ \beta_{G1} & \beta_{G2} & \cdots & \beta_{GG} \end{vmatrix} \tag{9.88}$$

$$\mathbf{\Gamma} = \begin{vmatrix} \gamma_{11} & \gamma_{12} & \cdots & \gamma_{1K} \\ \gamma_{21} & & & \gamma_{2K} \\ \vdots & & & \vdots \\ \gamma_{G1} & \gamma_{G2} & \cdots & \gamma_{GK} \end{vmatrix} \tag{9.89}$$

With this notation and system of equations, each of the solution methods can be developed into a general form that can be applied to any appropriate system of simultaneous linear equations.

The Reduced Form

The reduced form of the model is that obtained by expressing each endogenous variable in terms of the predetermined variables only. Assuming the **B** matrix is nonsingular, the reduced form of the model can be written

$$\mathbf{Y_j} = \mathbf{\Pi}\, \mathbf{X_j} + \mathbf{V_j} \tag{9.90}$$

where $\mathbf{\Pi} = G \times K$ matrix of reduced-form coefficients, and $\mathbf{V_j} =$ a column vector of G reduced-form disturbances. Applying ordinary least squares to equation 9.90 produces equation 9.91 as the solution for the matrix $\mathbf{\Pi}$.

$$\mathbf{\Pi} = -\mathbf{B}^{-1}\,\mathbf{\Gamma} \tag{9.91}$$

The disturbances, $\mathbf{V_j}$, are related to the original disturbances by

$$\mathbf{V_j} = \mathbf{B}^{-1}\,\mathbf{U_j} \tag{9.92}$$

Before any solutions are possible to the matrix 9.90, it is necessary to examine the form of the disturbance and certain other properties of the model.

Identification

The problems to be considered here are those of the solubility of the matrix 9.90, in terms of the βs and γs. As usual, in least-squares regression, some properties of the disturbances are assumed. First, it is assumed that the expected value of each disturbance term, U_{ij}, is zero.

$$\mathbf{E(U_j)} = \mathbf{0} \qquad \text{for all } \mathbf{j} \tag{9.93}$$

No a priori restrictions are made on the variance-covariance matrix which is therefore assumed to be

$$\mathbf{E(U_j,U_j')} = \mathbf{\Phi} = \begin{vmatrix} \sigma_{11} & \sigma_{12} & \cdots & \sigma_{1G} \\ \sigma_{21} & & & \sigma_{2G} \\ \vdots & & & \vdots \\ \sigma_{G1} & \sigma_{G2} & \cdots & \sigma_{GG} \end{vmatrix} \tag{9.94}$$

This formulation assumes that the variances and covariances are constant for all j, and makes no assumptions of the off-diagonal elements (the covariances) being zero. It should be noted that for each identity contained in the set of G models, one row and column of the matrix will be zero. (An identity is an equation which expresses an exact known relationship, not subject to errors of specification.) This, however, would cause $\mathbf{\Phi}$ to be singular. It is therefore assumed that the model has been solved initially to eliminate all identities, in which case $\mathbf{\Phi}$ is nonsingular.

In other words, no disturbance is assumed to be an exact linear function of any other disturbance(s).

In the simple example, equation 9.4 is an identity. Hence this identity must be absorbed into other equations in the system. In the example, this was done by substituting for Y in equation 9.3 in terms of C and I. The required simplification here is given by equation 9.12.

It was stated earlier that it is a necessary assumption that some of the βs and some of the γs are zero in each of the equations comprising the model. This is so because if all βs and all γs appeared in each equation, the system would be insoluble. This can be shown to be a necessary but not sufficient condition for identification.

Let G^a be the number of endogenous variables and K^b the number of predetermined variables which appear in the first equation of the model. Then define $G^{aa} = G - G^a$ (the number of endogenous variables *not* in the equation and $K^{bb} = K - K^b$ (the number of predetermined variables *not* in the equation).

The first equation of the set may be written

$$\beta_1 Y_j + \gamma_1 x_j = u_{1j} \tag{9.95}$$

The reduced form of equation 9.95 is equation 9.96, following the form of equation 9.90.

$$y_j = \pi x_j + v_j \tag{9.96}$$

Premultiplying equation 9.96 by β_1 and noting that $\beta_1 v_j = u_{1j}$ results in

$$\beta_1 y_j = \beta_1 \pi x_j + u_{1j} \tag{9.97}$$

Alternatively, equation 9.97 can be rewritten

$$\beta_1 y_j - \beta_1 \pi x_j = u_{1j} \tag{9.98}$$

Clearly, the coefficients of x_j in equation 9.98 must be identical to those in equation 9.95. Thus equation 9.99 follows.

$$-\gamma_1 = \beta_1 \pi \tag{9.99}$$

Without losing any generality, the numbering of the variables may be rearranged so as to put the nonzero coefficients at the beginning of each class. Thus the vectors β_1 and γ_1 can be partitioned.

$$\beta_1 = (\beta_{1a} 0_{aa}) = (\beta_{11} \ldots \beta_{1G^a} 0_{1G^a+1} \ldots 0_{1G}) \tag{9.100}$$

$$\gamma_1 = (\gamma_{1b} 0_{bb}) = (\gamma_{11} \ldots \gamma_{1K^b} 0_{1K^b+1} \ldots 0_{1K}) \tag{9.101}$$

Similarly, partitioning π into the first G^a and the remaining G^{aa} rows, and K^b and K^{bb} columns, equation 9.98 can be rewritten

$$-(\gamma_{1b} 0_{bb}) = (\beta_{1a} 0_{aa}) \begin{vmatrix} \pi_{ab} & \pi_{abb} \\ \pi_{aab} & \pi_{aabb} \end{vmatrix} \tag{9.102}$$

Multiplying out the right side of equation 9.102 produces

$$-\gamma_{1b} = \beta_{1a}\pi_{ab} \tag{9.103}$$

$$0_{bb} = \beta_{1a}\pi_{abb} \tag{9.104}$$

The coefficients, βs and γs, can then be identified if equation 9.104 can be solved to yield a unique vector β_{1a} in terms of the reduced-form coefficients π_{abb}, since equation 9.103 will yield the vector γ_{1b}. The conditions for identification can be determined by examining the rank and order of the matrix π_{abb}.

If the rank of π_{abb} is G^a, the set of equations 9.104 will have only the trivial solution of the zero vector. This is not permissible since it is assumed each relationship contains at least one endogenous variable. If the rank of π_{abb} is $G^a - 1$, the ratios of the G^a unknown β coefficients may be determined uniquely. This is all that is needed since one of the βs can be set arbitrarily to unity and the rest solved for. Since the matrix has G^a rows and K^{bb} columns, a necessary condition for its rank to be $G^a - 1$ is that equation 9.105 hold.

$$K^{bb} \geq G^a - 1 \tag{9.105}$$

This is the *order* condition for identification. In words, the number of predetermined variables excluded from the relation must be at least as many as the number of endogenous variables included minus one. Adding G^{aa} to both sides of equation 9.105 yields

$$G^{aa} + K^{bb} \geq G - 1 \tag{9.106}$$

In words, the total number of both predetermined and endogenous variables excluded from the relationship must be at least as many as the total number of endogenous variables minus one. It is also relevant to consider whether the condition of K^{bb} being greater than or equal in equation 9.105 to $(G^a - 1)$ has any effect on the identification conditions.

In the first case, $\rho(\hat{\pi}_{abb})$ will generally be equal to $G^a - 1$, so the ILS approach is feasible. If $K^{bb} > G^a - 1$, the ILS approach must be modified to ensure that $\rho(\hat{\pi}_{abb}) = G^a - 1$, or else a method of estimation must be used that does not require estimation of the reduced-form coefficients to estimate the structural coefficients. (Note, $\rho(N)$ denotes the rank of the matrix N.)

The two conditions for identification may be summarized.

$$K^{bb} \geq G^a - 1 \tag{9.107}$$

$$\rho(\pi_{abb}) = G^a - 1 \tag{9.108}$$

Having considered the necessary and sufficient conditions for identification, it becomes clear that more than one possible estimation method will be necessary. This is true because the conditions are not unique, but only one specific fulfillment of the conditions allows the use of the simplest estimation method.

Since there will usually be more than one endogenous variable in each equation, it is not possible to use ordinary least squares on each single equation. The extent and form of the biases introduced was dealt with in the previous section.

Indirect Least Squares

In the event that the structural relation for any equation is exactly identified, this method may be applied. That is, it may be applied when $K^{bb} = G^a - 1$. The technique involves the estimation for each relationship of the reduced-form equation using ordinary least squares. The estimates of the reduced-form coefficients, the $\hat{\pi}_{ij}$, can then be used to estimate the original coefficients. In general, the estimates of the $\hat{\pi}_{ij}$ will be the best linear unbiased estimates, but the derived structural coefficient estimates will not be. These derived coefficients will usually be biased, although they will also be consistent (i.e., the bias will approach zero as the sample size approaches infinity). If the structural disturbances, the u_{ij}, are normally distributed, the reduced-form disturbances will also be normally distributed. Thus, under these conditions, the least-squares estimators are the maximum-likelihood estimators.

Least-Variance Ratio

The Least-Variance Ratio (LVR) method and a second method, Least-Information Single Equation (LISE), yield the same estimating equations and so will be dealt with together. LVR (and LISE) is a method which is applicable in the case where the model is overidentified, that is, where $K^{bb} > G^a - 1$. In such a case, the rank of π_{abb} will usually be G^a, and hence the condition that $\rho(\pi_{abb}) = G^a - 1$ is not fulfilled. In this instance, an estimate of the vector β_{1a} in equation 9.109 will not be obtained.

$$0_{bb} = \beta_{1a}\hat{\pi}_{abb} \tag{9.109}$$

The technique was described in the simple example as being the attempt to minimize the ratio of the residual sum of squares of two equations in which all predetermined variables are assumed to enter the equation in the first instance, and only those structured into it in the second instance. Using the previous notation, it is assumed that γ_a is related to X_b and not to X_{bb}. Therefore, if Y_a is regressed on both X_b and

X, the best estimates of the coefficients will be those which minimize the ratio of the residuals of Y_a on X_b to Y_a on X.

It can be shown that the ratio to be minimized is

$$l = \frac{\beta_{1a} \, W_{aa}^b \, \beta_{1a}'}{\beta_{1a} \, W_{aa} \, \beta_{1a}'} \qquad (9.110)$$

where $W_{aa}^b = Y_a'Y_a - Y_a'X_b(X_b'X_b)^{-1}X_b'Y_a$
and $\quad W_{aa} = Y_a'Y_a - Y_a'X(X'X)^{-1}X'Y_a$

By differentiating l with respect to β_{1i} ($i=1,...,G^a$) and equating to zero, the problem is reduced to determining the solutions to the determinantal equation.

$$\left| \, W_{aa}^b - l W_{aa} \, \right| = 0 \qquad (9.111)$$

Solving for the smallest root of equation 9.111 (to achieve minimization), this value is substituted back in

$$(W_{aa}^b - l W_{aa})\hat{\beta}_{1a} = 0 \qquad (9.112)$$

Hence, the γs of equation 9.95 can be solved (using LVR and LISE) as shown by

$$\hat{\gamma}_{1b} = -\hat{\beta}_{1b} \, Y_a' \, X_b(X_b' \, X_b)^{-1} \qquad (9.113)$$

Two-stage Least Squares

The method of Two-stage Least Squares requires that in each relation one endogenous variable is isolated on the left side of the relation, and the least-squares estimates obtained of the coefficients of all the other variables in the equation. Since, in general, all the endogenous variables will be correlated with the error term, it is necessary to precede this estimation by an estimation of the other endogenous variables based on least-squares regressions of these variables on all the predetermined variables.

The set of equations may first be rewritten

$$Y_1 = -Y_2\beta_2' - X_b\gamma_{1b}' + U_1 \qquad (9.114)$$

The first stage comprises the regression of each of the y_js ($j=2,3,...,G$) on all of the Xs. Taking these one at a time, it may be seen that equations 9.115, 9.116, and so on result.

$$\hat{y}_2 = X(X'X)^{-1} \, X'y_2 \qquad (9.115)$$

$$\hat{y}_3 = X(X'X)^{-1} \, X'y_3 \qquad (9.116)$$

In general, these may be written

$$\hat{Y}_2 = X(X'X)^{-1} X'Y_2 \tag{9.117}$$

$$Y_2 = X(X'X)^{-1} X'Y_2 + V \tag{9.118}$$

where V = a matrix of reduced-form residuals for the $G^a - 1$ endogenous
variables

Hence equation 9.114 may be rewritten

$$y_1 = -(Y_2-V)\beta_2' - X_b\gamma_{1b}' + (U_j-V\beta_2') \tag{9.119}$$

This may be rewritten

$$y_1 = [(Y_2-V)X_b] \begin{vmatrix} \beta_2' \\ \gamma_{1b}' \end{vmatrix} + (U_1-V\beta_2') \tag{9.120}$$

Applying least squares to equation 9.120 to estimate the coefficient
matrix gives

$$\begin{vmatrix} \beta_2' \\ \gamma_{1b} \end{vmatrix} = -(A'A)^{-1} A'y_1 \tag{9.121}$$

where $A = [(Y_2-V)X_b]$

The term $(A'A)^{-1}$ can be rewritten to simplify equation 9.121. The value
$(Y_2-V)'(Y_2-V)$ is

$$(Y_2-V)'(Y_2-V) = Y_2'Y_2 - V'Y_2 - Y_2'V + V'V \tag{9.122}$$

However, equation 9.123 holds.

$$V'Y_2 = V'(\hat{Y}_2+V) = V'V \tag{9.123}$$

This follows, since it is a property of the least-squares estimation that
the residual is uncorrelated with the regression values, that is, $V'\hat{Y}_2 = 0$.
One can similarly express $Y_2'V$ as

$$Y_2'V = V'V \tag{9.124}$$

Equation 9.122 can then be simplified to

$$(Y_2-V)'(Y_2-V) = Y_2'Y_2 - V'V \tag{9.125}$$

The product $X_b'V$ is zero. This may be shown by considering, first, the
matrix $X'V$.

$$X'V = X'[Y_2-X(X'X)^{-1}X'Y_2] \tag{9.126}$$

This can be rewritten as equation 9.127 by expanding equation 9.126.

$$X'V = X'Y_2 - X'X(X'X)^{-1} X'Y_2 = 0 \tag{9.127}$$

Since $\mathbf{X}_b'\mathbf{V}$ is a submatrix of $\mathbf{X}'\mathbf{V}$, it also must be zero. Using these results, equation 9.121 can be rewritten

$$\begin{vmatrix} \hat{\beta}_2' \\ \hat{\gamma}_{1b}' \end{vmatrix} = - \begin{vmatrix} \mathbf{Y}_2'\mathbf{Y}_2 - \mathbf{V}'\mathbf{V} & \mathbf{Y}_2'\mathbf{X}_b \\ \mathbf{X}_b'\mathbf{Y}_2 & \mathbf{X}_b'\mathbf{X}_b \end{vmatrix}^{-1} \begin{vmatrix} \mathbf{Y}_2' - \mathbf{V}' \\ \mathbf{X}_b' \end{vmatrix} \mathbf{y}_1 \qquad (9.128)$$

The condition that the inverse matrix in equation 9.128 exists is the condition 9.105 for identification, that is, $K^{bb} \geq G^a - 1$.

Theil[3] has developed a generalization of this result which has some useful conceptual properties. If equation 9.128 is rewritten to include a multiplier k of each of $\mathbf{V}'$ and $\mathbf{V}'\mathbf{V}$, equation 9.129 results.

$$\begin{vmatrix} \hat{\beta}_2' \\ \hat{\gamma}_{1b}' \end{vmatrix} = - \begin{vmatrix} \mathbf{Y}_2'\mathbf{Y}_2 - k\mathbf{V}'\mathbf{V} & \mathbf{Y}_2'\mathbf{X}_b \\ \mathbf{X}_b'\mathbf{Y}_2 & \mathbf{X}_b'\mathbf{X}_b \end{vmatrix}^{-1} \begin{vmatrix} \mathbf{Y}_2' - k\mathbf{V}' \\ \mathbf{X}_b' \end{vmatrix} \mathbf{y}_1 \qquad (9.129)$$

If $k = 0$, this corresponds to ordinary least squares, since equation 9.129 then becomes

$$\begin{vmatrix} \hat{\beta}_2' \\ \hat{\gamma}_{1b}' \end{vmatrix} = - \begin{vmatrix} \mathbf{Y}_2'\mathbf{Y}_2 & \mathbf{Y}_2'\mathbf{X}_b \\ \mathbf{X}_b'\mathbf{Y}_2 & \mathbf{X}_b'\mathbf{X}_b \end{vmatrix}^{-1} \begin{vmatrix} \mathbf{Y}_2' \\ \mathbf{X}_b' \end{vmatrix} \mathbf{y}_1 \qquad (9.130)$$

Two-stage Least Squares is clearly the case where $k = 1$, and it can be shown that Limited Information corresponds to $k = \hat{l}$, where $\hat{l}$ is as defined in equation 9.110.

$\mathbf{A}$ may also be rewritten

$$\mathbf{A} = [(\mathbf{Y}_2 - \mathbf{V})\mathbf{X}_b] = \mathbf{X} \begin{vmatrix} (\mathbf{X}'\mathbf{X})^{-1} \mathbf{X}'\mathbf{Y}_2 & \mathbf{I} \\ & \mathbf{0} \end{vmatrix} \qquad (9.131)$$

where $\mathbf{I}$ is the unit matrix of order K^b and $\mathbf{0}$ is a zero matrix of order K^{bb} by K^b.

Before leaving Two-stage Least Squares, one further concept related to it should be examined. In demonstrating the equivalence of equation 9.129 to limited information in which $K = \hat{l}$, it is found that equations 9.132 and 9.133 hold,

$$\mathbf{W}_{aa}^b = \mathbf{Y}_2' \, \mathbf{B}_b \mathbf{Y}_a \qquad (9.132)$$

$$\mathbf{W}_{aa} = \mathbf{Y}_2' \, \mathbf{B}\mathbf{Y}_a \qquad (9.133)$$

where $\mathbf{B}$ and $\mathbf{B}_b$ are

$$\mathbf{B} = \mathbf{I} - \mathbf{X}(\mathbf{X}'\mathbf{X})^{-1}\mathbf{X}' \qquad (9.134)$$

$$\mathbf{B}_b = \mathbf{I} - \mathbf{X}_b(\mathbf{X}_b'\mathbf{X}_b)^{-1}\mathbf{X}_b' \qquad (9.135)$$

Define a quantity ϕ.

$$\phi = \beta_{1a} \, \mathbf{W}_{aa}^b \, \beta_{1a}' - \beta_{1a} \, \mathbf{W}_{aa} \, \beta_{1a}' \qquad (9.136)$$

It can be shown, using equations 9.132 and 9.133, that Two-stage Least Squares is the estimator of β_{1a} which minimizes ϕ. Also, in LISE, it was found that the procedure was to minimize the ratio l, where l was given by equation 9.110. Thus, LISE minimizes the *ratio* of the residual sums of squares and Two-stage Least Squares minimizes the *difference* between them.

Finally, it may be noted that a choice of methods among the values of k depends upon their properties and on computational considerations. The condition for consistency is

$$\lim_{n \to \infty} k = 1 \qquad (9.137)$$

Thus ordinary least squares is inconsistent for estimating simultaneous equations, since $k = 0$. Two-stage Least Squares is consistent, since $k = 1$, and so also is LISE, for $k = 1$ as n approaches ∞ for this method. It can further be shown, as would be expected from this, that Two-stage Least Squares has an advantage over LISE for small samples.

It should also be noted that the preceding methods all estimate the coefficients by utilizing one relation only out of the set. That is, only the first equation of the entire model has been utilized. A further estimation method, Three-stage Least Squares, is based on the idea of applying generalized least squares to the whole set of relations to estimate all coefficients simultaneously.

Three-stage Least Squares

This method has been developed by Zellner and Theil.[4] Consider again the first equation of Two-stage Least Squares, omitting the minus sign on the right.

$$\mathbf{Y}_1 = \mathbf{Y}_2 \boldsymbol{\beta}_2' + \mathbf{X}_b \boldsymbol{\gamma}_{1b}' + \mathbf{u}_1 \qquad (9.138)$$

This equation may be rewritten

$$\mathbf{y}_1 = \mathbf{Z}_1 \boldsymbol{\delta}_1 + \mathbf{u}_1 \qquad (9.139)$$

where $\mathbf{Z}_1$ and $\boldsymbol{\delta}_1$ are

$$\mathbf{Z}_1 = \mathbf{Y}_2 \, \mathbf{X}_b \qquad (9.140)$$

$$\boldsymbol{\delta}_1 = \begin{vmatrix} \boldsymbol{\beta}_2 \\ \boldsymbol{\gamma}_{1b}' \end{vmatrix} \qquad (9.141)$$

Premultiplying 9.139 by $\mathbf{X}'$ where $\mathbf{X}' = [\mathbf{X}_b \ \mathbf{X}_{bb}]$ results in

$$\mathbf{X}'\mathbf{y}_1 = \mathbf{X}'\mathbf{Z}_1 \boldsymbol{\delta}_1 + \mathbf{X}'\mathbf{u}_1 \qquad (9.142)$$

This is a system of K equations in $G^a - 1 + K^b$ unknowns, the matrix δ_1. If equation 9.139 is exactly identified, then $K^{bb} = G^a - 1$ and the number of unknown parameters is exactly equal to the number of equations. (There are K equations in $G^a - 1 + K^b$ unknowns. If $K^{bb} = G^a - 1$, there are $K^{bb} + K^b$ unknowns, i.e., K.) The matrix δ_1 may then be estimated using least squares.

$$\hat{\delta}_1 = (\mathbf{X}'\mathbf{Z}_1)^{-1} \mathbf{X}'\mathbf{y}_1 \tag{9.143}$$

This is in fact the indirect least-squares estimator. In the usual situation, however, the model is overidentified so that $K^{bb} > G^a - 1$, and there are more equations than unknowns. In this case, equation 9.142 is not suitable for the application of ordinary least squares. If, instead, generalized least squares is applied to equation 9.142, the solution is

$$\hat{\delta}_1 = [\mathbf{Z}_1'\mathbf{X}(\mathbf{X}'\mathbf{X})^{-1}\mathbf{X}'\mathbf{Z}_1]^{-1} \mathbf{Z}_1'\mathbf{X}(\mathbf{X}'\mathbf{X})^{-1} \mathbf{X}'\mathbf{y}_1 \tag{9.144}$$

As stated before, this method assumes that all the relations are used to estimate the coefficients simultaneously. Setting out all the equations in the form of equation 9.142 results in

$$\begin{vmatrix} \mathbf{X}'\mathbf{Y}_1 \\ \mathbf{X}'\mathbf{Y}_2 \\ \vdots \\ \mathbf{X}'\mathbf{Y}_G \end{vmatrix} = \begin{vmatrix} \mathbf{X}'\mathbf{Z}_1 & 0 & \cdots & 0 \\ 0 & \mathbf{X}'\mathbf{Z}_2 & \cdots & 0 \\ \vdots & & & \vdots \\ 0 & 0 & \cdots & \mathbf{X}'\mathbf{Z}_G \end{vmatrix} \begin{vmatrix} \delta_1 \\ \delta_2 \\ \vdots \\ \delta_G \end{vmatrix} + \begin{vmatrix} \mathbf{X}'\mathbf{u}_1 \\ \mathbf{X}'\mathbf{u}_2 \\ \vdots \\ \mathbf{X}'\mathbf{u}_G \end{vmatrix} \tag{9.145}$$

This may be written

$$\mathbf{A} = \mathbf{B}\delta + \mathbf{V} \tag{9.146}$$

where $\mathbf{A}$, $\mathbf{B}$, and $\mathbf{V}$ are defined by equations 9.145 and 9.146. Applying generalized least squares to 9.146 gives

$$\hat{\delta} = (\mathbf{B}'\mathbf{V}^{-1}\mathbf{B})^{-1} \mathbf{B}'\mathbf{V}^{-1}\mathbf{A} \tag{9.147}$$

The solution of this involves the unknown $\mathbf{V}$ matrix, comprising disturbance covariances δ_{ij}. Zellner and Theil[5] recommend replacing the δ_{ij} with their least-squares estimates, $\mathbf{S}_{ij}$.

Conclusions

Four alternative solutions to simultaneous equations have been examined. In general, it may be concluded that Indirect Least Squares is the most widely applicable method for an exactly identified model. In the case of overidentified models there are three alternative methods. Three-stage Least Squares may be expected to be more efficient than Two-stage Least

Squares and both have application to small samples. It is possible, however, that Three-stage Least Squares is liable to be seriously affected by specification error. In the case of large samples and an overidentified model, the Least-Variance Ratio or Limited-Information Single Equation method will be the most efficient.

Notes

1. J. Johnston, *Econometric Methods,* New York: McGraw-Hill Book Co., 1963, pp. 231–239.

2. Ibid, pp. 239–240.

3. Henri Theil, *Economic Forecasts and Policy*, 2d rev. ed., Amsterdam: North Holland Press, 1961, pp. 231–237.

4. A. Zellner and H. Theil, "Three-Stage Least Squares: Simultaneous Estimation of Simultaneous Equations," *Econometrica,* 1962, vol. 30, pp. 54–78.

5. Zellner and Theil, "Three-Stage Least Squares."

10 Canonical Correlation

Canonical correlation[1] is another technique that represents an outgrowth of simple linear-regression analysis. It is concerned with relationships between two sets of variables. The method comprises the derivation of a variate (a linear combination of the variables in a set) for each set such that the correlation between the sets is maximized.

There is an apocryphal story about the origin of the name "canonical" correlation. According to the story, the method was discovered originally by a mathematically inclined monk. Unable to invent a suitable name, he decided to canonize the method—hence, canonical correlation!

The method is designed for use when in effect not only the independent variables take a linear additive form, but also where there may be a set of two or more dependent variables in linear additive form. To date, applications of the method have been principally in economics for the estimation of demand and supply relationships, where both price and quantity may be expressed as a linear combination of variables. The method, however, assumes that the two sets of variables are distinct. In other words, it is not the same as simultaneous equations where endogenous variables are sometimes dependent variables and other times independent. Only one relationship is proposed in canonical correlation, that between two distinct sets of variables where each set contains a minimum of two variables. Waugh[2] applied canonical correlation to problems in economics with some success. In two such applications, he studied relations between the prices and rates of consumption of beef and pork and in the second between wheat and flour characteristics.

The method has also been used by Hotelling,[3] inter alia, in certain problems in psychology. Another application by Tintner[4] related certain price indices and certain production indices in economics. This application is used later to illustrate the method. It has not been used widely, but has some potential in various areas of planning and social science.

Canonical correlation is therefore applied in the situation where there is not only a set of variables to describe the causal agent, but also a set of variables needed to describe the result of changes in the causal agent. The idea is to build a linear model between these two sets of variables, where the coefficients in both sets are to be determined on the basis of obtaining the maximum correlation between the two sets of variables. Define a compound dependent variable as

$$U = a_1 y_1 + a_2 y_2 + a_3 y_3 + \dots + a_p y_p \qquad (10.1)$$

The causal agent, or set of independent variables, is also to be expressed as a linear combination.

$$V = b_1 x_1 + b_2 x_2 + b_3 x_3 + \dots + b_m x_m \tag{10.2}$$

The as and bs are then to be found simultaneously, so that the correlation between U and V is maximized. In the economic applications, the ys might be measures of supply, the xs of demand. Hence the problem is to see how to combine various measures of supply and demand so the relation between supply and demand is made to be as strong as possible, that is, the maximum correlation between the composite measures is achieved.

To help understand the concept further, consider an example from land-use planning. Within some definable land area, suppose a model is desired that predicts the amount of land that will be developed for various uses, such as residences, retail sales, services, primary employment (e.g., major industries), and transportation. Consider a theory of land use which postulates that residential development comes first and services to residents (such as transportation, retailing, services, government offices) and major employment (both primary and secondary) follow next. Finally, tertiary land uses, that is, those serving the second group of uses, are developed. Development of each type of land use will depend upon the amount of land available for that use and on the amount of land developed for uses that are higher in the hierarchy. This could be studied through canonical correlation, where the measures of land consumption for each type of land use form the ys of equation 10.1 and U is a composite measure of the total land consumption. Similarly, the xs of equation 10.2 could be measures of the available amount of land and existing development support (such as potable water, sewage disposal capacity), and V is a composite measure of land-use supply and development potential.

Now consider the mathematical problem of canonical correlation. Suppose there are two sets of variables, the first being a set of ys that are associated with the composite dependent variable and the second a set of xs associated with the composite independent variable. Two variates are to be found, where these are linear combinations of the two sets and the two variables are maximally correlated with each other.

The variables are first reformulated as standardized variables. Standardized variables were mentioned in the last two chapters. A rigorous derivation and examination of their properties is given here.

Standardized Variables

Definition and Derivation

A standardized variable is a variable that has a mean of zero and a variance of one. Any naturally occurring variable can be formed into a standardized variable by subtracting from it the mean value of the variable and dividing each resulting value by the standard deviation. This is shown in the following steps.

The original set of variables is $X_1, X_2, X_3, \ldots, X_p$. Each of these, X_j, has a deviate associated with it.

$$x_{ij} = X_{ij} - \bar{X}_j \tag{10.3}$$

(A deviate is a variate with a mean zero, that is, it is the variate minus the mean of all the observed values of the variate.) The variance of the X_js is

$$\sigma_j^2 = \frac{1}{n}\sum_{i=1}^{n}(X_{ij}-\bar{X}_j)^2 \tag{10.4}$$

Using equation 10.3, it can be seen that σ_j^2 can be rewritten

$$\sigma_j^2 = \frac{1}{n}\Sigma(x_{ij})^2 \tag{10.5}$$

Now define a new variable, z_j, such that equation 10.6 holds.

$$z_{ij} = x_{ij}/\sigma_j \tag{10.6}$$

Clearly, the standard deviation of the z_{ij}s is

$$\sigma_j' = \sqrt{\frac{1}{n}\Sigma(z_{ij})^2} \tag{10.7}$$

Writing equation 10.7 in terms of the original deviates and variance yields

$$\sigma_j' = \sqrt{\frac{1}{n}\Sigma\left(\frac{x_{ij}}{\sigma_j}\right)^2} \tag{10.8}$$

Since σ_j is the same for all i, it can be taken outside the summation.

$$\sigma_j' = \sqrt{\frac{\Sigma(x_{ij})^2}{\sigma_j^2 n}} \tag{10.9}$$

However, σ_j^2 is equal to $(1/n)\,\Sigma(x_{ij})^2$, so equation 10.9 reduces to

$$\sigma_j' = \pm 1 \tag{10.10}$$

Thus z_j is a standardized variable for the original variable X_j, where z_j is obtained by subtracting the mean value, $\bar{X}_j$ from each observation X_{ij} and dividing the result by σ_j, the standard deviation of the X_js.

Some Properties of Standardized Variables

The correlation between two variables x and y can be expressed

$$r_{yx} = \text{cov}(xy)/\sigma_x\sigma_y \tag{10.11}$$

Clearly, if $\sigma_x = \sigma_y = 1$, the case with standardized variables, equation 10.11 can be rewritten

$$r_{yx} = \text{cov}(xy) = \text{cov}(yx) = r_{xy} \qquad (10.12)$$

Now the covariance of two variables, x and y, is

$$\text{cov}(xy) = \frac{1}{n}\Sigma(x_i-\bar{x})(y_i-\bar{y}) \qquad (10.13)$$

Again, standardized variables have means equal to zero. Hence equation 10.13 can be written

$$\text{cov}(xy) = \frac{1}{n}\Sigma(x_i)(y_i) \qquad (10.14)$$

The correlations of equation 10.12 can then be written

$$r_{xy} = r_{yx} = \frac{1}{n}\Sigma x_i y_i \qquad (10.15)$$

Derivation of Canonical Variates

The problem is to determine two linear functions of the standardized variables as given by U and V.

$$U = a_1 z_1 + a_2 z_2 + a_3 z_3 + \dots + a_p z_p \qquad (10.16)$$

$$V = a_{p+1} z_{p+1} + a_{p+2} z_{p+2} + \dots + a_m z_m \qquad (10.17)$$

In these two equations, the as are to be determined so that U and V have maximum correlations with each other. Since the zs are standardized variables, and U and V are sums of standardized variables, it follows that $\bar{U}$ and $\bar{V}$ are both zero. Without loss of generality, U and V may be defined as standardized variables, that is, variables with unit variances. Assuming that the covariances of z_i, z_j all exist, the variance of U is

$$V(U) = \sum_i (a_1 z_{1i}+a_2 z_{2i}+a_3 z_{3i}+\dots+a_p z_{pi})^2 \qquad (10.18)$$

By using equation 10.15, equation 10.18 can be simplified to

$$V(U) = \sum_{s=1}^{p} \sum_{t=1}^{p} r_{st} a_s a_t \qquad (10.19)$$

where r_{st} is

$$r_{st} = \sum_{i=1}^{n} z_{si} z_{ti} \qquad (10.20)$$

By a similar process, the variance of V is

$$V(V) = \sum_{s=p+1}^{m} \sum_{t=p+1}^{m} r_{st} a_s a_t \tag{10.21}$$

The variances of U and V are constrained to be one, so equation 10.22 follows.

$$\sum_{s=1}^{p} \sum_{t=1}^{p} r_{st} a_s a_t = \sum_{s=p+1}^{m} \sum_{t=p+1}^{m} r_{st} a_s a_t = 1 \tag{10.22}$$

Subject to this constraint, the correlation between U and V is to be maximized; this correlation is

$$R = \text{cov}(U,V) \tag{10.23}$$

Given that U and V are standardized, equation 10.23 may be written

$$R = \sum_i U_i V_i \tag{10.24}$$

Expanding equation 10.24 in terms of equations 10.16 and 10.17 produces

$$R = \sum_i \{(a_1 z_{1i} + a_2 z_{2i} + \ldots a_p z_{pi})(a_{p+1} z_{p+1i} + a_{p+2} z_{p+2i} + \ldots + a_m z_{mi})\} \tag{10.25}$$

This may be simplified to

$$R = \sum_{s=1}^{p} \sum_{t=1}^{m} r_{st} a_s a_t \tag{10.26}$$

To maximize R under the constraints that the variances of each of U and V are unity, a function, F, may be constructed using Lagrange multipliers.

$$F = R - \tfrac{1}{2}\lambda V(U) - \tfrac{1}{2}\mu V(V) \tag{10.27}$$

Using the expressions for R (equation 10.26, $V(U)$ and $V(V)$ (equations 10.19 and 10.21, equation 10.27 can be rewritten

$$F = \sum_{s=1}^{p} \sum_{t=p+1}^{m} r_{st} a_s a_t - \tfrac{1}{2}\lambda \sum_{s=1}^{p} \sum_{t=1}^{p} r_{st} a_s a_t - \tfrac{1}{2}\mu \sum_{s=p+1}^{m} \sum_{t=p+1}^{m} r_{st} a_s a_t \tag{10.28}$$

F is to be minimized, so F is differentiated with respect to each a_s, and the differentials are equated to zero. From this, a set of equations of the form of equation 10.29 is obtained.

$$r_{1(p+1)}a_{p+1} + \ldots + r_{1m}a_m - \lambda a_1 - \lambda r_{12}a_2 - \ldots - \lambda r_{1p}a_p = 0$$

$$r_{p(p+1)}a_{p+1} + \ldots + r_{pm}a_m - \lambda r_{p1}a_1 - \lambda r_{p2}a_2 - \ldots - \lambda a_p = 0$$

$$r_{1(p+1)}a_1 + \ldots + r_{p(p+1)}a_p - \mu a_{p+1} - \ldots - \mu r_{(p+1)m}a_m = 0 \qquad (10.29)$$

$$r_{1m}a_1 + \ldots + r_{pm}a_p - \mu r_{(p+1)m}a_{p+1} - \ldots - \mu a_m = 0$$

The set of linear equations, equation 10.29, are very similar to the normal equations of multivariate linear regression. They represent a system of simultaneous linear equations in the values of a_s, for $s = 1,2,\ldots,m$.

To illustrate the solution process more clearly, suppose that $p = 3$ and $m = 5$. Then equation 10.29 may be rewritten

$$r_{14}a_4 + r_{15}a_5 - \lambda a_1 - \lambda r_{12}a_2 - \lambda r_{13}a_3 = 0$$

$$r_{24}a_4 + r_{25}a_5 - \lambda r_{21}a_1 - \lambda a_2 - \lambda r_{23}a_3 = 0$$

$$r_{34}a_4 + r_{35}a_5 - \lambda r_{31}a_1 - \lambda r_{32}a_2 - \lambda a_3 = 0 \qquad (10.30)$$

$$r_{14}a_1 + r_{24}a_2 + r_{34}a_3 - \mu a_4 - \mu r_{45}a_5 = 0$$

$$r_{15}a_1 + r_{25}a_2 + r_{35}a_3 - \mu r_{45}a_4 - \mu a_5 = 0$$

Multiplying the first $3(p)$ equations by $a_1, a_2, a_3 (\ldots,a_p)$ respectively and summing, the first three equations yield

$$-\lambda[a_1^2 + r_{21}a_1a_2 + r_{31}a_1a_3 + r_{12}a_2a_1 + a_2^2$$

$$+ r_{32}a_2a_3 + r_{13}a_3a_1 + r_{23}a_3a_1 + r_{23}a_3a_2 + a_3^2]$$

$$+ r_{14}a_4a_1 + r_{24}a_4a_2 + r_{34}a_4a_3$$

$$+ r_{15}a_5a_1 + r_{25}a_5a_2 + r_{35}a_5a_3 = 0 \qquad (10.31)$$

By gathering terms appropriately into summations, equation 10.31 may be rewritten

$$-\lambda \sum_{s=1}^{3}\sum_{t=1}^{3} r_{st}a_s a_t + \sum_{s=1}^{3}\sum_{t=4}^{5} r_{st}a_s a_t = 0 \qquad (10.32)$$

By multiplying the last $2(m-p)$ equations by $a_4, a_5, (\ldots,a_m)$ and summing, the last 2 equations from equation 10.30 can be written as

$$\sum_{s=1}^{3}\sum_{t=4}^{5} r_{st}a_s a_t - \mu \sum_{s=4}^{5}\sum_{t=4}^{5} r_{st}a_s a_t = 0 \qquad (10.33)$$

Substituting for the summations from equations 10.19, 10.21, and 10.26, equations 10.32 and 10.33 can be rewritten

$$-\lambda(V(U)) + R = 0 \tag{10.34}$$

$$R - \mu(V(V)) = 0 \tag{10.35}$$

Since V and U are standardized variables, their variances are 1 and equations 10.34 and 10.35 can be equated and written

$$-\lambda + R = R - \mu \tag{10.36}$$

Alternatively, one may write this

$$R = \lambda = \mu \tag{10.37}$$

Thus both Lagrange multipliers are equal to the canonical correlation coefficient, R. This fact may be used to rewrite the original equations 10.30, replacing μ with λ.

$$-\lambda \sum_{s=1}^{3} \sum_{t=1}^{3} r_{st} a_s + \sum_{s=1}^{3} \sum_{t=4}^{5} r_{st} a_t = 0$$
$$\sum_{s=1}^{3} \sum_{t=4}^{5} r_{st} a_s - \lambda \sum_{s=4}^{5} \sum_{t=4}^{5} r_{st} a_t = 0 \tag{10.38}$$

In the general case, this equation may be written

$$-\lambda \sum_{s=1}^{p} \sum_{t=1}^{p} r_{st} a_s + \sum_{s=1}^{p} \sum_{t=p+1}^{m} r_{st} a_t = 0$$
$$\sum_{s=1}^{p} \sum_{t=p+1}^{m} r_{st} a_s - \lambda \sum_{s=p+1}^{m} \sum_{t=p+1}^{m} r_{st} a_t = 0 \tag{10.39}$$

This can be solved readily by use of matrix algebra. The matrix of correlations may be represented as **R**, and the partitions of the correlation matrix as

$$\sum_{s=1}^{p} \sum_{t=1}^{p} r_{st} = \mathbf{R}_{11} \tag{10.40}$$

$$\sum_{s=1}^{p} \sum_{t=p+1}^{m} r_{st} = \mathbf{R}_{12} \tag{10.41}$$

$$\sum_{s=p+1}^{m} \sum_{t=p+1}^{m} r_{st} = \mathbf{R}_{22} \tag{10.42}$$

Hence the matrix, **R**, is being represented as equation 10.43, where the partitioning is between the set of variables belonging to U and that belonging to V.

$$R = \begin{vmatrix} R_{11} & R_{12} \\ R_{12} & R_{22} \end{vmatrix} \tag{10.43}$$

Similarly, the vector of coefficients, $\mathbf{a}$, can be partitioned into the sets for U and V.

$$\sum_{s=1}^{p} a_s = \mathbf{a}_1 \tag{10.44}$$

$$\sum_{t=p+1}^{m} a_t = \mathbf{a}_2 \tag{10.45}$$

Then the system of equations of 10.39 can be represented as

$$\begin{vmatrix} -\lambda R_{11} & R_{12} \\ R_{12} & -\lambda R_{22} \end{vmatrix} \begin{vmatrix} \mathbf{a}_1 \\ \mathbf{a}_2 \end{vmatrix} = 0 \tag{10.46}$$

For a nontrivial solution of this equation set, the matrix on the left must be singular, that is, the determinant of that matrix must be equal to zero. (A trivial solution would be one in which every element of the left or right matrix was zero. This result would not satisfy the condition that each of the variances, $V(U)$ and $V(V)$, are unity.) The solution then is

$$\begin{vmatrix} -\lambda R_{11} & R_{12} \\ R_{12} & -\lambda R_{22} \end{vmatrix} = 0 \tag{10.47}$$

Since, from equation 10.37, it was found that $R = \lambda$, and values of $a_s(s=1,2 \ldots, m)$ are to be chosen to maximize the canonical correlation, R, the largest root of the determinant 10.47 is used. Evaluating this root, a value for λ may be substituted in equation 10.46 and solutions obtained for $\mathbf{a}_1$ and $\mathbf{a}_2$. This allows the canonical variates, U and V, of the population to be evaluated such that U and V have maximum correlation.

It should be emphasized that this method does not necessarily generate results which can be readily interpreted in terms of the problem being addressed. In general, there is likely to be a problem of identification of the canonical variates. As an illustration of the application of this method, two examples are given.

Examples

An attempt is to be made to determine the relationship between certain price indices and some production indices using the technique of canonical correlation.[5] The variables are:

x_1 = production index of durable goods
x_2 = production index of nondurable goods

x_3 = production index of minerals
x_4 = index for agricultural products
x_5 = farm wholesale price index
x_6 = food wholesale price index
x_7 = other commodity wholesale price index

The task is to find two canonical variates, U and V, where equations 10.48 and 10.49 define U and V.

$$U = a_1 z_1 + a_2 z_2 + a_3 z_3 + a_4 z_4 \qquad (10.48)$$

$$V = a_5 z_5 + a_6 z_6 + a_7 z_7 \qquad (10.49)$$

The variances of each are to be constrained to unity, while their correlation is to be maximized. The correlation matrix is given by equation 10.50, using the data of Tintner,[6]

$$R = \left|
\begin{array}{cccc:ccc}
1.000 & 0.496 & 0.873 & 0.481 & -0.436 & -0.427 & -0.203 \\
 & 1.000 & 0.768 & 0.710 & 0.426 & 0.430 & 0.584 \\
 & & 1.000 & 0.712 & -0.038 & -0.044 & 0.139 \\
 & & & 1.000 & 0.261 & 0.267 & 0.378 \\
\hdashline
 & & & & 1.000 & 0.987 & 0.905 \\
 & & & & & 1.000 & 0.914 \\
 & & & & & & 1.000
\end{array}
\right| \qquad (10.50)$$

where the dashed lines represent the partitioning into $\mathbf{R}_{11}$, $\mathbf{R}_{12}$, and $\mathbf{R}_{22}$.

Equations 10.48 and 10.49 were set up in terms of the standardized variables, and the remaining calculations refer to these. The largest root of the determinant, equation 10.47, is found to be $\lambda = 0.88$ approximately. From this, U and V are solved to yield

$$U = 1.095 z_1 - 0.372 z_2 - 0.588 z_3 - 0.021 z_4 \qquad (10.51)$$

$$V = 1.000 z_5 - 0.011 z_6 - 0.215 z_7 \qquad (10.52)$$

It also follows, from the value of λ, that R is

$$R = \mu = \lambda = 0.88 \qquad (10.53)$$

Since the original variables are gross aggregations, and only a few indices are included here, the results must be interpreted with considerable caution. Some tentative conclusions can be entertained, however, as suggested by Tintner.[7]

Examining U suggests that in trying to estimate the mutual interdependence between production and prices, the largest weight has to be given to durable goods. This is in conformance with certain economic theories. The weight given to production of minerals is also high but negative. Nondurable goods are less important, and agricultural products play a very insignificant part. On the other hand, examining V, farm prices are the most important; wholesale food prices are almost negligible; and other commodity prices of some importance, but negative. For both U and V, the weighted differences, between indices of durable goods and minerals production for U, and between farm prices and other commodities for V, are the decisive factors. This, like the weights themselves, apparently conforms to theories of the business cycle.

Hence it may be seen that canonical correlation may be a useful method of data reduction, where a relationship can be hypothesized between two linear combinations of variables. It should be noted that the fitting procedure does not require any assumptions of causality in the relationship, nor are any distributional assumptions made. There is also no reference made to any assumed error in the relationship.

The second example is drawn from transportation planning. The study concerns freight transportation and attempts to seek relationships between shipments and trucks. Shipments can be described in terms of a number of characteristics, such as weight, volume, needs for special packing, special handling, and distance to be shipped. Trucks can be described in terms of capacity, vehicle type, ownership, and time spent loading and unloading.

Using data from two shipping firms, the canonical correlates of equations 10.54 and 10.55 were sought,

$$U = a_1 x_1 + a_2 x_2 + a_3 x_3 + a_4 x_4 \qquad (10.54)$$

where x_1 = shipment volume
   x_2 = shipment weight
   x_3 = special packing
and x_4 = special handling

$$V = a_5 x_5 + a_6 x_6 + a_7 x_7 + a_8 x_8 + a_9 x_9$$
$$+ a_{10} x_{10} + a_{11} x_{11} + a_{12} x_{12} \quad (10.55)$$

where x_5 = truck capacity
   x_6 = truck type
   x_7 = inbound or outbound
   x_8 = ownership
   x_9 = stops before this shipment picked up
   x_{10} = stops after shipment is picked up
   x_{11} = time truck at firm
and x_{12} = loading/unloading time

The analysis proceeded as described earlier in the chapter, resulting in the solution

$$R = \mu = \lambda = 0.519 \qquad (10.56)$$

Table 10–1
Coefficients for Shipment-Truck Canonical Correlates

Coefficient	Value
a_1	0.803
a_2	0.275
a_3	−0.052
a_4	0.013
a_5	−0.735
a_6	1.228
a_7	−0.235
a_8	0.470
a_9	−0.152
a_{10}	−0.297
a_{11}	−0.199
a_{12}	0.223

The coefficients for this solution are shown in table 10–1. First, it must be noted that the solution is not that good, since the value of R is relatively low. Furthermore, the canonical correlates are not easily interpreted. The first canonical correlate, U, is made up almost entirely of the effects of the first two variables—shipment volume and weight. The primary variables in the second correlate, V, are truck capacity, truck type, and ownership. Most other variables have relatively small effects on the correlate. Strangely, truck capacity has a negative sign, which is contrary to a priori expectations. This may have arisen from relatively high correlations between truck capacity and truck type. Apart from this, the canonical correlation suggests that shipment volume and weight correlate best with truck type, truck capacity, and ownership. For the last variable, high values indicated common carrier and lower values indicated ownership by shipper or receiver. Hence it appears that large shipments (by weight or volume) are more likely to be shipped by common carrier or by other means than vehicles owned by shipper or receiver.

This second example illustrates one of the common problems of difficult interpretability of canonical correlations results. Nevertheless, the method may provide useful insights into the analysis of various problems and data.

Notes

1. The method was introduced by Hotelling in H. Hotelling, "Relations between Two Sets of Variables," *Biometrika*, 1936, vol. 28, pp. 321 ff. See also S.S. Wilks, *Mathematical Statistics*, Princeton, N.J.:

Princeton University Press, 1943; M.G. Kendall, *The Advanced Theory of Statistics,* London: Charles Griffin and Co., 1946; Gerhard Tintner, *Econometrics,* New York: John Wiley & Sons, 1952, pp. 114–121.

2. F.V. Waugh, *"Regressions between Sets of Variables,"* *Econometrica,* 1942, vol. 10, pp. 290 ff.

3. Hotelling, "Relations between Two Sets of Variables."

4. Gerhard Tintner, "Some Applications of Multivariate Analysis to Economic Data," *Journal of the American Statistical Association,* 1946, vol. 41, pp. 487 ff.

5. Tintner, *Econometrics,* pp. 118–121.

6. Ibid, pp. 120–121.

7. Ibid.

11 Rudiments of Factor Analysis

Basic Concept

In common with the techniques described in chapters 9 and 10, factor analysis is a procedure with roots in the basic principles of multivariate linear-regression analysis. One of the problems noted in chapter 8, with respect to multivariate linear regression, is that multicollinearity of two or more independent variables causes misestimation of the coefficients of such variables. Frequently, exclusion of one of the multicollinear variables will result in a less than satisfactory relationship, while inclusion leads to counterintuitive values for the coefficients. The analyst is thus placed on the horns of a dilemma. Factor analysis and, more particularly, principal components or principal factors (a special case of factor analysis) provide a means to resolve this dilemma.

In very broad terms, factor analysis is a method for reformulating a set of natural or observed independent variables into a new set (usually fewer in number, but necessarily not more in number) of independent variables, such that the latter set has certain desired properties specified by the analyst. Kendall[1] has suggested a difference between factor analysis and principal-components analysis. He suggests that principal-components analysis is the search through data to try to find factors or components that "may reduce the dimensions of variation" and may "be given a possible meaning."[2] On the other hand, he suggests that factor analysis starts with the hypothesis of a model and tests it against the data to see if it fits. This distinction seems useful and is pursued in the latter part of the chapter. (It is notable that other authors, e.g., Harman[3] and Tintner,[4] do not appear to make use of this distinction.) The chapter concentrates on principal-components analysis, where among other things the latter set of variables has the property of zero correlations between all the new variables. Orthogonal factor analysis is also dealt with later. In all versions of factor analysis, the variables (factors) in the new set are formed as linear combinations of all the variables in the original set.

The method was originally developed in mathematical psychology as a means to reduce a large number of measures (probably collinear) obtained in psychological measurement to a small number of largely uncorrelated salient characteristics, capable of describing character traits or other concepts in psychology. For readers familiar with psychological measurement, factor analysis was originally developed as the principal means of constructing multidimensional scales.[5] Although the method

237

was criticized by many psychologists, it still remains one of the most tractable and understandable methods of undertaking multidimensional scaling.[6]

Some examples may be useful to illustrate the purpose and procedure of factor analysis. Consider, first, a situation in marketing cars. A sample of people have been asked to rate some of the aspects of the finish and comfort of a number of car models. The ratings were made by allocating points on each attribute to each model, where the points may be selected from zero to 100. Fourteen attributes are used, as shown in table 11–1. After obtaining the scores of each person, one could obtain the means for each model on each attribute (see figure 11–1).

There is a great deal of useful information in figure 11–1. However, it is very hard to absorb the information and even harder to gain an impression of which model is superior. Furthermore, many of the attributes may be related, either in the minds of the raters or in the technological design. For example, the efficiency and effectiveness of the air conditioning and heating are likely to be related, while these same items are probably quite unrelated to the exterior finish. Applying factor analysis to these ratings will lead to greater insights into the images of the various models and will also show how the various attributes link together. The grouping of attributes is shown by the coefficients of the original variables in the composition of the factors. Those with large (negative or positive) coefficients make a major contribution to the factor, while those with small (near zero) coefficients make little contribution to the factor.

Suppose a factor analysis of the ratings has yielded the groupings of attributes shown in table 11–2, where the three factors account for more

Table 11–1
Attributes Used to Rate Various Car Models

Attributes

Color
Seat width
Seat support
Interior finish (panels, fascia)
Exterior finish (trims, etc.)
Radio/tape player
Window controls
Air conditioning
Heating
Interior noise
Carpeting
Entry/exit space
Safety belts
Seat adjustment

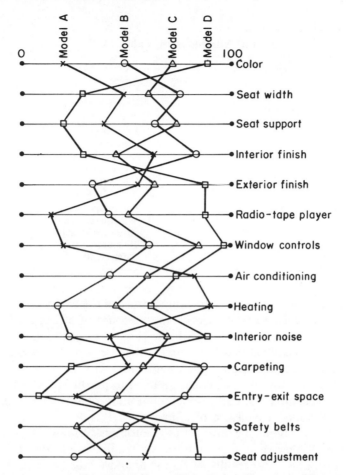

Figure 11-1. Raw Attribute Scores for Four Car Models

than 95% of the variance of the original attribute ratings. Factor 1 might be termed exterior design, factor 2 interior comfort, and factor 3 environmental controls. For each factor, an average score for each car model could be obtained and plotted, as shown in figure 11-2. The information provided by figure 11-2 is at once much easier to comprehend, and has also shown how certain items are correlated to produce a major effect, that is, the three factors.

In a similar manner, a number of measures may be made of the performance of a locomotive. These might include such items as the brake horsepower at given rpm (revolutions per minute), the rate of fuel consumption delivered at the same rpm, the acceleration rate from 0 mph (miles per hour), the operating temperature, the quantity of unburnt

Table 11–2
Factor Groupings for Attributes of Table 11–1

Factor	Attributes
1	Color Exterior finish Entry/exit space
2	Seat width Seat support Interior finish Interior noise Carpeting Seat adjustment
3	Radio-tape player Window controls Air conditioning Heating Safety belts

hydrocarbons in the exhaust, the temperature of the exhaust gases, and so on. It may be desired to develop some measure of efficiency for such a locomotive. Again, it is highly probable that many of the measures listed above are strongly related and would obscure a simple model of efficiency. Furthermore, one is again confronted with an overabundance of information. Factor analysis could be used to produce a smaller number of salient measures that can be combined more readily in a simple linear model to produce a measure of efficiency.

Finally, it should be noted that factor analysis is not a modeling technique. Rather, it is a procedure for manipulating data prior to developing models. The results of factor analysis can be used directly in some forms of decision making or appraisal, but such use does not conform to the usual definitions of modeling.

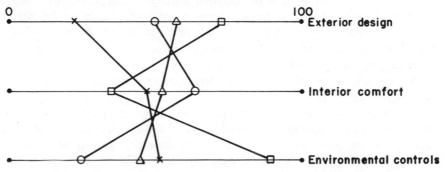

Figure 11–2. Factor Ratings of Four Automobile Models

Derivation of Principal Factors

Consider the situation in which measures, such as those suggested in the previous section, are obtained on a number of attributes of a phenomenon. These measures are denoted X_{1i}, X_{2i}, X_{3i}, ..., X_{Mi}, where the subscripts 1 through M refer to the attributes 1 through M, and the subscript i refers to the observation number. It is assumed that the measures are developed from a sample of data. It is desired to find a set of $N(N \leqslant M)$ factors, denoted F_{1i}, F_{2i}, ..., F_{Ni}, that may be used for analysis in place of the original attribute measures.

It would be desirable to express the original variables completely in terms of a set of factors, where N is less than M. In practice, this is not usually possible.[7] Instead, one must make a trade-off—a set of factors are developed where N is less than M, but at the sacrifice of not explaining all of the variance in the original set of M variables. To this end, the factors are selected by finding first the factor that explains the largest possible proportion of the variance in the original variables. The second factor is chosen to be uncorrelated with the first and to explain the largest portion of the remaining (residual) variance. Each succeeding factor is chosen in the same manner, being uncorrelated with any of the previously defined factors and capturing the maximum amount of residual variance.

The factors are defined to be linear transforms of the original variables. More complex transforms could be considered, but Kendall[8] suggests this would overcomplicate the theory. Instead, he suggests that any nonlinear relationships should be transformed to linear ones before the principal-factors analysis is undertaken.

It is clear that the original variables, the X_{ji}s, will contribute in varying degrees to the definition of the principal factors. Some variables may be measured in such a way as to produce values of thousands of units, while others may be measured in ones and twos. The variance of the former may be several orders of magnitude greater than that of the latter, with the result that the effects of the latter variables are swamped by the former ones. To remove this problem, the variables are standardized first, as described in chapter 10. Rewriting the standardized versions of the X_{ij}s as z_{1i}, z_{2i}, ..., z_{Mi}, the factors are defined as

$$z_{ji} = a_{j1}F_{1i} + a_{j2}F_{2i} + ... + a_{jM}F_{Ni} \tag{11.1}$$

where i = subscript referring to a specific observation
and j = a subscript referring to one of the original variates, where
$j = 1,2, ..., M$

Hence, the problem is to find a matrix of coefficients, A, such that the factors are uncorrelated with each other and where the factors are defined in descending order of the variance captured from the original variables. The factors are also defined to be standardized variables, having zero means and unit standard deviations.

Defining the Constraints

The correlation between two standardized variables is equal to the sum of the cross products of the two variables, divided by the sample size. Thus if there are n observations in the sample, the condition for uncorrelated factors is

$$\frac{1}{n}\Sigma_i F_{si}F_{ti} = 0 \qquad (s \neq t) \qquad (11.2)$$

It is also useful to keep in mind that all the sums of squares of the standardized variables are equal to n.

$$\Sigma_i F_{si}^2 = n \qquad \forall\, s \qquad (11.3)$$

In finding the principal factors by determining the values of the matrix **A**, the original correlations among the variates, z_{ji}, are to be preserved.

$$r_{st} = \frac{1}{n}\Sigma_i z_{si} z_{ti} \qquad (11.4)$$

To obtain a solution for the matrix **A**, it is useful to express equation 11.4 in terms of the principal factors and the coefficients to be found. By so doing, it is possible to formulate a solution process for the matrix. Equation 11.4 can be rewritten

$$r_{st} = a_{s1}a_{t1} + a_{s2}a_{t2} + \ldots + a_{sN}a_{tN} \qquad (11.5)$$

This may be shown in a few steps. First, z_{si} and z_{ti} may each be written in the form of equation 11.1.

$$z_{si} = a_{s1}F_{1i} + a_{s2}F_{2i} + \ldots + a_{sN}F_{Ni} \qquad (11.6)$$

$$z_{ti} = a_{t1}F_{1i} + a_{t2}F_{2i} + \ldots + a_{tN}F_{Ni} \qquad (11.7)$$

By cross multiplying these two equations and summing the results over all n, equation 11.4 becomes

$$r_{st} = \frac{1}{n}\Sigma_i [a_{si}a_{ti}F_{1i}^2 + a_{s2}a_{t2}F_{2i}^2 + \ldots + a_{sN}a_{tN}F_{Ni}^2$$

$$+ a_{s1}a_{t2}F_{1i}F_{2i} + a_{s1}a_{t3}F_{1i}F_{3i} + \ldots + a_{s1}a_{tN}F_{1i}F_{Ni}$$

$$+ \ldots + a_{sN}a_{tN-1}F_{Ni}F_{N-1i}] \qquad (11.8)$$

Applying the summation to each term of equation 11.8, it may be noted that the summation applies only to the products and squares of the terms F_{ki}. Furthermore, using equations 11.2 and 11.3, it may be seen that all cross-product terms of the form $a_{sp}a_{tr}\sum_{i}F_{pi}F_{ri}$ will vanish by virtue of equation 11.2. Similarly, the terms of the form of $a_{sk}a_{tk}\sum_{i}F_{ki}^2$ reduce to $na_{sk}a_{tk}$, by virtue of equation 11.3. Hence equation 11.8 reduces to equation 11.5.

Principal Factor Problem Restated

As stated above, the problem is to find a set of factors in decreasing order of the variance they capture from the original set of variables. It is necessary therefore to determine the variance captured by each factor.

The variance of each variate, z_j, may be written in terms of the factors and their coefficients by substituting for z_{ji} in equation 11.9.

$$V(z_j) = \frac{1}{n}\sum_i(z_{ji})^2 \tag{11.9}$$

In substituting equation 11.1 into equation 11.9, the cross products will vanish, as before, and terms in the square of a factor become simply the square of the factor coefficient. Hence equation 11.9 becomes

$$n V(z_j) = a_{j1}^2 + a_{j2}^2 + a_{j3}^2 + \dots + a_{jN}^2 \tag{11.10}$$

Consider now the set of M equations of this form. These will be of the form shown by

$$n V(z_1) = a_{11}^2 + a_{12}^2 + a_{13}^2 + \dots + a_{1N}^2 \tag{11.11}$$

$$n V(z_2) = a_{21}^2 + a_{22}^2 + a_{23}^2 + \dots + a_{2N}^2 \tag{11.12}$$

$$n V(z_M) = a_{M1}^2 + a_{M2}^2 + a_{M3}^2 + \dots + a_{MN}^2 \tag{11.13}$$

The contribution of the first factor to the variance of all the z_js is given by summing the first terms of each of the equations of the form of equations 11.11 through 11.13. This variance contribution is

$$S_1 = a_{11}^2 + a_{21}^2 + a_{31}^2 + \dots + a_{M1}^2 \tag{11.14}$$

where S_1 = the variance captured by principal factor 1

For factor 1, equation 11.14 is to be maximized, subject to the constraint, detailed earlier, that the correlations among the original variates be preserved. For simplicity, equation 11.14 can be rewritten in terms of a summation, and the maximization problem is

$$\max S_1 = \sum_m a_{m1}^2 \tag{11.15}$$

$$\text{subject to } \sum_s\sum_t r_{st} = \sum_s\sum_t\sum_n a_{sn}a_{tn}$$

Using Lagrange multipliers, this problem can be written as a single equation.

$$\min T = S_1 - \sum_s \sum_t \mu_{st} r_{st} \qquad (11.16)$$

where T = Lagrange function to be minimized
and μ_{st} = Lagrange multipliers

Equation 11.16 can be written in terms of the coefficients and Lagrange multipliers alone, and solved by the normal process of setting the first differentials with respect to the unknowns equal to zero. Equation 11.17 shows the value of T in terms of the coefficients.

$$T = \sum_m a_{m1}^2 - \sum_s \sum_t \sum_n a_{sn} a_{tn} \mu_{st} \qquad (11.17)$$

To evaluate the first partial differentials, it is useful to expand equation 11.17.

$$T = a_{11}^2 + a_{21}^2 + a_{31}^2 + \ldots + a_{M1}^2 - [\mu_{11} a_{11}^2$$

$$+ \mu_{12} a_{11} a_{21} + \mu_{13} a_{11} a_{31} + \ldots + \mu_{1N} a_{11} a_{N1} + \mu_{11} a_{12}^2$$

$$+ \mu_{12} a_{12} a_{22} + \ldots + \mu_{NN} a_{N1}^2 + \ldots + \mu_{NN} a_{NN}^2] \qquad (11.18)$$

Differentiating with respect to the first coefficient, a_{11}, results in

$$\frac{\partial T}{\partial a_{11}} = 2a_{11} - \sum_t \mu_{1t} a_{t1} - \sum_s \mu_{1s} a_{s1} \qquad (11.19)$$

Since s and t are being used as dummy subscripts over the set of original variates, M, the last two terms of equation 11.19 are identical. Hence equation 11.19 can be written

$$\frac{\partial T}{\partial a_{11}} = 2[a_{11} - \sum_t \mu_{1t} a_{t1}] \qquad (11.20)$$

This may be generalized to the sth variable and the kth factor.

$$\frac{\partial T}{\partial a_{sk}} = 2[a_{sk} - \sum_t \mu_{st} a_{tk}] \qquad (k=1) \qquad (11.21)$$

$$\frac{\partial T}{\partial a_{sk}} = -2 \sum_t \mu_{st} a_{tk} \qquad (k \neq 1) \qquad (11.22)$$

In these equations, it is assumed that the Lagrange multipliers are symmetrical, that is, $\mu_{st} = \mu_{ts}$. However, the coefficients are not assumed to be symmetrical, that is, $a_{sk} \neq a_{ks}$. This is why there are two equations for the generalization of equation 11.20. Clearly, there will only be terms in the differential equation from S_1 for $k = 1$.

Equations 11.21 and 11.22 represent a system of equations for $s = 1,\ldots, N$ and $k = 1,\ldots, M$. To minimize T both equations are set to zero. It is necessary to manipulate the system of equations to produce a solution for the coefficients, as shown in the next few steps. (The reader may proceed directly to equation 11.35 for the solution, if he or she is not interested in exploring the process of manipulation that yields the solution.)

Multiplying the equations of the system by a_{j1} yields

$$a_{s1}^2 - \sum_t \mu_{st} a_{t1} a_{s1} = 0 \tag{11.23}$$

$$-\sum_t \mu_{st} a_{tk} a_{s1} = 0 \quad (k=2,3,\ldots,M) \tag{11.24}$$

Summing these equations over all j $(=1,\ldots,N)$ produces

$$\sum_s a_{s1}^2 - \sum_s \sum_t \mu_{st} a_{t1} a_{s1} = 0 \tag{11.25}$$

$$-\sum_s \sum_t \mu_{st} a_{tk} a_{s1} = 0 \quad (k=2,3,\ldots,M) \tag{11.26}$$

Returning for a moment to equation 11.21, it can be seen that equation 11.27 holds.

$$a_{s1} = \sum_t \mu_{st} a_{t1} \tag{11.27}$$

Setting the first term of equation 11.25 to a constant, λ_1, and using equation 11.27, equations 11.25 and 11.26 can be manipulated to produce equations 11.28 and 11.29, noting that s and t are interchangeable.

$$\lambda_1 - \sum_t a_{t1}^2 = 0 \tag{11.28}$$

$$-\sum_s a_{s1} a_{sk} = 0 \quad (k=2,3,\ldots,M) \tag{11.29}$$

Observe that equations 11.28 and 11.29 together represent a system of M equations.

$$\left. \begin{array}{l} \lambda_1 - \sum_t a_{t1}^2 = 0 \\ -\sum_s a_{s1} a_{s2} = 0 \\ -\sum_s a_{s1} a_{s3} = 0 \\ \quad \vdots \\ -\sum_s a_{s1} a_{sM} = 0 \end{array} \right\} \tag{11.30}$$

Multiplying each of these equations by the appropriate value of a_{tk} ($k = 1,2,3,\ldots,M$) produces the following system of equations.

$$\left. \begin{array}{ll} \lambda_1 a_{t1} - \sum_t a_{t1}^2 a_{t1} = 0 \\ -\sum_s a_{s1} a_{s2} a_{t2} & = 0 \\ -\sum_s a_{s1} a_{s3} a_{t3} & = 0 \\ \quad \vdots \\ -\sum_s a_{s1} a_{sM} a_{tM} & = 0 \end{array} \right\} \tag{11.31}$$

Summing this set of equations produces equation 11.32, where t and s are assumed to be interchangeable.

$$\lambda_1 a_{t1} + \sum_s \sum_k a_{s1} a_{sk} a_{tk} = 0 \tag{11.32}$$

Now consider equation 11.5 again; it can be written

$$r_{st} = \sum_k a_{sk} a_{tk} \tag{11.33}$$

Using this, equation 11.32 may be rewritten

$$\lambda_1 a_{t1} - \sum_s a_{s1} r_{st} = 0 \qquad (t-1,2,3,...,N) \tag{11.34}$$

This equation produces the following system of equations.

$$\left.\begin{array}{l}
a_{11} + a_{21} r_{12} + a_{31} r_{13} + ... + a_{N1} r_{1N} = \lambda_1 a_{11} \\[8pt]
a_{11} r_{21} + a_{21} + a_{31} r_{23} + ... + a_{N1} r_{2N} = \lambda_1 a_{21} \\[8pt]
\vdots \qquad\qquad\qquad\qquad\qquad\quad \vdots \\[8pt]
a_{11} r_{N1} + a_{21} r_{N2} + a_{31} r_{N3} + ... + a_{N1} = \lambda_1 a_{N1}
\end{array}\right\} \tag{11.35}$$

(It should be noted that r_{11}, r_{22}, etc., are all unity and that this property has been used in expanding equation 11.34.) Equation 11.35 can be constructed as a homogeneous system of linear equations by moving the terms $\lambda_1 a_{t1}$ across the equal signs and expressing the system as one in which the r_{st}s are known and the a_{sk}s are unknowns.

$$\left.\begin{array}{l}
(1-\lambda_1) a_{11} + r_{12} a_{21} + r_{13} a_{31} + ... + r_{1N} a_{N1} = 0 \\[8pt]
r_{21} a_{11} + (1-\lambda_1) a_{21} + r_{23} a_{31} + ... + r_{2N} a_{N1} = 0 \\[8pt]
\vdots \qquad\qquad\qquad\qquad\qquad\qquad \vdots \\[8pt]
r_{N1} a_{11} + r_{N2} a_{21} + r_{N3} a_{31} + ... + (1-\lambda_1) a_{N1} = 0
\end{array}\right\} \tag{11.36}$$

This system can only have a nontrivial solution if its determinant (shown in equation 11.37) is equal to zero.

$$\begin{vmatrix}
(1-\lambda_1) & r_{12} & r_{13} & ... & r_{1N} \\
r_{21} & (1-\lambda_1) & r_{23} & ... & r_{2N} \\
\vdots & \vdots & \vdots & & \vdots \\
r_{N1} & r_{N2} & r_{N3} & ... & (1-\lambda_1)
\end{vmatrix} = 0 \tag{11.37}$$

This may be solved by applying the normalizing condition given by equation 11.10 and noting that, since z_j is a standardized variable, $V(z_j)$ is equal to 1.

$$a_{j1}^2 + a_{j2}^2 + a_{j3}^2 + ... + a_{jN}^2 = 1 \tag{11.38}$$

From equation 11.28, the largest root of the determinant (equation 11.37) is associated with the first principal component, that is, the component found by maximizing S_1. By determining this largest root, the values of the coefficients for that principal component can be obtained from equation 11.35. This gives the first principal component which will account for the largest portion of the variance of the original standardized variables.

In a similar manner, the second largest root can be found and the coefficients for this value of λ_1 can be computed to produce the second principal component. This process is continued until some cutoff point determined by the analyst. One possibility is to continue until some predetermined percentage of the variance has been accounted for by the principal components, for example, 75%, 80%, or 90%. Alternatively, those components may be selected for which the eigenvalue is greater than unity, or some other selected value.

An Example

A useful example is provided by the same data used for the illustration of canonical correlation in chapter 10. The data and results are taken from Tintner.[9] The problem concerns data on a set of production indices and a set of price indices for comparison with the canonical-correlation problem; the two sets of indices were factor-analyzed separately.

Considering first the production indices, the correlation matrix for the variables is extracted from equation 10.50.

$$R = \begin{vmatrix} 1.000 & 0.496 & 0.873 & 0.481 \\ & 1.000 & 0.768 & 0.710 \\ & & 1.000 & 0.712 \\ & & & 1.000 \end{vmatrix} \qquad (11.39)$$

The system of equations, given by equation 11.35, becomes

$$\left. \begin{aligned} 1.000a_{11} + 0.496a_{21} + 0.873a_{31} + 0.481a_{41} &= \lambda_1 a_{11} \\ 0.496a_{11} + 1.000a_{21} + 0.768a_{31} + 0.710a_{41} &= \lambda_1 a_{21} \\ 0.873a_{11} + 0.768a_{21} + 1.000a_{31} + 0.712a_{41} &= \lambda_1 a_{31} \\ 0.481a_{11} + 0.710a_{21} + 0.712a_{31} + 1.000a_{41} &= \lambda_1 a_{41} \end{aligned} \right\} \quad (11.40)$$

The determinant, equation 11.36, is therefore given by

$$\begin{vmatrix} 1.000-\lambda & 0.496 & 0.873 & 0.481 \\ 0.496 & 1.000-\lambda & 0.768 & 0.710 \\ 0.873 & 0.768 & 1.000-\lambda & 0.712 \\ 0.481 & 0.710 & 0.712 & 1.000-\lambda \end{vmatrix} = 0 \qquad (11.41)$$

The largest root of this determinant is found to be given by $\lambda_1 = 3.033$. Substituting this value in equation 11.40 and solving for the unknown coefficients produces the values shown in table 11–3.

This solution shows that all four of the production indices make a fairly substantial contribution to the first principal component. However, it is important to find out what proportion of the original variance is captured by this first component. Referring to equation 11.14, this contribution is the sum of the squared values of the coefficients of the production indices divided by their total variance. The total variance of four standardized variables is the value 4. Hence the proportion of variance captured by the first principal component is 3.033 (from table 11–3) divided by 4, or 0.758. Thus the first principal component accounts for slightly more than three-quarters of the original variance. Further components could be sought but this seems less than necessary, considering both the amount of variance accounted for and the high loadings of all four variables on this first component.

A similar set of computations may be carried out on the set of three price indices used in the original example. In this case, the largest root of the determinantal equation is given by $\lambda_1 = 2.874$. This produces the values given in table 11–4 for the coefficients. Again, all the variables load heavily on the first principal component, which captures 95.8% of the original variance (as deduced from the sum of squared coefficient values in table 11–4). No further components need be sought in this case.

It should be noted that the results of the factor analysis are quite different from those of the canonical correlation. The two methods seek to develop quite different relationships, so that similarity should not be expected.

Table 11–3
Solutions for the First Principal Component for Production Indices

Coefficient	Value from Largest Root	Squared Coefficients
a_{11}	0.817	0.667
a_{21}	0.888	0.789
a_{31}	0.952	0.906
a_{41}	0.819	0.671
Sum		3.033

Table 11–4
Solutions for the First Principal Component for Price Indices

Coefficient	Value from Largest Root	Squared Coefficients
a_{51}	0.988	0.976
a_{61}	0.990	0.980
a_{71}	0.958	0.918
Sum		2.874

Interpretation

In conducting factor analysis, one is clearly striving to obtain a simpler structure of the variables. It is not sufficient, however, to obtain just a mathematical solution. The factor-analysis solution should also be conceptually interpretable. In other words, the components or factors obtained should be named sensibly and in such a way as to convey information to both the analyst and the audience. Frequently, this is one of the major difficulties encountered in factor analysis.

A procedure that may assist with interpretation is rotation. The idea behind this procedure is to simplify interpretation by rotating the factors about the origin to increase the larger variable loadings and decrease the small ones. This may be illustrated by considering a simple two-factor situation. Suppose the initial solution is given by figure 11–3, where F_1 and F_2 represent the two factors and A_1 through A_8 are the original attributes (variables). As shown, some problems of interpretation are likely since several attributes cluster around a line at about 45° to each of the factors, thus indicating coefficient values of about 0.5 for those variables. This situation could be improved by rotating the factors with respect to the attributes, as shown in figure 11–4. In this case, the attributes have either smaller or larger angles with the factors, thus aiding interpretability.

A number of methods of rotation have been developed, two of which are specifically for orthogonal factors, namely, quartimax and varimax. The varimax rotation is the most widely used and is the only one described in this text. Other methods may be found in texts such as Harman.[10]

Varimax Method of Rotation

The Varimax method attempts to simplify the structure of the factors by reducing the number of variables with nonzero coefficients in each factor, or by increasing the values of the large coefficients as much as possible.

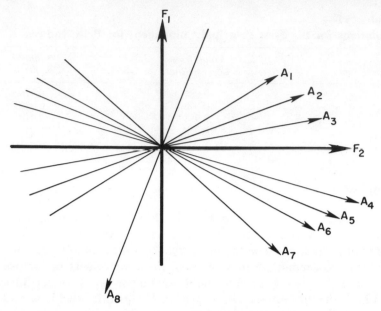

Figure 11–3. Two-factor Solution before Rotation

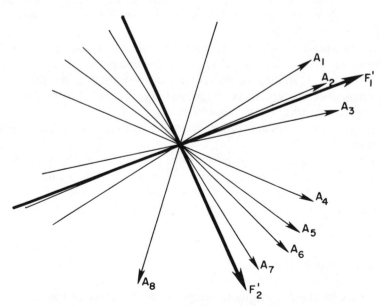

Figure 11–4. Two-factor Solution after Rotation

The initial coefficients are denoted a_{sk}, where s refers to the original variable and k to the factor. Rotation involves finding a new set of coefficients, denoted b_{sk}. It is necessary to determine a matrix of transformation values, τ_{st}, where the original orthogonality of the factors must be preserved.

Before developing the Varimax method, certain constraints can be defined. After any form of rotation, the variances of the original variables must remain unchanged, where the variances were given in equation 11.10. After rotation, the variance is given by the sum of the squared rotated coefficients divided by n. The necessary constraint is

$$n V(z_s) = \sum_k a_{sk}^2 = \sum_k b_{sk}^2 = h_s^2 \qquad (11.42)$$

The square of the variance must also remain constant for each variable in the original set. This condition is shown by

$$(\sum_k b_{sk}^2)^2 = \sum_k b_{sk}^4 + 2 \sum_{k<j} b_{sk}^2 b_{sj}^2 = g_s^4 \qquad (11.43)$$

where g_s = a constant for variable s

To see what happens when rotation is performed, sum equation 11.43 over all the original variables, thus producing

$$\sum_s \sum_k b_{sk}^4 + 2 \sum_s \sum_{k<j} b_{sk}^2 b_{sj}^2 = G^4 \qquad (11.44)$$

where G = a constant for the set of original variables

It follows, from equation 11.44, which could also be written in terms of the initial coefficient estimates, a_{sk}, that for each b_{sk} that is larger than the original a_{sk}, there must be a b_{sj} whose value is less than the original a_{sj}.

Having established the constraints of equations 11.42 and 11.43 and noting the implication of equation 11.44, it is possible to proceed with the Varimax rotation method. The procedure is based on the definition of the simplicity of a factor, where simplicity is intended to indicate interpretability. A factor with maximum simplicity would be one where variables have coefficients of ± 1 or 0 only. A factor with minimum simplicity would have coefficients of ± 0.5 only. Based on these notions, simplicity is defined by equation 11.45, which represents the variance of the squared loadings on factor k.

$$\sigma_k^2 = \frac{1}{n} \sum_s (b_{sk}^2)^2 - \frac{1}{n^2} (\sum_s b_{sk}^2)^2 \qquad (11.45)$$

This variance is maximized when all b_{sk} values are either 1 or 0, and is minimized when all b_{sk} values are identical and equal to 0.5.

Since all the factors are orthogonal initially and must remain so, it is not possible to obtain an acceptable solution by maximizing the simplicity of each factor in isolation. Rather, the simplicity of the vector of factors must be simplified simultaneously. This may be done by maximizing the sum of the simplicities of the factors.

$$\sigma^2 = \frac{1}{n} \sum_k \sum_s (b_{sk}^2)^2 - \frac{1}{n^2} \sum_k (\sum_s b_{sk}^2)^2 \tag{11.46}$$

Equation 11.46 is found, however, to result in a biased solution, since the factors are likely to have rapidly decreasing variances in terms of the original variables. Thus using equation 11.46 will tend to provide good interpretability of the first factor, but with increasingly poor interpretability of the subsequent factors. To correct this, the factor simplicities should be standardized, as described next. It may be recalled that the variance of each original variable was set to a constant, as given in equation 11.42. If each b_{sk} is divided by the appropriate value of h_s, so that the coefficient now represents the proportion of the original variation captured by factor k, the simplicity may be rewritten in standardized form.

$$\sigma'^2 = \frac{1}{n} \sum_k \sum_s (b_{sk}/h_s)^4 - \frac{1}{n^2} \sum_k (\sum_s b_{sk}^2/h_s^2)^2 \tag{11.47}$$

For convenience, multiply equation 11.47 throughout by n^2, and write $n^2 \sigma'^2$ as V.

$$V = n \sum_k \sum_s (b_{sk}/h_s)^4 - \sum_k (\sum_s b_{sk}^2/h_s^2)^2 \tag{11.48}$$

The problem is now to express the values b_{sk}/h_s in terms of a_{sk}/h_s and the angle of rotation, ϕ, for a maximum of V.

Consider a simple situation in which there are only two factors, k and j. Any orthogonal transformation in the plane of two factors through an angle ϕ must satisfy equations 11.49 and 11.50.

$$b_{sk} = a_{sk} \cos \phi + a_{sj} \sin \phi \tag{11.49}$$

$$b_{sj} = -a_{sk} \sin \phi + a_{sj} \cos \phi \tag{11.50}$$

This can be illustrated geometrically as follows. Consider a point P_i with coordinates (a_{s1}, a_{s2}) in the original frame of reference, and (b_{s1}, b_{s2}) in a new frame of reference, as shown in figure 11–5. P_i is to be expressed in terms of the rotated factors M_1 and M_2, while F_1 and F_2 were the original unrotated factors. From the geometry of figure 11–5, equation 11.51 can be determined.

$$OR_i^2 + R_i P_i^2 = OS_i^2 + S_i P_i^2 \tag{11.51}$$

Similarly, the projections of the original coordinates onto M_1 and M_2 can be determined.

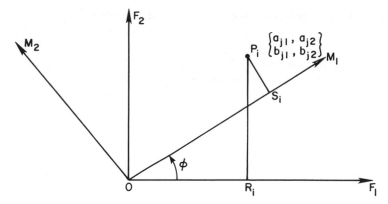

Figure 11–5. Geometry of a Two-factor Rotation

$$OR_i \cos \phi + R_iP_i \sin \phi = OS_i \tag{11.52}$$

$$-OR_i \sin \phi + R_iP_i \cos \phi = S_iP_i \tag{11.53}$$

Now the original coordinates of P_i with respect to F_1 and F_2 define OR_i as being a_{sk} and R_iP_i as a_{sj}. Similarly, S_iP_i is b_{sj} and OS_j is b_{sk}. Therefore equations 11.52 and 11.53 may be rewritten as equations 11.54 and 11.55, which are identical to equations 11.49 and 11.50.

$$b_{sk} = a_{sk} \cos \phi + a_{sj} \sin \phi \tag{11.54}$$

$$b_{sj} = -a_{sk} \sin \phi + a_{sj} \cos \phi \tag{11.55}$$

Thus equations 11.54 and 11.55 can be substituted into equation 11.48 to define V in terms of the known values of the as and the one unknown, ϕ. The idea is then to maximize V in that equation by selecting the value of ϕ. The solution for this is mathematically cumbersome and is not detailed here. The result may be summarized as

$$\tan 4\phi = \frac{D - 2AB/n}{C - (A^2 - B^2)/n} \tag{11.56}$$

where $A = \sum\limits_{s}(a^2_{sk}/h^2_k - a^2_{sj}/h^2_j)$

$B = \sum\limits_{s}(2a_{sk}a_{sj}/h_kh_j)$

$C = \sum\limits_{s}(a^2_{sk}/h^2_k - a^2_{sj}/h^2_j - 2a_{sk}a_{sj}/h_kh_j)$

and $D = 2\sum\limits_{s}(a^2_{sk}/h^2_k - a^2_{sj}/h^2_j)(2a_{sk}a_{sj}/h_kh_j)$

There are four possible solutions to equation 11.56 in terms of the angle ϕ, and the maximum solution is determined by examining the sign of the second differential of V with respect to ϕ. Solving for ϕ, this value is substituted in equations 11.54 and 11.55 to determine the values of b_{sk} and b_{sj}.

This solution process can be extended readily (but with some mathematical complexity) to any number of factors, but the process is still to find a single value of ϕ in each plane of the solution space. Thus, a three-factor solution will generate two angles and an n-factor solution $(n-1)$ angles.

An Example

A survey was conducted on people traveling to work. Questions were asked on their ratings of 12 attributes of the transportation system, namely, total travel time, ability to carry parcels, safety from accidents, ability to estimate time of arrival, waiting time, ability to carry out other tasks en route, protection against weather, independence from schedules, individual privacy, cost, availability of a seat, and ability to control the temperature and ventilation. These attributes were rated for six different means of travel. The correlation matrix is shown in table 11–5. By substituting these values into equation 11.37, the first several roots can be determined. The first was determined to be $\lambda_1 = 5.016$, which accounted for 41.8% of the variance of the original variables and produced the set of coefficients shown in table 11–6. Five additional factors were found, which accounted together for 80.7% of the original variance. The complete set of factor loadings are shown in table 11–7.

In the initial configuration (table 11–7) there are some serious problems of interpretability. No variables load heavily on factors 4, 5, and 6. Total travel time loads equally on factors 1 and 2, while cost has heavy loadings on both factors 2 and 3. For interpretation, factor 1 is made up of total time, parcels, waiting time, errands, weather protection, independence, privacy, seats, and control of temperature. This appears to be some composite of time, comfort, and convenience. Similarly, factor 2 is made up of total travel time, safety, estimate time, and cost; while factor 3 comprises cost only. (The reader is referred to the beginning of this section for a full description of each of the variables, described cryptically here.)

Varimax rotation was applied to this factor-analytic solution, in an attempt to improve its interpretability. The transformation matrix for this rotation is shown in table 11–8. Applying this rotation produces the factor loadings shown in table 11–9. In this instance, the interpretability of the factors has been enhanced considerably. All the factors have one or more variables with a high loading and only one variable—privacy—has equal

Table 11–5
Correlation Matrix for the Second Example

Attribute	Total Time	Parcels	Safety	Estimate Time	Wait Time	Errands	Weather Protection	Inde-pendence	Privacy	Cost	Seats	Control Temp.
Total time	1.00	.423	.093	.452	.355	.336	.335	.268	.313	.151	.270	.305
Parcels		1.00	-.131	.251	.367	.563	.599	.548	.602	-.075	.613	.636
Safety			1.00	.255	-.024	-.123	-.155	-.091	-.095	.084	-.173	-.182
Estimate time				1.00	.371	.249	.216	.229	.240	.092	.182	.173
Wait time					1.00	.467	.385	.354	.394	.151	.311	.379
Errands						1.00	.489	.587	.580	-.064	.399	.543
Weather protection							1.00	.495	.539	-.012	.549	.596
Independence								1.00	.615	-.232	.429	.527
Privacy									1.00	-.127	.534	.644
Cost										1.00	-.032	-.048
Seats											1.00	.645
Control temperature												1.00

Table 11-6
Initial Loadings for the First Factor

Coefficient	Value from Largest Root	Squared Loadings
a_{11}	.527	.278
a_{21}	.821	.674
a_{31}	−.151	.022
a_{41}	.400	.160
a_{51}	.594	.353
a_{61}	.757	.573
a_{71}	.761	.579
a_{81}	.740	.548
a_{91}	.803	.645
a_{101}	−.065	.004
a_{111}	.727	.529
a_{121}	.809	.654
Sum		5.019

Table 11-7
Initial Factor Loadings for Six Factors

	Factor 1	Factor 2	Factor 3	Factor 4	Factor 5	Factor 6
Total time	.527	.523	.033	.073	−.489	.342
Parcels	.821	−.072	.018	.192	−.099	.102
Safety	−.151	.611	−.493	.358	.442	.083
Estimate time	.400	.667	−.249	−.053	−.248	−.354
Wait time	.594	.342	.162	−.487	.184	−.296
Errands	.757	−.025	−.071	−.339	.139	.268
Weather protection	.761	−.079	.140	.147	.011	−.051
Independence	.740	−.162	−.312	−.171	.112	.158
Privacy	.803	−.118	−.127	.013	.163	.063
Cost	−.065	.529	.737	.065	.266	.175
Seats	.727	−.167	.176	.394	−.003	−.293
Control temperature	.809	−.179	.120	.183	.108	−.050

Table 11-8
Transformation Matrix for Varimax Rotation

	Factor 1	Factor 2	Factor 3	Factor 4	Factor 5	Factor 6
Factor 1	.720	−.031	.274	.263	.576	−.064
Factor 2	−.224	.425	.529	.469	−.106	.508
Factor 3	.197	.792	−.050	.043	−.220	−.530
Factor 4	.565	.001	−.430	.142	−.517	.457
Factor 5	.058	.370	−.051	−.704	.347	.491
Factor 6	−.262	.233	−.675	.440	.470	.084

Table 11–9
Rotated Factor Loadings

	Factor 1	Factor 2	Factor 3	Factor 4	Factor 5	Factor 6
Total time	.192	.132	.182	.887	.194	.036
Parcels	.687	−.054	.039	.323	.391	−.051
Safety	−.137	.058	.074	.044	−.035	.969
Estimate time	.138	−.100	.748	.440	−.010	.269
Wait time	.196	.254	.736	−.019	.446	−.107
Errands	.283	.023	.156	.162	.803	−.087
Weather protection	.691	.046	.130	.148	.320	−.096
Independence	.376	−.261	.094	.098	.714	.026
Privacy	.580	−.101	.108	.076	.582	.047
Cost	−.013	.950	.066	.098	−.115	.057
Seats	.895	−.023	.131	.035	.055	−.071
Control temperature	.769	.023	.070	.051	.378	−.074

highest loadings on more than one factor. In this case, the factors may be named fairly readily. Factor 1 could be called comfort (comprising parcels, weather protection, privacy, seats, and control temperature). Factor 2 is cost, since this is the only variable loading on it. Factor 3 is excess time, comprising estimate time and waiting time; and factor 4 is total time, since this is the only variable loading heavily on it. Factor 5, consisting of errands, independence, and privacy, could be called convenience; and factor 6 is safety. This example illustrates very clearly how the use of rotation can improve the results of factor analysis. The final factor loadings are generally very well-conditioned, exhibiting very low or very high loadings only. In other words, the simplicity is very much higher.

To demonstrate the effect of rotation on the factor simplicities, the individual simplicities are shown in table 11–10. It can be seen that all

Table 11–10
Factor Simplicities before and after Rotation

Factor	Simplicity before Rotation	Simplicity after Rotation
1	.056	.069
2	.024	.061
3	.023	.040
4	.005	.046
5	.006	.042
6	.002	.066
Total	.116	.324

factor simplicities have increased and that the total simplicity has been increased almost threefold. This bears out the clear improvements gained in interpretability of the factors.

General Factor Analysis

In the beginning of the chapter, a distinction was noted between factor analysis and principal-components analysis. It was suggested that factor analysis involves testing a hypothesized model against a set of data. The basic difference can be demonstrated by examining the initial problem formulation for factor analysis as opposed to that of principal components given on p. 241, and repeated here in equation 11.57.

$$z_{ji} = \sum_k a_{jk} F_{ki} \qquad (11.57)$$

In factor analysis, a somewhat different problem is assumed.

$$z_{ji} = \sum_k a_{jk} F_{ki} + b_j S_{ji} + c_j \epsilon_{ji} \qquad (11.58)$$

This model formulation distinguishes two types of factors and an error term. The F_k are factors that may appear in more than one original variable. The S_j is a factor specific to z_j, and ϵ_j is an error term. All these elements are assumed, as before, to have unit variance and zero means and are also assumed to be independent.

The formulation of equation 11.58 is general enough to include equation 11.57 as a special case. Clearly, if the number of factors equals the number of original variables, the z_js can be expressed in terms of the F_js only, without the need for specific factors or residual errors. However, if many fewer factors are used than there are original variables, there is likely to be residual variation that can be ascribed to a specific factor and an error term for each variable. The coefficient c_j is used to ensure that the error term has unit variance.

Communality, Reliability, and Specificity

In equation 11.58, the F_k are called common factors. If such a factor occurs in all variables, it is termed a general factor.

Consider now the variances and covariances of the original variables. The covariance of z_{ji} with z_{mi} is

$$\text{cov}(z_j, z_m) = \sum_k a_{jk} a_{mk} + b_j b_m \text{ cov}(S_j, S_m)$$

$$+ c_j c_m \text{ cov}(\epsilon_j, \epsilon_m) \qquad (j \neq m) \quad (11.59)$$

Assuming that specific factors and errors are uncorrelated across original variables, the last two terms of equation 11.59 vanish, resulting in

$$\text{cov}(z_j, z_m) = \sum_k a_{jk} a_{mk} \qquad (j \neq k) \qquad (11.60)$$

Similarly, the variance of z_j may be written

$$V(z_j) = \sum_k (a_{jk})^2 + b_j^2 + c_j^2 \qquad (11.61)$$

This latter equation is also equal to unity, since z_j is a standardized variable. Therefore equation 11.62 is true.

$$\sum_k (a_{jk})^2 + b_j^2 + c_j^2 = 1 \qquad (11.62)$$

It is interesting to note that the coefficients b_j and c_j appear only in the variance and not in the covariance.

The various terms in these equations are also given specific names.[11] The error term, ϵ_j, is called the unreliability factor, and c_j^2 is called the unreliability. Similarly, the sum of the other two terms in equation 11.62, namely, $\sum_k (a_{jk})^2 + b_j^2$, is called the reliability. Because b_j is the coefficient of the specific factor, b_j^2 is called the specificity and $\sum a_{jk}^2$ is called the communality and is generally written as h_j^2. The value of the complement of h_j^2, $(1-h_j^2)$ is called the uniqueness. It is interesting to note that the communality represents the amount of the variance accounted for by the factors.

Solving the General Factor-Analysis Problem

As noted at the outset of the chapter, it is not intended to provide an exhaustive treatment of general factor analysis. However, it is useful to consider the basic procedure by which the problem is solved. Assume there are no specific factors but there is a residual error. Equations 11.60 and 11.61 can be generalized to

$$\text{cov}(z_j, z_m) = \sum_k a_{jk} a_{mk} + \delta_{jm} V(\epsilon_j) \qquad (11.63)$$

where $\delta_{jm} = 1$ if $j = m$, and is zero otherwise.

The covariances on the left side of equation 11.63 are known from the data, and equation 11.63 could be solved (as it was for the principal-components case) if it were not for the terms $\delta_{jm} V(\epsilon_j)$. These terms exist on the diagonal of the matrix only and indicate that the estimates off the diagonal are all correct, while those on the diagonal are biased, when the principal-components solution is used.

Hence, the principal-components solution procedure could be used here, except that the diagonal terms would be

$$\sum_k a_{jk}^2 + V(\epsilon_j) = 1 \tag{11.64}$$

In contrast, the principal-components solution assumes that equation 11.65 holds.

$$\sum_k a_{jk}^2 = 1 \tag{11.65}$$

Thus the principal-components solution will overestimate the value of $\sum_k a_{jk}^2$. It is necessary, therefore, to find a means of removing the error variances from the diagonal of the matrix. It will be recalled that $\sum_k a_{jk}^2$ is the communality. This is clearly the value that is required on the diagonal of the matrix. Hence the matrix to be solved is

$$
\begin{vmatrix}
h_1^2 & \mathrm{cov}(z_1,z_2) & \cdots \mathrm{cov}(z_1,z_M) \\
\mathrm{cov}(z_2,z_1) & h_2^2 & \cdots \mathrm{cov}(z_2,z_M) \\
\vdots & \vdots & \vdots \\
\mathrm{cov}(z_M,z_1) & \mathrm{cov}(z_M,z_2) & h_M^2
\end{vmatrix}
$$

$$
=
\begin{vmatrix}
\sum_k a_{1k}^2 & \sum_k a_{1k} a_{2k} & \cdots \sum_k a_{1k} a_{Mk} \\
\sum_k a_{2k} a_{1k} & \sum_k a_{2k}^2 & \cdots \sum_k a_{2k} a_{Mk} \\
\vdots & \vdots & \vdots \\
\sum_k a_{Mk} a_{1k} & \sum_k a_{Mk} a_{2k} & \cdots \sum_k a_{Mk}^2
\end{vmatrix}
\tag{11.66}
$$

This problem cannot be solved uniquely because there are insufficient conditions for solution. There are a number of different procedures for obtaining estimates of the h_j^2s, which are detailed in some of the texts on factor analysis.[12] In most instances, an iterative solution is used, in which initial estimates are made for the h_j^2s and are refined by subsequent solutions of the factor-analysis problem.

The addition of the error term to the problem provides the distinction noted by Kendall and quoted at the beginning of the chapter. It is clear that the principal-components procedure is a search for simpler structure, while the general factor-analysis procedure assumes a residual error, which is a function of a hypothesis structure tested on the data.

Notes

1. Maurice G. Kendall, *A Course in Multivariate Analysis*, London: Charles Griffin and Co., 1965.
2. Kendall, *Multivariate Analysis*, p. 37.

3. H.H. Harman, *Modern Factor Analysis,* Chicago: University of Chicago Press, 1960; R.J. Rummel, *Applied Factor Analysis*, Evanston, Ill.: Northwestern University Press, 1973.

4. Gerhard Tintner, *Econometrics,* New York: John Wiley & Sons, 1952.

5. Harman, *Modern Factor Analysis.*

6. Frank S. Koppelman and John R. Hauser, "Consumer Travel Choice Behavior: An Empirical Analysis of Destination Choice for Non-Grocery Trips," Working Paper No. 414–09, Evanston, Ill.: The Transportation Center, Northwestern University, July 1977.

7. Kendall, *Multivariate Analysis,* p. 10.

8. Ibid.

9. Tintner, *Econometrics,* pp. 110–114.

10. Harman, *Modern Factor Analysis.*

11. Kendall, *Multivariate Analysis,* p. 41.

12. See, for example, Harman, *Modern Factor Analysis,* and Rummel, *Applied Factor Analysis.*

12

Introduction to Discriminant Analysis

Setting up the Problem

Discriminant analysis was originally developed for use in the biological sciences. Two derivations of it exist, one by Neyman and Pearson,[1] the other by Fisher.[2] The derivation used here is by Fisher, but the assumptions used by both methods are identical. A brief condensed derivation of discriminant analysis is given here.

The basic hypothesis of discriminant analysis runs as follows: it is assumed that a population is made up of two distinct subpopulations. It is further assumed that it is possible to find a linear function of certain measures or attributes of the population that will allow an observer to discriminate between the two subpopulations. Originally, the technique was devised to assist biologists in identifying subspecies. In this context, suppose a researcher has two subspecies of a particular plant species, say, a daffodil. In general appearance the two subspecies are so alike that one cannot with any certainty state which is which. However, the length, width, and thickness of the leaves and the maximum diameter of the bulb can be measured. It can be hypothesized that a linear combination of these measures can be devised which will allow the analyst to discriminate between the two subspecies with the least possible likelihood of error.

This technique appears to have many possible applications in the social sciences and engineering. An application frequently used pertains to the choice of transportaation mode, where the two populations are considered as being auto users and transit users, and a linear combination of system and user characteristics is sought as a basis of discriminating between the two populations. Using this as an example, the frequency distributions of the two populations—auto users and transit users—can be plotted against some function z. Now over the range z_1 to z_2 (see figure 12–1), it is not certain whether an individual with a z value in that range is an auto or a transit user. Suppose it has to be decided whether each person in this total population is an auto or a transit user. How can one proceed so as to make as few mistakes as possible? To put it in a slightly different way, how can one minimize the number of misclassified people?

Two important things should be noted here, namely, that each member has to be assigned to one or the other subpopulation, and that the decision as to how to divide the population has already been made, that is, into auto and transit users. In other words, discriminant analysis is not

263

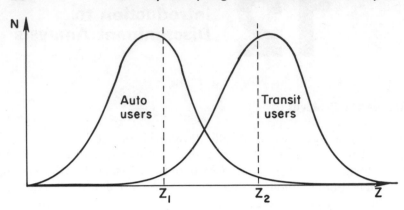

Figure 12–1. Frequency Distribution of Auto versus Transit Users

designed as a procedure for seeking population groupings, like cluster analysis.

It should be clear from this that discriminant analysis has many potential uses beyond the original biological ones. Whenever it is desired to find relationships that would permit classifying human populations into groups, discriminant analysis may be an applicable method. Likewise, engineering often calls for classification of various physical substances into groups having distinct properties. Whenever such grouping cannot be made on obvious grounds, discriminant analysis may be used to find relevant compound measures to permit appropriate classification.

The problem is one of determining the function z that will best permit the discrimination between members of the two populations; z is defined as a function of a number of factors $x_1, x_2, x_3, ..., x_k$, and the subpopulation is designated by subscript i and the members of the subpopulations by subscript j.

$$z_{ij} = a_1 x_{1ij} + a_2 x_{2ij} + ... + a_k x_{kij} \qquad (12.1)$$

For convenience, equation 12.1 may be abbreviated to

$$z_{ij} = \sum_{p=1}^{k} a_p x_{pij} \qquad (12.2)$$

where i = 1,2
$\quad\quad\ j$ = 1,2,...,n
and $\quad a_p$ = the weighting coefficient of the pth factor, x_{pij}

The task is to determine the values of the weighting coefficients, the a_ps, such that one can best discriminate between the two subpopulations. To set about determining these coefficients, it is necessary to define what is

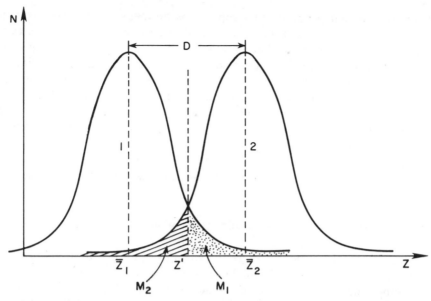

Figure 12–2. Definition of Discrimination between Two Subpopulations

meant by discrimination between two subpopulations. Two alternative definitions could be postulated. Consider figure 12–2. One may postulate that the rule sought is to state that all members of the population with a value of z less than z' are to be classified as being in subpopulation 1, while those greater than z' are classified in subpopulation 2. Clearly, those members of population 2 who fall in the shaded area, M_2, will be misclassified by this rule, as will those of population 1 who fall in the shaded area, M_1.

Neyman and Pearson[3] suggested that discrimination be treated as an attempt to minimize the total number of misclassifications, that is, the sum of M_1 and M_2. The task is to state this mathematically so as to define the coefficients of the discriminant function, z.

Fisher[4] suggested, alternatively, that discrimination could be defined as achieving the maximum separation of the two subpopulations. This is, of course, equivalent in effect to minimizing misclassifications, but it generates a different mathematical statement of the problem and is the derivation used in this text. Of course, with Fisher's definition of discrimination, some care is needed in selecting the measure of separation of the two subpopulations. If one were to measure this as the distance, D, between the two measures, $\bar{z}_1$ and $\bar{z}_2$, one could clearly increase D simply by multiplying the z function by some factor greater than one. This, however, would not increase the separation in any meaningful way.

Therefore, Fisher proposed that separation be measured as the distance between the means, D, relative to the within-subpopulation variances. This would make it clear that scaling the discriminant function has no effect upon the separation of the two subpopulations. Finally, the simple distance between the two means may not be the best measure, since one may experience sign problems. (A priori, one may not know which subpopulation has the smaller z values. Hence, by measuring from the second subpopulation mean, say, from figure 12–2, the distance would be negative and the largest negative distance would be desired.) This can be overcome by maximizing the square of the distance between the two subpopulations, where this value is also the between-population variance.

Fisher's Derivation of the Discriminant Function

Regardless of which method is used, two important assumptions have to be made: the distributions of the factors are multivariate normal, and the two subpopulations are homoscedastic with respect to the factors (i.e., the covariance matrices for the factors are the same for both populations). The method of choosing the weighting coefficients can thus be stated as maximizing the between-population variance relative to the within-population variance. The between-population variance is simply $(\bar{z}_1 - \bar{z}_2)^2$, where $\bar{z}_1$ and $\bar{z}_2$ are the means of the two subpopulations, as shown in figure 12–2.

The within-population variances are $\sum_j (z_{1j} - \bar{z}_1)^2$ and $\sum_j (z_{2j} - \bar{z}_2)^2$. Considering the first subpopulation, the value $(z_{1j} - \bar{z}_1)$ can be written in terms of the variables, x_p, and the weighting coefficients, a_p.

$$(z_{1j} - \bar{z}_1) = \sum_p a_p x_{p1j} - \sum_p a_p \bar{x}_{p1} \tag{12.3}$$

This can be simplified to

$$(z_{1j} - \bar{z}_1) = \sum_p a_p (x_{p1j} - \bar{x}_{p1}) \tag{12.4}$$

Thus the square of this value, used in determining the first within-population variance, can be written

$$(z_{1j} - \bar{z}_1)^2 = \sum_p a_p (x_{p1j} - \bar{x}_{p1}) \sum_p a_p (x_{p1j} - \bar{x}_{p1}) \tag{12.5}$$

The within-population variance for population 1 is

$$\sum_j (z_{1j} - \bar{z}_1)^2 = \sum_j [\sum_p a_p (x_{p1j} - \bar{x}_{p1}) \sum_p a_p (x_{p1j} - \bar{x}_{p1})] \tag{12.6}$$

This may be written

$$\sum_j (z_{1j} - \bar{z}_1)^2 = \sum_p \sum_q a_p a_q C_{pq} \tag{12.7}$$

where $\mathbf{C}_{pq}$ is the matrix shown in equation 12.8.

$$\mathbf{C}_{pq} = \begin{vmatrix} C_{11} & C_{12} & C_{13} & \cdots & C_{1k} \\ C_{21} & C_{22} & C_{23} & \cdots & C_{2k} \\ \cdots & \cdots & \cdots & & \cdots \\ \cdots & \cdots & \cdots & & \cdots \\ \cdots & \cdots & \cdots & & \cdots \\ C_{k1} & C_{k2} & C_{k3} & \cdots & C_{kk} \end{vmatrix} \qquad (12.8)$$

Definitions of the matrix elements are

$$C_{11} = \sum_j (x_{11j}-\bar{x}_{11})^2 \qquad (12.9)$$

$$C_{12} = \sum_j (x_{11j}-\bar{x}_{11})(x_{21j}-\bar{x}_{21}) \text{ etc.} \qquad (12.10)$$

$$C_{12} = C_{21} = \sum_j (x_{11j}-\bar{x}_{11})(x_{21j}-\bar{x}_{21}) \qquad (12.11)$$

Also, according to the assumption of homoscedasticity, the within-population variance of population 1 is the same as that of population 2, that is, $\mathbf{C}_{pq}$ is common to both populations.

Equation 12.7 may be shown to hold by the following procedure. First, equation 12.6 is expanded in terms of the summations over p.

$$\sum_j (z_{1j}-\bar{z}_1)^2 = \sum_j \Big[a_1^2 (x_{11j}-\bar{x}_{11})^2$$
$$+ a_1 a_2 (x_{11j}-\bar{x}_{11})(x_{21j}-\bar{x}_{21}) + \cdots$$
$$+ a_k^2 (x_{k1j}-\bar{x}_{k1})^2 \Big] \qquad (12.12)$$

The summation over j can then be applied to each element of equation 12.12.

$$\sum_j (z_{1j}-\bar{z}_1)^2 = a_1^2 \sum_j (x_{11j}-\bar{x}_{11})^2 + a_1 a_2 \sum_j (x_{11j}-\bar{x}_{11})(x_{21j}-\bar{x}_2)$$
$$+ \cdots + a_k^2 \sum_j (x_{k1j}-\bar{x}_{k1})^2 \qquad (12.13)$$

This equation can be simplified by noting that the summed terms in equation 12.13 are variances or covariances. For example, the covariance of the two variables, u and v, is

$$n[\text{cov}(u,v)] = \sum_j (u_j-\bar{u})(v_j-\bar{v}) \qquad (12.14)$$

Likewise, the variance of a variable, u, is represented by

$$n[V(u)] = \sum_j (u_j-\bar{u})^2 \qquad (12.15)$$

Hence each of the terms in equation 12.13 is a variance or covariance. Denoting this in terms of the matrix defined by equation 12.8 leads to equation 12.7.

Having stated the within-subpopulation variances in terms of the original variables and the to-be-defined weighting coefficients, the problem of determining the values of the coefficients can now be addressed. First, it is useful to state the problem mathematically. The objective is to maximize a function, G.

$$G = \frac{(\bar{z}_1 - \bar{z}_2)^2}{\sum_j (z_{1j} - \bar{z}_1)^2} \qquad (12.16)$$

Equation 12.16 can be rewritten in terms of the x_ps and a_ps as

$$G = \frac{\left[\sum_p a_p \bar{x}_{p1} - \sum_p a_p \bar{x}_{p2}\right]^2}{\sum_p \sum_q a_p a_q C_{pq}} \qquad (12.17)$$

Equation 12.17 may be written more usefully as equation 12.18, by gathering the terms in the numerator.

$$G = \frac{\left[\sum_p a_p (\bar{x}_{p1} - \bar{x}_{p2})\right]^2}{\sum_p \sum_q a_p a_q C_{pq}} \qquad (12.18)$$

For simplicity, $(\bar{z}_1 - \bar{z}_2)^2$ may be written as D^2 and $\sum_j (z_{1j} - \bar{z}_1)^2$ as S.

$$G = \frac{D^2}{S} \qquad (12.19)$$

To maximize G, by choosing the a_ps, G is differentiated with respect to each a_p (equation 12.20) and each partial differential is set to zero.

$$\frac{\partial G}{\partial a_p} = \frac{1}{S^2}\left[S\, 2D \frac{\partial D}{\partial a_p} - D^2 \frac{\partial S}{\partial a_p}\right] = 0 \qquad \forall p \qquad (12.20)$$

Equation 12.20 may be simplified and written

$$\frac{D}{S^2}\left[2S \frac{\partial D}{\partial a_p} - D \frac{\partial S}{\partial a_p}\right] = 0 \qquad \forall p \qquad (12.21)$$

Rearranging equation 12.21 leads to

$$\frac{S}{D} \cdot \frac{\partial D}{\partial a_p} = \frac{1}{2} \cdot \frac{\partial S}{\partial a_p} \qquad \forall p \qquad (12.22)$$

Since $\frac{S}{D}$ is a constant for all $\frac{\partial G}{\partial a_p}$, equation 12.22 can be written

$$\frac{\partial S}{\partial a_p}\, \alpha = \frac{\partial D}{\partial a_p} \qquad \forall p \qquad (12.23)$$

To see how equation 12.23 is solved for the coefficients, it is helpful to use a simple example. Consider therefore a function with three variables.

$$z_{1j} = a_1 x_{11} + a_2 x_{12} + a_3 x_{13} \tag{12.24}$$

The covariance of x_1 and x_2 is C_{12}, etc. S can then be expressed

$$S = \sum_{p=1}^{3} \sum_{q=1}^{3} a_p a_q C_{pq} \tag{12.25}$$

Expanding the summation over q yields

$$S = \sum_{p=1}^{3} a_p \, [a_1 C_{p1} + a_2 C_{p2} + a_3 C_{p3}] \tag{12.26}$$

Now the sumation over p may be expanded to show those terms that will appear in each differentiation of S.

$$S = a_1 [a_1 C_{11} + a_2 C_{12} + a_3 C_{13}] + a_2 \, [a_1 C_{21} + a_2 C_{22} + a_3 C_{23}]$$
$$+ a_3 \, [a_1 C_{31} + a_2 C_{32} + a_3 C_{33}] \tag{12.27}$$

Differentiating equation 12.27 with respect to a_1 results in

$$\frac{\partial S}{\partial a_1} = 2a_1 C_{11} + a_2 C_{12} + a_3 C_{13} + a_2 C_{21} + a_3 C_{31} \tag{12.28}$$

By gathering terms, equation 12.28 can be simplified to

$$\frac{\partial S}{\partial a_1} = 2[a_1 C_{11} + a_2 C_{12} + a_3 C_{13}] \tag{12.29}$$

Equation 12.29 can then be written

$$\frac{\partial S}{\partial a_1} = 2 \sum_{q=1}^{3} a_q C_{1q} \tag{12.30}$$

Similarly, D can be expressed

$$D = a_1 \, (\bar{x}_{11} - \bar{x}_{21}) + a_2 \, (\bar{x}_{12} - \bar{x}_{22}) + a_3 \, (\bar{x}_{13} - \bar{x}_{23}) \tag{12.31}$$

Differentiating D with respect to a_1 results in

$$\frac{\partial D}{\partial a_1} = (\bar{x}_{11} - \bar{x}_{21}) = d_1 \tag{12.32}$$

Returning to the general solution, it may therefore be stated that equation 12.33 may be rewritten as the set of equations 12.34.

$$\frac{\partial S}{\partial a_p} \, \alpha = \frac{\partial D}{\partial a_p} \tag{12.33}$$

$$\sum_q a_q C_{pq}\, \alpha = d_p \qquad p = 1,2,\ldots,k \tag{12.34}$$

The a_ps are weighting coefficients, so they may be scaled as desired. Without loss of generality, therefore, α may be set to one, and equation 12.34 can be rewritten

$$\sum_q a_q C_{pq} = d_p \qquad p = 1,2,\ldots,k \tag{12.35}$$

This system of equations may be rewritten in matrix terms.

$$\mathbf{AC = D} \tag{12.36}$$

Because $\mathbf{A}$, the matrix of weighting coefficients, is the matrix to be determined, equation 12.36 can be rearranged to yield

$$\mathbf{A = C^{-1}D} \tag{12.37}$$

In terms of the original variables and coefficients, equation 12.37 can be rewritten

$$\sum_q a_q = \sum_q \sum_p \mathbf{C}^{pq}\, (\bar{x}_{p1} - \bar{x}_{p2}) \tag{12.38}$$

where $\mathbf{C}^{pq}$ denotes the inverse of $\mathbf{C}_{pq}$

Therefore, z_{ij} is expressed

$$z_{ij} = \sum_q \sum_p \mathbf{C}^{pq}\, (\bar{x}_{p1} - \bar{x}_{p2}) x_{qij} \tag{12.39}$$

Since the true means, variances, and covariances of the populations are not usually known, maximum-likelihood estimates of these values from the sample population observations must be used, as shown in equation 12.40 for estimates of $\hat{a}_q$. A similar expression arises for z_{ij}.

$$\sum_q \hat{a}_q = \sum_q \sum_p \hat{\mathbf{C}}^{pq}(\hat{\bar{x}}_{p1} - \hat{\bar{x}}_{p2}) \tag{12.40}$$

Statistical Tests of the Function

Next it has to be determined whether the discriminator is significant. Kendall[5] suggests three ways in which this can be accomplished.

First, there may be a real difference between the populations, but they are so close together that a discriminator is not very effective. This is measured by the errors of misclassification which, though the minimum, may still be large. Second, there may be a real difference between the populations but the sample is not large enough to produce a very reliable discriminator; this is really a matter of setting confidence limits to the function or to its coefficients. This is discussed a little later. Third, it may be that the parent populations are identical and that a discriminant function is illusory, for example, that all auto users will use transit and all transit users will use autos.

For the purposes of the use of discriminant analysis in this context, the latter is unlikely to occur. In general, the use of this technique is only suggested when separate populations can be identified. So this point is not considered any further here.

However, the other two questions of significance are very relevant. Considering the first point, the populations may be interpreted as being so close together in two ways. Either the wrong factors were chosen or not all of the significant factors to build the discriminant function. (This is the same as one of the problems that may arise in regression.) Or the assumptions of rationality and consistency implicit in this modeling technique are so tenuous that no set of factors can discriminate effectively between the two populations. In either case the significance of the observed discrimination can be measured in terms of the probability of the observed discrimination occurring at random. This can be done using the variance ratio, which is derived below. Because the variance ratio involves measures of both the distance between the populations and the numbers in the sample, it can also be used as an indicator of the reliability of the discriminator due to sample size.

One may also wish to test the significance of each factor in the discriminant function. One way in which the standard errors of the coefficients of each factor can be derived, which can be used for this significance test, is to refer to the equivalent multiple-regression problem. This is presented in relation to Kendall's[6] derivation.

Kendall's Regression Analogy[7]

Initially, it is necessary to propose a dummy variate, y, where for each member of population 1, y takes the value $n_2/(n_1+n_1)$, and for each member of population 2, y takes the value $-n_1$ (n_1+n_2), where n_1 and n_2 are the total numbers of observations in populations 1 and 2, respectively. The sum of the y values is

$$\sum_{i=1}^{n_1+n_2} y_i = n_1 \frac{n_2}{n_1 + n_2} + n_2 \frac{-n_1}{n_1 + n_2} \qquad (12.41)$$

It can be seen readily that equation 12.41 reduces to

$$\sum_i y_i = 0 \qquad (12.42)$$

Equation 12.42 states that y_i has a zero mean. The sum of squares of y_i

about the mean is

$$\sum_{i=1}^{n_1+n_2} (y_i-\bar{y})^2 = \sum_{i=1}^{n_1+n_2} y_i^2 \tag{12.43}$$

Since there are n_1 values of $n_2/(n_1+n_2)$ and n_2 values of $-n_1/(n_1+n_2)$ for y_i, equation 12.43 can be rewritten

$$\Sigma y_i^2 = n_1 \frac{n_2^2}{(n_1+n_2)^2} + n_2 \frac{n_1^2}{(n_1+n_2)^2} \tag{12.44}$$

Gathering terms and simplifying produces

$$\Sigma y_i^2 = \frac{n_1 n_2}{n_1 + n_2} \tag{12.45}$$

Setting up the regression for y_i produces

$$y_i = \sum_{p=1}^{k} a_p (x_{pi}-\bar{x}_p) + \epsilon_i \tag{12.46}$$

where x_p is defined as

$$\bar{x}_p = \frac{n_1 \bar{x}_{p1} + n_2 \bar{x}_{p2}}{n_1 + n_2} \tag{12.47}$$

In other words, an attempt is made to determine a function of the x_ps ($p=1,2,\dots,k$) such that the dichotomy of the population can be explained. This is shown in graphical terms in figure 12–3, where an attempt is made to fit the straight line on the graph of y against z, where z is a function of the attributes $x_1, x_2, \dots, x_k$. (It should be noted that there is some question as to the validity of a regression in which the dependent variable is a dummy variable. Kendall[8] notes this and cites Bartlett[9] where an attempt has been made to examine this question.)

Following the normal regression procedure, the error term is

$$\epsilon_j = y_j - \sum_p a_p(x_{pj}-\bar{x}_p) \tag{12.48}$$

The squared error term is

$$\epsilon_j^2 = [y_j - \sum_q a_p (x_{pj}-\bar{x}_p)]^2 \tag{12.49}$$

The quantity to be minimized, the sum of the squared errors, is

$$\sum_j \epsilon_j^2 = \sum_j [y_j-\sum_p a_p(x_{pj}-x_p)]^2 \tag{12.50}$$

Note that this regression is constrained by its formulation to pass through the origin, so there is no constant term. Also note that observations for each factor, x_p, are combined for the two populations, x_{p1j} and x_{p2j}, into a new matrix of observations $x'_{pj}(j=1$ to $n_1+n_2)$. Differentiating

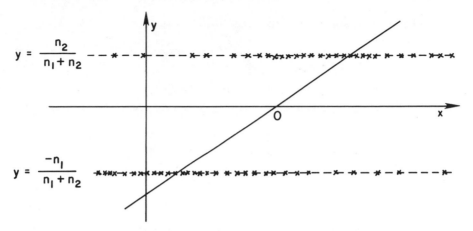

Figure 12–3. Illustration of Dichotomy of Population

equation 12.50 with respect to a_p and setting the expression equal to zero results in

$$\frac{\partial \Sigma \epsilon_j^2}{\partial a_p} = 2\sum_j [y_j - \sum_p a_p(x_{pj} - \bar{x}_p)] (x_{pj} - \bar{x}_p) = 0 \tag{12.51}$$

Equation 12.51 can be rearranged

$$\sum_j [y_j(x_{pj} - \bar{x}_p)] = \sum_j [(x_{pj} - \bar{x}_p)\sum_p a_p(x_{pj} - \bar{x}_p)] \tag{12.52}$$

For a particular a_p, say a_q, equation 12.52 should be written

$$\sum_j [y_j(x_{qj} - \bar{x}_q)] = \sum_j [(x_{qj} - \bar{x}_q)\sum_p a_p (x_{pj} - \bar{x}_p)] \tag{12.53}$$

It can then be shown that equation 12.53 may be solved for a_p.

$$a_p = \frac{n_1 n_2}{(n_1 + n_2)^2} [1 - \sum_p a_p(\bar{x}_{p1} - \bar{x}_{p2})] \sum_p C^{pq}(\bar{x}_{q1} - \bar{x}_{q2}) \tag{12.54}$$

The reduction of equation 12.53 to equation 12.54 involves a number of steps of manipulation, which are shown next. (The reader who is not interested in having equation 12.54 proved may proceed to the bottom of page 275.

Consider, first, the summation on the left of the equals sign in equation 12.53. By partitioning the data into the two subpopulations, this may be rewritten

$$\sum_j [y_j (x_{qj} - \bar{x}_q)] = \frac{n_2}{n_1 + n_2} \sum_{j=1}^{n_1} (x_{qj} - \bar{x}_q) + \frac{-n_1}{n_1 + n_2} \sum_{j=n_1+1}^{n_1+n_2} (x_{qj} - \bar{x}_q) \tag{12.55}$$

Carrying out the summation produces

$$\sum_j [y_j (x_{qj} - \bar{x}_q)] = \frac{n_2}{n_1 + n_2} [n_1 \bar{x}_{q1} - n_1 \bar{x}_q] - \frac{n_1}{n_1 + n_2} [n_2 \bar{x}_{q2} - n_2 \bar{x}_q] \tag{12.56}$$

Hence equation 12.55 can be simplified to

$$\sum_j [y_j(x_{qj}-\bar{x}_q)] = \frac{n_1 n_2}{n_1+n_2} (\bar{x}_{q1}-\bar{x}_{q2}) \qquad (12.57)$$

The right side of equation 12.53 can be written

$$\sum_j [(x_{qj}-\bar{x}_q)\sum_p a_p(x_{pj}-\bar{x}_p)] = \sum_p [a_p\sum_j (x_{pj}-\bar{x}_p)(x_{qj}-\bar{x}_q)] \qquad (12.58)$$

The summation over j may then be expanded, as shown in equation 12.59, by partitioning into the two subpopulations.

$$\sum_{j=1}^{n_1+n_2} (x_{qj}-\bar{x}_q)(x_{pj}-\bar{x}_p) =$$

$$\sum_{j=1}^{n_1} (x_{qj}-\bar{x}_{q1})(x_{pj}-\bar{x}_{p1})$$

$$+ \sum_{j=1+n_1}^{n_1+n_2} (x_{qj}-\bar{x}_{q2})(x_{pj}-\bar{x}_{p2})$$

$$+ \sum_{j=1}^{n_1} (\bar{x}_{q1}-\bar{x}_q)(\bar{x}_{p1}-\bar{x}_p)$$

$$+ \sum_{j=1+n_1}^{n_1+n_2} (\bar{x}_{q2}-\bar{x}_q)(\bar{x}_{p2}-\bar{x}_p) \qquad (12.59)$$

The first two terms are summations of cross products for each population. The covariance matrix was previously defined as $\mathbf{C}_{pq}$, so the first two terms of equation 12.59 can be written

$$\sum_{j=1}^{n_1} (x_{qj}-\bar{x}_{q1})(x_{pj}-\bar{x}_{p1}) + \sum_{j=1+n_1}^{n_1+n_2} (x_{qj}-\bar{x}_{q2})(x_{pj}-\bar{x}_{p2}) = (n_1+n_2)\mathbf{C}_{pq} \qquad (12.60)$$

The second half of the equation can be expanded using the definition of $\bar{x}_p$, from equation 12.47. In the first of the two terms, this definition results in

$$\sum_{j=1}^{n_1} (\bar{x}_{q1}-\bar{x}_q)(\bar{x}_{p1}-\bar{x}_p)$$

$$= \frac{n_1}{(n_1+n_2)^2} [(\bar{x}_{q1}n_1+\bar{x}_{q1}n_2-\bar{x}_{q1}n_1-\bar{x}_{q2}n_2)$$

$$(\bar{x}_{p1}n_1+\bar{x}_{p1}n_2-\bar{x}_{p1}n_1-\bar{x}_{p2}n_2)] \qquad (12.61)$$

Collecting the terms in equation 12.61 leads to

$$\sum_{j=1}^{n_1} (\bar{x}_{q1}-\bar{x}_q)(\bar{x}_{p1}-\bar{x}_p) = \frac{n_1 n_2^2}{(n_1+n_2)^2}(\bar{x}_{q1}-\bar{x}_{q2})(\bar{x}_{p1}-\bar{x}_{p2}) \qquad (12.62)$$

Equation 12.63, for the last term of equation 12.59, is developed through similar manipulations.

$$\sum_{j=1+n_1}^{n_1+n_2} (\bar{x}_{q2}-\bar{x}_q)(\bar{x}_{p2}-\bar{x}_p) = \frac{n_2 n_1^2}{(n_1+n_2)^2}(\bar{x}_{q2}-\bar{x}_{q1})(\bar{x}_{p2}-\bar{x}_{p1}) \quad (12.63)$$

Substituting all these expressions into equation 12.58 results in

$$\sum_j \{(x_{qj}-\bar{x}_q)\sum_p a_p(x_{pj}-\bar{x}_p)\} =$$
$$\sum_p a_p \Big\{(n_1+n_2)\mathbf{C}_{pq} + \frac{n_1 n_2}{(n_1+n_2)^2}[n_2(\bar{x}_{q1}-\bar{x}_{q2})(\bar{x}_{p1}-\bar{x}_{p2})$$
$$+ n_1(\bar{x}_{q2}-\bar{x}_{q1})(\bar{x}_{p2}-\bar{x}_{p1})]\Big\} \quad (12.64)$$

Gathering the terms within the brackets allows equation 12.64 to be simplified to

$$\sum_j \{(x_{qj}-\bar{x}_q)\sum_p a_p(x_{pj}-\bar{x}_p)\} = \sum_p a_p \Big\{(n_1+n_2)\mathbf{C}_{pq}$$
$$+ \frac{n_1 n_2}{(n_1+n_2)}(\bar{x}_{q1}-\bar{x}_{q2})(\bar{x}_{p1}-\bar{x}_{p2})\Big\} \quad (12.65)$$

Substituting equations 12.65 and 12.57 into equation 12.53 produces

$$\frac{n_1 n_2}{(n_1+n_2)}(\bar{x}_{q1}-\bar{x}_{q2}) = \sum_p a_p \Big\{(n_1+n_2)\mathbf{C}_{pq}$$
$$+ \frac{n_1 n_2}{n_1+n_2}(\bar{x}_{q1}-\bar{x}_{q2})(\bar{x}_{p1}-\bar{x}_{p2})\Big\} \quad (12.66)$$

Equation 12.66 can be rearranged to

$$\frac{n_1 n_2}{n_1+n_2}(\bar{x}_{q1}-\bar{x}_{q2})[1-\sum_p a_p(\bar{x}_{p1}-\bar{x}_{p2})] = (n_1+n_2)\sum_p a_p \mathbf{C}_{pq} \quad (12.67)$$

Hence equation 12.54 follows. Returning to the original solutions for the weighting coefficients, as given in equations 12.34 and 12.38, it can be seen that the original solution yields

$$a_p = \alpha \sum_p \mathbf{C}^{pq}(\bar{x}_{q1}-\bar{x}_{q2}) \quad (12.68)$$

Comparing equation 12.68 with equation 12.54 suggests the two are equivalent if α is

$$\alpha = \frac{n_1 n_2}{(n_1+n_2)^2}[1-\sum_p a_p(\bar{x}_{p1}-\bar{x}_{p2})] \quad (12.69)$$

It may also be noted that, as given by equation 12.4, the summation term in equation 12.69 is simply equal to $(\bar{z}_1-\bar{z}_2)$. Hence equation 12.69

can be written

$$\alpha = \frac{n_1 n_2}{(n_1+n_2)^2}[1-(\bar{z}_1-\bar{z}_2)] \tag{12.70}$$

It was already shown that the variance of y about its mean is given by equation 12.45. From the regression solution, it is also found that the regression and residual sum of squares are defined as

$$\text{SSR} = \frac{n_1 n_2}{n_1+n_2} \sum_{p=1}^{k} a_p(\bar{x}_{p1}\ \bar{x}_{p2}) \tag{12.71}$$

$$\text{SSE} = \frac{n_1 n_2}{n_1+n_2}[1-\Sigma a_p(\bar{x}_{p1}-\bar{x}_{p2})] \tag{12.72}$$

These expressions were determined from the analysis-of-variance table shown in table 12–1.

Significance Tests

Kendall[10] outlines the reasoning behind the fact that even though a dummy regression was performed the usual significance tests for a null hypothesis can still be applied to this set of mean squares. Thus the overall significance of the discrimination can be tested using the F test and F is

$$F = \frac{\text{MSR}}{\text{MSE}} = \frac{[\Sigma a_p(\bar{x}_{p1}-\bar{x}_{p2})](n_1+n_2-k-1)}{[1-\Sigma a_p(\bar{x}_{p1}-\bar{x}_{p2})]k} \tag{12.73}$$

where k = the number of coefficients in the discriminant function

Similarly, the coefficients may be tested using a two-tailed t test, in the standard format for a regression, where t is

$$t = \frac{a_q}{\text{s.e. of } a_q} \tag{12.74}$$

Table 12–1
Analysis of Variance Table

Source	Sum of Squares	d.f.
Regression	$\frac{n_1 n_2}{n_1+n_2}\Sigma_p a_p(x_{p1}-x_{p2})$	k
Residual	$\frac{n_1 n_2}{n_1+n_2}[1+\Sigma a_p(x_{p1}-x_{p2})]$	n_1+n_2-k-1
Total	$n_1 n_2/(n_1+n_2)$	n_1+n_2-1

where s.e. of a_q = standard error of a_q and is computed as

$$\text{s.e. } a_q = \sqrt{\text{MSE } C'^{qq}} \tag{12.75}$$

C'^{qq} is an element of the $\mathbf{C}'^{pq}$ matrix (the inverted common covariance matrix) that would be used in the regression analysis, not the inverted covariance matrix from the discriminant analysis. The difference between these two is that the regression matrix is built using the entire matrix of observations x_{pi}, while the discriminant matrix is the result of pooling the matrices for the separate populations.

The derivation and testing of the discriminant function depend on the assumptions of normality and homoscedasticity. It would be appropriate to develop means of testing whether these assumptions are valid in any particular case. Homoscedasticity can be tested by examining the covariance matrices of the two populations. It is suggested that corresponding elements of the matrices should not vary by more than a factor or two in more than 5% of cases. If this is true of the covariance matrix, the assumption of homoscedasticity would appear to be acceptable. (This test is somewhat arbitrary, and there are considerable grounds for discussion here. No clear test appears to have been devised.)

Normality can be tested for by examining the shape of the frequency distributions of each factor for the two populations, and also the shape of the final discriminant function distributions. The percentage of misclassifications can be compared with the theoretical misclassification. This would be done by calculating the area of overlap of two normal distributions of appropriate dimensions.

One may also wish to compare the discriminant functions obtained by using various sets of factors. A convenient measure is the correlation coefficient, R. However, it also has to be determined if the differences are significant. This could be tested using Fisher's transformation which is

$$z = 1/2[\ln(1+R) - \ln(1-R)] \tag{12.76}$$

It is not clear, though, whether this applies to discriminant functions. An alternative could be the "jackknife method."[11] This involves taking 10 different random sets of the calibration data, using 90% of the data at a time (that is, taking 10 different 90%s). The standard error of the values of R is then determined from the 10 sets. From this, multiples of the standard error can be used to yield significance levels. (Twice the standard error is exceeded only 5% of the time, 2.3 times only 1% of the time, etc.)

In general, the approach to discriminant analysis just discussed is open to question. The pseudoregression measures produced by the analogy frequently give very poor values. At least one reason for this is the fact that a regression on a dummy variable where there is a considerable spread of values of the independent variables must yield generally poor measures of fit. This apart, the validity of such regression procedures is open to question, as discussed in chapter 15. Unfortunately, no other basis has yet been put forward for testing the results of discriminant analysis.

Applications to Multiple Populations

Extensions of this theory are possible for the case where the analyst deals with more than two clearly defined populations. Initially, three populations are considered, but it will be readily apparent how this could be extended to m populations.

Obviously, the simplest extension is to define a separate discriminant function between each neighboring pair of populations (assuming that only two overlap). This can even be done if they overlap so that the tails of several distributions overlap. By direct extension of the earlier result a discriminator can be specified for each population region R_t such that, in the three-population case, R_1 is defined as

$$\sum_p \sum_q C^{pq} (\bar{x}_{p1} - \bar{x}_{p2})x_q \geqslant \beta_{12} \tag{12.77}$$

$$\sum_p \sum_q C^{pq} (\bar{x}_{p1} - \bar{x}_{p3})x_q \geqslant \beta_{13} \tag{12.78}$$

where β_{12} and β_{13} = constants

Similarly, R_2 is defined as

$$\sum_p \sum_q C^{pq}(\bar{x}_{p2} - \bar{x}_{p1})x_q \geqslant \beta_{21} \tag{12.79}$$

$$\sum_p \sum_q C^{pq}(\bar{x}_{p2} - \bar{x}_{p3})x_q \geqslant \beta_{23} \tag{12.80}$$

where β_{21} and β_{23} = constants

R_3 would be defined by two further boundary conditions, β_{31} and β_{32}. The problem is not considered here any further. More detailed discussions of this issue can be found in Kendall[12] and Rao.[13]

Probability of Classification

The next aspect of discriminant analysis to be discussed is the probability of an individual observation being classified in one or the other population. Strictly speaking, this notion is a violation of the original concepts of discriminant analysis. It has, however, received such widespread use that

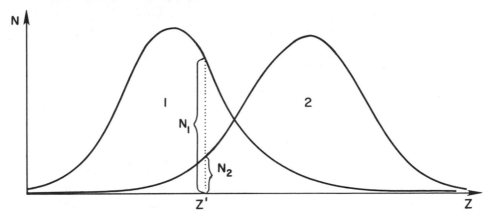

Figure 12–4. Probability of Classification

it would be a serious error of omission not to include it in the text. The reader should bear in mind that this is a very questionable usage of discriminant functions.

If the frequency distributions of values of z for members of both populations are considered, the best estimate of the probability of an observation being classified in population 2 with a given z value, z', is the ratio of the ordinate of the population 2 distribution to the sum of the ordinates of the two population distributions, as illustrated in figure 12–4.

If these two frequency distributions are represented as $F_1(z')$ and $F_2(z')$ (see figure 12–4) and if the probability of being classified in population 2 at z' is $p_2(z')$, this probability can be expressed

$$p_2(z') = \frac{F_2(z')}{F_1(z')+F_2(z')} = \frac{1}{1+F_1(z')/F_2(z')} \tag{12.81}$$

It was assumed that the two distributions are multivariate normal and homoscedastic. For convenience, $(\bar{z}_1-\bar{z}_2)$ is denoted as $2D$ and the origin is shifted to the point where it is midway between the two means, that is, Z equals $z + t$ where t causes this shift. If the variance of each distribution is σ^2, the two frequency distributions can be expressed as

$$F_1(z) = \frac{1}{\sigma\sqrt{2\pi}} \exp\left[-\frac{(Z-D)^2}{2\sigma^2}\right] \tag{12.82}$$

$$F_2(z) = \frac{1}{\sigma\sqrt{2\pi}} \exp\left[-\frac{(Z+D)^2}{2\sigma^2}\right] \tag{12.83}$$

Substituting equations 12.82 and 12.83 into equation 12.81 shows that

$p_2(z)$ is

$$p_2(z) = \frac{1}{1 + \exp\{-[(Z-D)^2 - (Z+D)^2]/2\sigma^2\}} \qquad (12.84)$$

Expanding the terms in the brackets allows equation 12.84 to be simplified to

$$p_2(z) = \frac{1}{1 + \exp(-2DZ/\sigma^2)} \qquad (12.85)$$

It can be shown that when the constant term in the equation for $a_q, \alpha,$ is set to unity, as done here, then G in Fisher's[14] maximization criterion used to derive this equation takes the value of the distance between the means, $2D$, which is equal to the variance.

$$\sigma^2 = 2D \qquad (12.86)$$

This is shown by considering equations 12.22 and 12.23. Substituting for α in equation 12.22 results in

$$\alpha = \frac{D}{2S} \qquad (12.87)$$

Equating α to unity produces

$$S = D/2 \qquad (12.88)$$

G was defined by equation 12.19. Substituting equation 12.88 into equation 12.19 produces

$$G = 2D \qquad (12.89)$$

Equation 12.85 can therefore be rewritten

$$p_2(z) = \frac{1}{1 + \exp(-Z)} \qquad (12.90)$$

Since Z was set equal to $(z+t)$, equation 12.90 can be rewritten in terms of the original discriminant function, z.

$$p_2(z) = \frac{1}{1 + \exp[-(z+t)]} \qquad (12.91)$$

Multiplying throughout by $\exp(z+t)/\exp(z+t)$ produces

$$p_2(z) = \frac{\exp(z+t)}{1 + \exp(z+t)} \qquad (12.92)$$

In the situation where there are different numbers of observations in each population, n_1 and n_2, as will usually be the case, the ordinate of

$F_2(z)$ will be inflated by n_2/n_1. This means that equation 12.83 should be written

$$F_2(z) = \frac{n_2}{n_1} \frac{1}{\sigma\sqrt{2\pi}} \exp[-(z+D)^2/2\sigma^2]$$ (12.93)

The expression for $F_1(z)$ will be as before. Therefore the probability of being classified in population 2 can be expressed

$$p_2(z) = \frac{1}{1 + \left(\dfrac{n_1}{n_2}\right) \exp[-2DZ/\sigma^2]}$$ (12.94)

By the same process used to convert equation 12.85 to equation 12.92, equation 12.94 may be rewritten

$$p_2(z) = \frac{(n_2/n_1)\exp(z+t)}{1 + \dfrac{n_2}{n_1} \exp(z+t)}$$ (12.95)

If it is intended to use results based on the analysis of a sample to predict probabilities of classification into one or the other population, the analyst can proceed in two ways.

1. One can find the value of t and predict probabilities directly from the above relationship. The value of t is the distance from the origin to the point midway between the means of the two distributions.

2. One can calculate the value of $F_2(z)/[F_1(z)+F_2(z)]$ from the actual frequency distribution of z yielded by the sample data for equal subranges of z values covering the area of the overlap of the distributions. These values are then used as probability estimates which can be fitted to a generalized version of the probability function.

$$p_2(z) = F_2(z)/[F_1(z) + F_2(z)] = \frac{\exp k(z+c)}{1+\exp k(z+c)}$$ (12.96)

This can be fitted by linear regression after the reduction of this expression to a linear format (see chapter 15). First, equation 12.96 may be written

$$p = \frac{\exp k(z+c)}{1+\exp k(z+c)}$$ (12.97)

Rearranging equation 12.97 produces

$$\frac{p}{(1-p)} = \exp k(z+c)$$ (12.98)

Taking logarithms of both sides of equation 12.98 produces the linear model

$$\ln(p/(1-p)) = k(z+c)$$ (12.99)

The closeness of the value of k to 1 and c to t in this regression is indicative of the closeness of the distributions to normality for the z range considered. This is an additional means of checking the validity of the analysis. It should also be noted that statistical goodness-of-fit measures for the above regression are only applicable to the model in the form of equation 12.99 and should not be interpreted as applying to equation 12.97.

An Example

The data for this example relate to travel to work in a large city (Chicago).[15] For 159 people, information was obtained on the costs of their travel to work, the time the trip took, mode of travel used, income, age and sex of the respondent, and information on the availability of a car for the work trip. The problem tackled is to see if a discriminant function can be used to separate transit and auto users. A total of 61 subjects used autos and 98 used transit.

Car availability was entered as a dummy variable, while income, sex, and age were entered as values. Time and cost were entered as differences between transit and auto. The results of the discriminant analysis are shown in table 12–2. For this model, the F score, with 7 and 151 degrees of freedom, was 8.08 which is significant beyond the 99% confidence level. The value of D^2 was 1.564. The total number of misclassifications was 37 out of the 159 respondents, where 11 car users were assigned to transit and 26 transit users to car, giving an overall misprediction of 15 too many car users and 15 too few transit users.

Examining the results depicted in table 12–2, it may be seen that all but two of the coefficients are significantly different from zero at a 95% confidence level. Car users are found to be associated with positive values

Table 12–2
Results of Discriminant Analysis on Chicago Data

Variable	Coefficient	t Score
Travel	.0330	2.48
Travel cost	.0050	3.15
Car availability 1	.0631	0.36
Car availability 2	−.5039	2.98
Income	−.8167	3.36
Sex	−.2713	1.81
Age	−.0203	2.39
Constant	1.179	—

of the discriminant function and transit users with negative values. It appears therefore that car users are associated with positive travel-time and travel-cost differences. For classification purposes, since these differences are expressed as transit minus auto times and costs, auto drivers are more likely, ceteris paribus, to be associated with shorter travel times and costs than for the transit trip. Similarly, transit users are likely, ceteris paribus, to be older and have higher incomes than car users and are more likely to be female (sex was entered as a 1,2 variable, such that 1 indicated male and 2 female). Transit users are also likely to have more favorable times and costs by their chosen mode than by car.

The interpretation of this example illustrates the conceptual difference between the correct use of discriminant analysis as a classification tool and the incorrect use of discriminant analysis as a probabilistic model. As the correct use, one may conclude that transit users are most likely to be female, older, with higher incomes, not to have a car available for the work trip, and to enjoy both cost and time advantages relative to car. Hence, given data on a new individual, one would classify that person as a transit or car user by evaluating their value of the discriminant function. This is the correct classification procedure and makes sense in those terms.

Interpreted as a probabilistic model of choice, the model would suggest—using the form of equation 12.92—that relatively little can be done to change mode choices since so much of the discriminant value is derived from socioeconomic variables. One would also conclude that little can be done to change auto users to transit users through cost changes, but time changes (which are much more difficult to achieve) might achieve something of a shift. The results of this model should be compared with those reported in chapters 14 and 15.

In summary, based on the value of D^2, F, and the number of misclassifications, one may conclude that auto users and transit users can be discriminated between relatively well by using this fitted discriminant function. Further, the discriminant function seems to make reasonable, intuitive sense.

It is instructive to see how discriminant analysis is used as a classification procedure. Suppose data are obtained on three additional people, as shown in table 12–3. (Note that age and income were coded as normalized variables.) The values of the discriminant function for each of these individuals are also shown in table 12–3. The discriminant function was defined in table 12–2 so that zero represents the boundary between the two populations, with negative values indicating a transit user and positive values a car user. Hence the discriminant function suggests that individual 1 is a car user, while individuals 2 and 3 are transit users. This is intuitively satisfying. The time and cost differences for individual 1 show

Table 12–3
Additional Commuters to Be Classified

Individual	Time Differ- ence	Cost Differ- ence	Car Avail- ability 1	Car Avail- ability 2	Income	Sex	Age	Z
1	10	5	0	1	0.5	1	0.5	0.340
2	−20	−10	0	0	0.95	2	3.2	−0.914
3	−15	25	1	0	0.76	2	2.2	−0.336

that for this person the car is quicker and cheaper. Hence the car is a logical choice and the classification is reasonable. Similarly, individual 2 has time and cost advantages on transit and is classified as a transit user. Individual 3 has a time advantage by transit and a cost disadvantage. While the cost disadvantage is larger, the relative coefficient sizes of the two variables mean the travel time advantage has far more effect on the classification of this person than the cost disadvantage. Transit classification is therefore intuitively right.

Notes

1. J. Neyman and E.S. Pearson, "On the Problem of the Most Efficient Test in Statistical Hypothesis," *Philosophical Transactions Acta,* vol. 231, pp. 289–337.

2. Ronald A. Fisher, "The Use of Multiple Measurements in Taxonomic Problems," *Annals of Eugenics,* 1936, vol. 7, no. 2, pp. 179–188.

3. Neyman and Pearson, "Statistical Hypothesis."

4. Fisher, "Multiple Measurements."

5. Maurice G. Kendall, *A Course in Multivariate Analysis*, London: Charles Griffin and Co., 1965, pp. 144–,8.

6. Ibid.

7. Ibid.

8. Ibid.

9. M.S. Bartlett, "The Standard Errors of Discriminant Function Coefficients," *Journal of the Royal Statistical Society,* 1939, vol. 6, p. 169.

10. Kendall, *Multivariate Analysis,* pp. 161–162.

11. David A. Quarmby, "Choice of Travel Mode for the Journey to Work: Some Findings," *Journal of Transport Economics and Policy,* 1967, vol. 1, no. 3, pp. 301–314.

12. Maurice G. Kendall, *Advanced Theory of Statistics,* vol. 2, London: Charles Griffin and Co., 1946.

13. C.R. Rao, *Advanced Statistical Methods in Biometric Research,* New York: John Wiley & Sons, 1952.

14. Fisher, "Multiple Measurements."

15. Thomas E. Lisco, "The Value of Commuter's Travel Time: A Study in Urban Transportation," unpublished Ph.D. dissertation, Department of Economics, University of Chicago, 1967.

13 Maximum-Likelihood Estimation

The Concept of Likelihood

The concept of likelihood is derived from the ideas of probability and experimental outcomes.[1] One could say that the outcome of a particular experiment may have different probabilities based on certain assumptions about the process that has been experimented upon. In simple terms, consider an experiment to measure the level of carbon monoxide in an urban area. Suppose the experiment is conducted by measuring the carbon monoxide level at a number of specific points selected randomly in the urban area. Because of local variations in concentration, dispersion, proximity of sources, and so forth, one will obtain a range of different values from these different points. Before or after undertaking the experiment, one may postulate a general level of carbon monoxide in the urban area. For example, the experiment may have generated the data of table 13–1.

One may postulate that the average level of carbon monoxide in the area is 20 parts per million (ppm). Now one would like to know how *likely* it is that these values could have been obtained if the average urban level is 20 ppm. This can be determined by calculating the likelihood of the values of table 13–1, given the assumption the average level is 20 ppm. One may also calculate the *likelihood* of these values, given some other assumed average level, say, 25 ppm. As is seen later, additional assumptions must be made to estimate the value of the likelihood. For the moment, however, it is appropriate to consider further the meaning of likelihood.

For any assumed average level of carbon monoxide, it is possible to define a probability that each of the values in table 13–1 could have occurred. The likelihood is then defined as the joint probability under the assumed average value. Suppose the probability of measuring 26 ppm at location 1 under the assumed value of 20 ppm is denoted $p_{(1|20)}$, the value of 13 ppm at location 2 denoted $p_{(2|20)}$, and so forth. Then the likelihood of the values of table 13–1 is

$$L(\overline{CO}=20) = p_{(1|20)}p_{(2|20)}p_{(3|20)} \cdots p_{(7|20)} \tag{13.1}$$

where $L(\overline{CO}=20)$ = likelihood of the measurements being obtained if the average level of carbon monoxide is 20 ppm

Similarly, equation 13.2 defines the likelihood for the assumption of a mean level of 25 ppm.

$$L(\overline{CO}=25) = p_{(1|25)}p_{(2|25)}p_{(3|25)} \cdots p_{(7|25)} \tag{13.2}$$

287

Table 13–1
Observations on Carbon Monoxide Level
(parts per million)

Location	Carbon Monoxide Level
1	26
2	13
3	18
4	32
5	23
6	25
7	17

One might wish now to consider which of the assumptions is more likely: an average level of 20 ppm or 25 ppm. The more likely value is clearly going to be the one that yields the higher value of the likelihood. This notion leads to the extension of the concept of likelihood to *hypothesis testing*.

To test a hypothesis, one may calculate the likelihood under the hypothesis and under some alternative hypothesis. Clearly, if the likelihood under the first hypothesis is larger, one would be inclined to accept that hypothesis in preference to the alternative hypothesis. This notion leads to several very important considerations.

First, it is clearly of considerable importance to construct two hypotheses for such tests to determine which hypothesis is the more likely and hence the more acceptable. Second, the choice of the alternative hypothesis is as important as that of the principal hypothesis. It is clear that if, in the example, one chose the alternative hypothesis as 200 ppm against the principal hypothesis of 20 ppm, the principal hypothesis would be accepted. However, if one chose 21 or 19 ppm, it is no longer clear that the principal hypothesis would be accepted. The test is now more stringent.

Third, it is not sufficient simply to determine which of the two likelihoods is larger. One may raise the question of how much larger one likelihood must be to consider that the difference could not have been caused by chance. This is, of course, the principle underlying most statistical hypothesis testing, much of which has been utilized in earlier portions of the book. It is worthwhile exploring these considerations in more detail here since they lead to the notion of maximum-likelihood estimation and its accompanying statistical tests, which are the concern of subsequent chapters.

Calculating Likelihoods

Before developing the notions of likelihood and hypothesis testing further, it is worthwhile to see how one may calculate likelihoods. Return to the data of table 13–1. It was noted that additional assumptions must be made to compute the likelihood. Essentially, it is necessary to assume there is some underlying distribution of values of carbon monoxide, from which the samples of table 13–1 were drawn. Suppose the distribution is assumed to be normal with a standard deviation of 6 ppm. Then the probability of obtaining one measurement, x, of carbon monoxide is

$$p_{(1,\bar{x})} = \frac{1}{\sigma\sqrt{2\pi}} \exp\left[-\frac{(x_1-\bar{x})^2}{2\sigma^2}\right] \tag{13.3}$$

where σ = standard deviation of the distribution
and $\bar{x}$ = a hypothesized mean value of carbon monoxide

Assuming the value of $\bar{x}$ to be 20 ppm, σ to be 6 ppm, and the values of table 13–1, the likelihood can be estimated from

$$L(x_1, x_2, x_3, \ldots, x_7 \,|\, \bar{x}=20)$$

$$= \frac{1}{\sigma\sqrt{2\pi}} \exp\left[-\frac{(x_1-\bar{x})^2}{2\sigma^2}\right] \frac{1}{\sigma\sqrt{2\pi}} \exp\left[-\frac{(x_2-\bar{x})^2}{2\sigma^2}\right]$$

$$\cdots \frac{1}{\sigma\sqrt{2\pi}} \exp\left[-\frac{(x_7-\bar{x})^2}{2\sigma^2}\right] \tag{13.4}$$

By gathering the terms, this may be simplified to

$$L(x_1, x_2, x_3, \ldots, x_7 \,|\, \bar{x}=20) = \left[\frac{1}{\sigma\sqrt{2\pi}}\right]^7 \exp\left[-\frac{1}{2\sigma^2}\sum_{i=1}^{7}(x_i-\bar{x})^2\right] \tag{13.5}$$

Using the values listed above, the value of equation 13.5 is found to be 1.2435×10^{-10}. In a similar manner, one can calculate the likelihood under the assumption that $\bar{x}$ is 25 ppm, rather than 20 ppm. This yields an estimate of 7.2322×10^{-11} which is smaller than the previous estimate. So one may conclude that it is less likely that the values of table 13–1 would have been obtained if the mean value was 25 ppm than if it was 20 ppm. The unresolved question here is whether the difference in the two values is significant. It is also important to note that, in this example, if one rejects the hypothesis of a mean value of 25 ppm, one accepts (or does not reject) the hypothesis of 20 ppm. However, this hypothesis may still be wrong, simply because both hypotheses chosen were wrong. This issue is resolved in the next section.

Two further points are worth noting here. It has been shown that one additional assumption is needed to compute the likelihood, that is, the distribution of the phenomenon being measured. Second, the likelihood is shown to be a very small value. This follows since the likelihood is a joint probability obtained by multiplying together all the individual probabilities for each observation. Since probabilities must lie between zero and one, it follows that the likelihood will be very small, decreasing in size as the sample size increases.

Choosing Hypotheses

It is useful to observe, from the previous example, that the maximum value the likelihood could take under the first hypothesis ($\bar{x}=20$ ppm) would occur when the measurements obtained were all equal to 20 ppm. In other words, this would occur if all the entries in table 13–1 were 20. This follows from the general expression for the likelihood.

$$L(x_1,x_2,x_3,\ldots,x_n\,|\,\bar{x}=\bar{x}_0)= \left[\frac{1}{\sigma\sqrt{2\pi}}\right]^n \exp\left[-\frac{1}{2\sigma^2}\sum_{i=1}^{n}(x_i-\bar{x}_0)^2\right] \quad (13.6)$$

In the example, this would give a value of 5.7453×10^{-9}. Having calculated this value, three pieces of information have been obtained:

$$H_0:\bar{x} = 20 \qquad L_0 = 12.435 \times 10^{-11}$$
$$H_1:\bar{x} = 25 \qquad L_1 = 7.2322 \times 10^{-11}$$
$$H^*:x_i = 20 \qquad i = 1,2, \ldots n, \bar{x} = 20;$$
$$L^* = 574.53 \times 10^{-11}$$

Based on these pieces of information, it might be possible to develop a decision rule that says if L_0 differs from L^* by more than a certain amount, H_0 (hypothesis nought) should be rejected. However, it would normally be the case if H_0 is rejected, H_1 is accepted. This would clearly be wrong, since H_1 yields a lower likelihood value than H_0.

Two Types of Errors

From this example, it is possible to see that hypothesis testing based on likelihood is subject to two types of errors. The first type, also referred to as type I error, is rejecting a hypothesis when it is true. Type II error is incurred by accepting a hypothesis when it is not true. In any situation, there is a finite probability of incurring either error. These probabilities are

$$p(\text{type I}) = \alpha = p(\text{reject } H_0\,|\,H=H_0) \qquad (13.7)$$

$$p(\text{type II}) = \beta = p(\text{accept } H_0\,|\,H\neq H_0) \qquad (13.8)$$

The two types of error occur in many situations. An interesting illustration is found in the law courts based on the English system of law. A defendant is presumed innocent unless proved guilty beyond a reasonable doubt. Type I error arises when the defendant is found guilty though in fact innocent. Similarly, type II error arises when the defendant is found innocent though in fact guilty. Under the system of law, an attempt is made to stay as far away as possible from punishing an innocent person, that is, to minimize the probability of making a type I error. This must clearly be done at the expense of incurring a greater probability of a type II error. This illustration also makes it clear that as the probability of one type of error is reduced, that of the other must increase. In statistics, the only way to improve the situation is to increase the sample size, a procedure that simultaneously reduces the probability of both errors. For a given sample size, however, it still holds that reducing α increases β and vice versa.

As is seen later in the chapter, statistics uses the reverse of the law courts procedure. In the law courts, one attempts to minimize α for given β. In statistics, one attempts to minimize β for given α. The reason for this is that it is easier to calculate α than β. Thus one can set up the necessary rule to minimize β, but can calculate the value of α.

Definition of a Rejection Region

It is crucial to define an appropriate criterion for rejection. This has been done formally by Neyman and Pearson.[2] The formal proofs and theorems are not developed here. Instead, consideration is given to the concepts and important results stemming from the theory.

Suppose the likelihood test is to be conducted on some parameter or parameter set, denoted θ. Suppose further there is some parameter space that can be defined to include all the possible values of θ, where this parameter space is denoted Ω. For example, if θ was a variance of a set of observations, Ω is all positive values from zero to infinity. If θ was a mean of a set of observed carbon monoxide levels, Ω might be all nonnegative values.

Two types of hypothesis tests may be conducted. The first is called a *simple hypothesis,* which involves testing one point in Ω. This might be stated as follows:

$$H_0 : \theta = \theta_0$$

$$H_1 : \theta \text{ in } \Omega \neq \theta_0$$

The second is a *composite hypothesis*. This involves testing a set of points in the parameter space, Ω. Suppose one defines a region of the parameter space as ω; then the hypothesis might be that θ lies in ω, as follows:

$$H_0 : \theta \text{ in } \omega$$

$$H_1 : \theta \text{ in } \Omega - \omega$$

These two examples of hypotheses have introduced a new and important concept. The alternative hypothesis, H_1, has been defined in both cases to represent all possible values of the parameter, θ, not included in H_0. This resolves the earlier dilemma of how to choose the alternative hypothesis, except that a new problem has been generated of how to estimate likelihoods for H_1. There is still an unresolved problem of how to decide when L_0 is sufficiently much larger than L_1 to accept H_0 and reject H_1 or when L_0 is sufficiently close to or smaller than L_1 that one rejects H_0 and accepts H_1.

The Neyman-Pearson theory suggests that an appropriate strategy is to minimize the type II error, β, subject to a given level of type I error, α. Under H_0, α is the probability that the set of observations lies in the rejection region when H_0 is in fact true. (Note that H_0 is usually termed the null hypothesis.) To minimize the type II error for H_0 is equivalent to maximizing the probability of rejecting H_1, when H_0 and H_1 have been set up as discussed previously. Mathematically, this is given by

$$\max p\left[(x_1, x_2, x_3, \ldots, x_n) \text{ in } R \,|\, H_1\right] \qquad (13.9)$$

where $R =$ the rejection region

Equation 13.9 is also equal to minimizing the probability that the values $(x_1, x_2, x_3, \ldots, x_n)$ are in the acceptance region for H_1. The type II error is that of accepting H_0 when it is not true, that is, when H_1 is true. Hence, equation 13.9 defines the type II error for H_0, the null hypothesis. The rule put forward by Neyman and Pearson is to minimize β, subject to a given value of α. This may be written

$$\max p\left[(x_1, x_2, x_3, \ldots, x_n) \text{ in } R \,|\, H_1\right] = \beta$$

$$\text{subject to } p\left[(x_1, x_2, x_3, \ldots, x_n) \text{ in } R \,|\, H_0\right] \leqslant \alpha \qquad (13.10)$$

It can then be shown[3] that this is equivalent to defining the rejection region in terms of the ratio of the likelihoods under H_0 and H_1, such that the ratio must be less than or equal to some chosen value, k, which is less than one and is determined by selecting the value of α;

$$\text{reject } H_0 \text{ when } \frac{L_0}{L_1} \leqslant k \qquad (13.11)$$

where $L_0 =$ likelihood of the observations
$\qquad (x_1, x_2, x_3, \ldots, x_n)$ under H_0
and $\quad L_1 =$ likelihood of the observations
$\qquad (x_1, x_2, x_3, \ldots, x_n)$ under H_1

In the situation defined by the earlier example, using carbon-monoxide observations, it is relatively simple to apply this procedure. If H_0 were that $\bar{x} = 20$ ppm and H_1 that $\bar{x} = 25$ ppm, the two likelihoods and hence their ratios can be calculated readily. However, this situation assumes a simple hypothesis where $\Omega \neq \theta_0$ is defined by a single point. This would not normally be the case. Assuming that $\Omega \neq \theta_0$ is a set of possible values, the question is: which one should be used to calculate the likelihood, L_1? Intuitively, it would seem reasonable to find the maximum value of the likelihood in the region defined by $\Omega \neq \theta_0$. Thus one would compare the likelihood for θ_0 with the maximum that can be determined for any other value of θ than θ_0. The same principle may be applied when a composite hypothesis is to be tested. In this case, the likelihoods would each be chosen as the maxima under the respective hypotheses, H_0 and H_1.

$$\text{reject } H_0 \text{ if } \frac{\max L_0 \ (\theta \text{ in } \omega)}{\max L_1 \ (\theta \text{ in } \Omega-\omega)} \leq k \qquad (13.12)$$

Suppose H_0 is true. Then the maximum value of L_1 will be less than the maximum value of L_0 with a probability of $(100-\alpha)\%$. It has already been specified that k must be less than one. Under the condition that max L_0 is greater than max L_1, H_0 cannot be rejected for any value of k less than one. Therefore, H_0 can only be rejected with a probability less than α, as desired. It will also follow, in this case, that the ratio of max L_0 to the maximum likelihood in the entire parameter space, Ω, will be equal to one with a probability of $(100-\alpha)\%$. Therefore, if k is less than one, equation 13.12 is equivalent to

$$\text{reject } H_0 \text{ if } \lambda = \frac{\max L_0 \ (\theta \text{ in } \omega)}{\max L \ (\theta \text{ in } \Omega)} \leq k \qquad (13.13)$$

Now if H_0 is not true, there is a probability of $(100-\alpha)\%$ that the maximum value of L_0 (θ in ω) will be less than the maximum value of L_1 (θ in $\Omega-\omega$). Hence, under this condition, equations 13.12 and 13.13 are still equivalent. Since it is much easier to calculate the maxima of equation 13.13 than the maxima of equation 13.12, equation 13.13 represents the most desirable version of the *likelihood-ratio test*. It should be noted also that this final form of the test has removed the need to specify a rejection region. Thus what had seemed to be the major stumbling block—how to define an appropriate rejection region—has been eliminated entirely. It only remains to calculate values of k for various levels of α to be able to conduct the likelihood-ratio test. This is done by making certain assumptions about the shape (that is, distribution) of the likelihood function in the parameter space, Ω.

The use of likelihood ratios in hypothesis testing is the basis of statistical tests, such as the t test and the F test, in which likelihood ratios are set up for the specific tests to be done based on specified distributional assumptions. The mathematical expressions which result from these assumptions form the basis for the derivation of the values of statistics, such as t, F, χ^2, etc.

Maximum-Likelihood Estimation

The preceding sections defined the concept of likelihood and developed the notions of likelihood-ratio tests and the behavior of the likelihood function. These ideas form the basis for maximum-likelihood estimation. Consider the numerator of equation 13.13, shown here in expanded form as

$$\max L[(x_1, x_2, x_3, \ldots, x_n) \,|\, \theta \text{ in } \omega] \tag{13.14}$$

This equation suggests that if H_0 were true, θ determined at the maximum of the likelihood in ω is the best estimate for the observations $(x_1, x_2, x_3, \ldots, x_n)$. Now if H_0 is true, θ determined at the maximum of the likelihood in Ω is the same as θ determined at the maximum of the likelihood in ω. Therefore the best estimate of θ is obtained by maximizing equation 13.15.

$$\max L[(x_1, x_2, x_3, \ldots, x_n) \,|\, \theta \text{ in } \Omega] \tag{13.15}$$

To find the maximum value of the likelihood, the standard procedure would be to differentiate the likelihood with respect to θ and equate this to zero.

$$\frac{dL[x_1, x_2, x_3, \ldots, x_n) \,|\, \theta]}{d\theta} = 0 \tag{13.16}$$

(It is, of course, important to ascertain that the solution of this is a maximum and not a minimum or a point of inflection.)

It is pertinent to ask, at this point, why maximum-likelihood estimates are needed for parameter values. To see this, it is necessary to consider the desired properties of parameter estimates. In general, estimates of parameters should have four properties: consistency, lack of bias, efficiency, and sufficiency.

A *consistent estimator* has an accuracy which increases with the size of the sample used to compute the estimate. Thus an infinite sample would yield the exact value of the parameter. An *unbiased estimator* has a sampling distribution with a mean of the true value of the parameter. In other words, if several samples are taken and estimates made of the parameter from each separate sample, the mean of the estimates will be the true value, or not statistically significantly different from the true value, of the parameter if the estimates are all unbiased. An *efficient estimator* exhibits a small variance for the sampling distribution of the unbiased estimator. In fact, the efficient estimator has the smallest variance of any estimator. A *sufficient estimator* is somewhat more complex and cannot be adequately described in a few words.[4] Briefly, a sufficient estimator uses all the available information from a sample for the estimation of some parameter, θ. If a sufficient estimator exists, no nonsufficient estimators need be considered.

Thus the properties of estimators may be summarized as involving a tendency to exact estimation as sample size increases, lack of bias in the estimation, minimizing the sampling distribution variance, and using all the information provided by a sample.

Maximum-likelihood estimates are consistent, efficient, and sufficient (if sufficient estimates exist), but not always unbiased.[5] Hence maximum-likelihood estimates fulfill three of the four properties of estimates listed above; the fourth—lack of bias—is usually approximated since the bias decreases rapidly with increasing sample size.

Maximum-likelihood estimates possess another property which has important implications for testing the adequacy of the estimates. This property is that the distribution of a maximum-likelihood estimate (MLE), $\hat{\theta}$, approaches that of a normal distribution with a mean of θ (where θ is the true value) and a variance given by equation 13.17 as the sample size approaches infinity.

$$\sigma_{\hat{\theta}}^2 = E[(\partial^2 \ln L/\partial\theta^2)]^{-1} \qquad (13.17)$$

This is known as the *asymptotic normality property*.[6]

To recap, the MLE of a parameter, θ, is one particular estimator that has a number of desirable properties and may frequently be the best estimator that can be found. Put in colloquial terms, the MLE of a parameter is the most likely value, given the available sample data. As the sample size increases, the most likely value becomes ever closer to the true value of the parameter. A few simple examples will illustrate maximum-likelihood estimation.

Example 1

Suppose N independent observations have been drawn from a normal distribution, with mean $\bar{x}$ and variance σ^2. The problem is to estimate the values of $\bar{x}$ and σ^2 from the sample, using maximum-likelihood procedures. The likelihood of obtaining the observations $(x_1, x_2, x_3, \ldots, x_N)$ if $\bar{x}$ and σ^2 are the values of the mean and variance, respectively, is shown by

$$L(x_1, x_2, \ldots, x_N) \mid \bar{x}, \sigma^2) = \left[\frac{1}{\sigma\sqrt{2\pi}}\right]^N \exp\left[-\frac{1}{2\sigma^2}\sum_{i=1}^{N}(x_i - \bar{x})^2\right] \quad (13.18)$$

Though $\partial L/\partial \bar{x}$ and $\partial L/\partial \sigma^2$ can be estimated readily, it is computation-ally simpler to estimate $\partial \ln L/\partial \bar{x}$ and $\partial \ln L/\partial \sigma^2$, which are identical to the former when the partial derivatives are equated to zero. The log of the likelihood is

$$\ln L(x_1, x_2, \ldots, x_N \mid \bar{x}, \sigma^2) = N \ln \frac{1}{\sigma \sqrt{2\pi}} - \frac{1}{2\sigma^2} \Sigma (x_i - \bar{x})^2 \quad (13.19)$$

Taking the partial derivative of equation 13.19 with respect to $\bar{x}$ yields

$$\frac{\partial \ln L}{\partial \bar{x}} = \frac{2}{2\sigma^2} \Sigma (x_i - \bar{x}) = 0 \quad (13.20)$$

Similarly, the partial derivative with respect to σ^2 is

$$\frac{\partial \ln L}{\partial \sigma^2} = \frac{1}{2} N\sigma \sqrt{2\pi} \frac{1}{\sqrt{2\pi}} - \frac{1}{2} (-\sigma^{-4}) \Sigma (x_i - \bar{x})^2 \quad (13.21)$$

Setting equation 13.21 equal to zero yields

$$\frac{N}{2\sigma^2} + \frac{1}{2\sigma^4} \Sigma (x_i - \bar{x})^2 = 0 \quad (13.22)$$

Rearranging equations 13.20 and 13.22 yields

$$\frac{1}{\sigma^2} \Sigma (x_i - \bar{x}) = 0 \quad (13.23)$$

$$\frac{1}{2\sigma^4} \Sigma (x_i - \bar{x})^2 - \frac{N}{2\sigma^2} = 0 \quad (13.24)$$

These equations can be simplified further, first by rejecting the trivial solution that $1/\sigma^2$ is zero. One may divide equation 13.23 by $1/\sigma^2$ and equation 13.24 by $1/2\sigma^4$ to yield

$$\Sigma x_i - \Sigma \bar{x} = 0 \quad (13.25)$$

$$\Sigma (x_i - \bar{x})^2 - N\sigma^2 = 0 \quad (13.26)$$

Again rejecting the trivial solutions of equations 13.25 and 13.26 gives the maximum-likelihood estimates of $\bar{x}$ and σ^2, $\hat{\bar{x}}$ and $\hat{\sigma}^2$.

$$\hat{\bar{x}} = \frac{1}{N} \Sigma x_i \quad (13.27)$$

$$\hat{\sigma}^2 = \frac{1}{N} \Sigma (x_i - \bar{x})^2 \quad (13.28)$$

These are familiar estimating equations and demonstrate that the estimates of $\hat{\bar{x}}$ and $\hat{\sigma}^2$ obtained are maximum-likelihood estimators of the mean and variance. It is in fact true that the estimate $\hat{\sigma}^2$ is slightly biased for small N so the unbiased estimator used for small samples is

$$\hat{\sigma}^2 = \frac{1}{(N-1)} \sum (x_i - \bar{x})^2 \tag{13.29}$$

Example 2

This example is drawn from regression analysis. Suppose, as hypothesized, the error terms ϵ_i are independent and drawn from a normal distribution. The likelihood function for the Y_is, Y_1, Y_2, Y_3,..., Y_n is

$$L(Y_1, Y_2, Y_3, ..., Y_N | \epsilon_i) = \left[\frac{1}{\sigma\sqrt{2\pi}} \right]^n \exp\left[-\frac{1}{2\sigma^2} \sum \epsilon_i^2 \right] \tag{13.30}$$

This follows because ϵ_i is assumed to have a zero mean. Taking logs as before, the log likelihood is

$$\ln L = n \ln\left[\frac{1}{\sigma\sqrt{2\pi}} \right] - \frac{1}{2\sigma^2} \sum \epsilon_i^2 \tag{13.31}$$

For a bivariate regression between Y and X, $\sum \epsilon_i^2$ is

$$\sum \epsilon_i^2 = \sum (Y_i - a_1 X_i - a_0)^2 \tag{13.32}$$

Substituting equation 13.32 in equation 13.31 yields

$$\ln L = n \ln\left[\frac{1}{\sigma\sqrt{2\pi}} \right] - \frac{1}{2\sigma^2} \sum (Y_i - a_1 X_i - a_0)^2 \tag{13.33}$$

To obtain maximum-likelihood estimates of a_1 and a_0, the log likelihood is differentiated with respect to each parameter, and the differentials are equated to zero.

$$\frac{\partial \ln L}{\partial a_0} = -\frac{1}{2\sigma^2} 2\sum [-(Y_i - a_1 X_i - a_0)] = 0 \tag{13.34}$$

$$\frac{\partial \ln L}{\partial a_1} = -\frac{1}{2\sigma^2} 2\sum [(-X_i)(Y_i - a_1 X_i - a_0)] = 0 \tag{13.35}$$

Assuming that σ^2 is not equal to zero, these can be solved as shown by

$$\sum (Y_i - a_1 X_i - a_0) = 0 \tag{13.36}$$

$$\sum X_i (Y_i - a_1 X_i - a_0) = 0 \tag{13.37}$$

These are the identical estimating equations to those derived on the basis of least squares. Hence provided that ϵ_i is a random normal variable with mean zero and variance σ^2, the least-squares estimators are the maximum-likelihood estimators. This is one of the reasons why least squares is used as the basis of estimation rather than, say, minimizing $|\Sigma\epsilon_i|$.

In each of the above examples, the second derivatives should also have been found to ensure that a maximum value of the log of the likelihood had in fact been found. The reader can satisfy herself or himself rather easily that a maximum was found in each case.

In these examples the likelihood functions are all linear in the unknown parameters, so the solution for the maximum is obtained easily by a simple algebraic solution. This will not, however, always be the case. The likelihood functions for some parameters may be nonlinear in the unknowns, thus requiring some form of iterative search procedure to find the maximum. In such situations, it is essential to check the second partial derivatives to determine if one is climbing toward the maximum, away from it, or has landed on a point of inflection.

Finally, it is important to note that the maximum-likelihood estimator of a parameter is only one possible estimator. There are in fact a number of other possible estimators that may fulfill more or less of the desired properties of estimators. It must be remembered that MLEs are not always unbiased, so other estimators may be desirable when an MLE is suspected of being seriously biased, a situation which is more likely to occur with very small samples.

Notes

1. The concept was originally introduced by Gauss and developed by R.A. Fisher. See Sylvain Ehrenfeld and Sebastian Littauer, *Introduction to Statistical Method,* New York: McGraw-Hill Book Co., 1964, p. 253.

2. Ehrenfeld and Littauer, *Statistical Method,* pp. 256–269.

3. Ibid, pp. 259–261.

4. Ibid, pp. 345–357.

5. Ibid, pp. 359–362.

6. Ibid, p. 362.

14 Probit Analysis

The Concept

As discussed in chapter 12, there are many instances in engineering and the social sciences where a simple linear model is not appropriate for the situation. Discriminant analysis provides a method for classification, which may have a number of potential uses. Another model of interest provides estimates of a limited dependent variable, such as a probability. A probability must have a value that lies between zero and one. Clearly, a linear model cannot be used unless all the independent variables are limited in their values and the combined function of the independent variables is limited to values between one and zero. Such a situation is generally unlikely. Suppose one has a situation in which one wishes to estimate the probability of some occurrence as a function of a set of independent variables that may take values over a virtually unlimited range. Suppose further one cannot observe probabilities, but can only observe whether or not an event occurs. In this situation, a model process is needed that will permit calibration of a model for estimating a limited dependent variable from observations on the occurrence of an event. Two such processes are probit analysis, described in this chapter, and logit analysis, described in chapter 15.

Some examples of the type of situation envisaged here might be useful. In civil engineering, for example, one may wish to express the probability of the collapse of a concrete slab as a function of certain parameters describing the slab (dimensions, mix of concrete materials, curing time and temperature) and of parameters of the loads carried (tension, compression, torsion). For any slab, observations can be made only on whether or not it collapses under certain conditions. The conditions may be repeated on a number of slabs and observation made of how many collapse or they may be varied over a number of slabs and observations made on which slabs collapse.

Similarly, in economics, one may wish to model the probability of a household purchasing a consumer durable, such as a refrigerator. The purchase may be expressed as a function of the real disposable household income, the age and size of the existing refrigerator, the number of cars owned, the type of dwelling, and the size of the family. Again, a probability of purchase cannot be observed. Rather, one can observe a household as its characteristics change and observe when a new refrigerator is bought, or one may observe a number of households with various characteristics and note which makes a purchase within a specified time.

In both cases, a model is desired that can provide estimates of probabilities of an event occurring as a function of continuous and unlimited independent variables (parameters of the slab and its load or parameters of the household and its current refrigerator) and will produce this model, not from observations of the probabilities, but from observations of the occurrence or nonoccurrence of the event.

Probit analysis, like discriminant analysis, is a modeling technique with foundations in bioassay and was developed by D.J. Finney.[1] It arose as a statistical technique for dealing with the situation in which one wishes to attempt to place a mathematical explanation on the response of a subject to some level of stimulus. Much of this work is associated with toxicology, as is the nomenclature.

Consider the responses of a number of subjects to a given stimulus. A frequency distribution of their responses can be drawn if they respond by showing varying response levels. Or if the response is a quantal one, the analyst could consider the size of the stimulus necessary to produce a given response and construct an analogous distribution of the frequency of response. This might appear as shown in figure 14–1.

It can be shown that with most skew distributions like this one, a transformation of the independent variable will usually change the distribution to one that closely approximates a normal distribution. Finney[2] shows that in many cases in bioassay, the transformation required is

$$x_i = \ln(\lambda_i) \tag{14.1}$$

where λ_i = the value of the original stimulus
and x_i = the transformed stimulus.

This gives the familiar normal distribution shown in figure 14–2.

These distributions represent the number of subjects at each given value of X who respond in a certain specified way to the stimulus. One

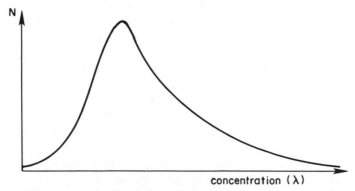

Figure 14–1. A Typical Stimulus-Response Curve

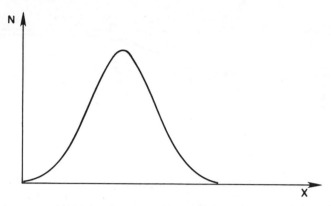

Figure 14–2. A Logarithmic Transform of the Stimulus-Response Curve

may also be interested in the percentage who have responded at any given stimulus level. This can be done by plotting a cumulative curve. The skew distribution results in a sigmoid curve of the form shown in figure 14–3. Similarly, the normalized distribution of figure 14–2 would result in the normal sigmoid curve when cumulated, as shown in figure 14–4.

Returning to the intensity of stimulus, consider a situation in which the stimulus is varied and a quantal response is observed. Consider a threshold or tolerance corresponding to the stimulus which marks the difference between nonresponse and response in each individual. The distribution of tolerance can be expressed

$$dP = f(\lambda)d\lambda \qquad (14.2)$$

where dP = a proportion of the population consisting of individuals
whose tolerances lie between λ and $(\lambda + d\lambda)$
$f(\lambda)$ = the probability-density function of λ

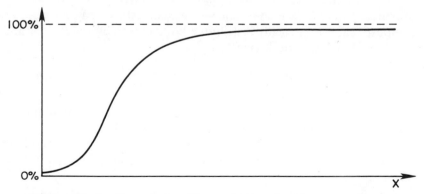

Figure 14–3. Cumulative Skewed Stimulus-Response Curve

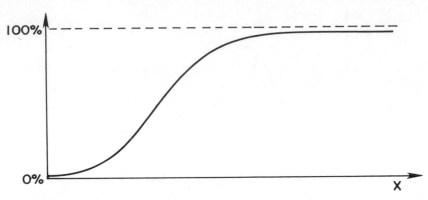

Figure 14–4. Cumulative Normal Stimulus-Response Curve

If stimulus λ_0 is administered, the proportion of the population who respond will be those whose tolerance level is less than λ_0.

$$P = \int_0^{\lambda_0} f(\lambda)d\lambda \tag{14.3}$$

If stimulus values can range from 0 to $+\infty$, it follows that the integration over all possible values of the stimulus is 1 (i.e., a response will occur by $+\infty$).

$$\int_0^{\infty} f(\lambda)d\lambda = 1 \tag{14.4}$$

Suppose the stimulus measure, λ, is normalized by some general function of the form

$$x_i = \phi(\lambda_i) \tag{14.5}$$

It is desirable to express equation 14.3 in terms of the normalized stimulus, to define a model that will permit estimation and calibration. Suppose that x_i is normally distributed with a mean of μ and a variance of σ^2. Equation 14.2 may be written

$$dP = \frac{1}{\sigma\sqrt{2\pi}} \exp\left[-\tfrac{1}{2}\frac{(x-\mu)^2}{\sigma^2}\right]dx \tag{14.6}$$

This defines dP, and one may then develop an expression for P in terms of the unknowns μ and σ^2. It should be noted, however, that while the stimulus was assumed to be nonnegative, the transformed stimulus, x_i, may range over any values—both negative and positive. Using the variable x_i, equation 14.3 may be written

$$P_i = \frac{1}{\sigma\sqrt{2\pi}} \int_{-\infty}^{x_i} \exp\left[-\tfrac{1}{2}\frac{(x_i-\mu)^2}{\sigma^2}\right]dx \tag{14.7}$$

It is not easy to obtain a direct solution for μ and σ^2 from equation 14.7. As a result, a transformation of equation 14.7 is defined, known as the *probit transformation*. Consider a standard normal deviate, u (with mean zero and variance of one). The probability of a response at some value Y_i is

$$P = \frac{1}{\sqrt{2\pi}} \int_{-\infty}^{Y_i} \exp[-\tfrac{1}{2}u^2]du \qquad (14.8)$$

The value of P in equation 14.8 can be made identical to that of equation 14.7 by selecting the value of Y appropriately. To find the value of Y that is needed, it is necessary only to integrate equations 14.7 and 14.8 and equate them to each other. Doing this yields

$$Y_i = \frac{1}{\sigma}(x_i - \mu) \qquad (14.9)$$

Thus the value of Y_i that will yield the identical probability to that of equation 14.7 is a function of σ, μ, and the stimulus value, x_i. Y_i is therefore the standard normal deviate that corresponds to a probability P_i obtained from a normalized, cumulative stimulus-response curve. Finney[3] coined the name "probit" for the value Y_i, derived from the notion that it is the equivalent deviate for a probability value.

Unfortunately, since the values of P cannot be observed, there is no direct and simple method to estimate the values of μ and σ from equation 14.9. If probabilities could be observed, it would be a simple matter to look up the corresponding values of Y_i from a table of standard normal deviates and carry out a regression between the values of Y_i and observed values of x_i. Equation 14.9 can be rewritten for this purpose;

$$Y_i = \alpha + \beta x_i \qquad (14.10)$$

where $\alpha = -\mu/\sigma$
and $\beta = 1/\sigma$.

However, since P_i cannot be observed, some other type of solution process is required for equation 14.9 or 14.10. Finney[4] outlines two methods: a graphical method and a computational method. When Finney originally developed probit analysis in the 1940s, high-speed computers had not yet appeared, the computational method was very tedious, and the graphical method represented the only really satisfactory procedure. With the advent of modern high-speed computers in the last two decades, the graphical methods have been almost totally superseded by computational methods. Thus the concern of this chapter is restricted to the computational method. The interested reader will find an extensive treatment of the graphical methods in earlier editions of Finney's book.[5]

Before proceeding to the method, it is important to note two factors about the model proposed so far. First, the model proposed is termed a *binary* probit model. In this use, binary refers to the fact that the outcome from applying the stimulus is binary in nature, that is, only two outcomes are possible—an event either occurs or does not occur. Second, the proposed model is a *bivariate* model. That is, the stimulus for the event is considered to be a single variable (one independent variable) in the same way that simple linear-regression analysis also considers one independent variable. The probit model is bivariate in the sense that equation 14.10 involves a relationship between two variables, Y_i and x_i. As dealt with later in the chapter, this restriction can be relaxed rather easily, although the calibration (or fitting) of the probit model becomes more tedious when it becomes multivariate.

Computational Estimation of the Binary, Bivariate Probit Model

To estimate the parameters of the probit model, that is, α and β in equation 14.10, it is necessary to find a procedure that can yield estimates of α and β that fit the observed data. The computational method developed by Finney[6] is based upon finding the maximum-likelihood estimates (MLEs) of α and β.[7] To do this it is necessary to develop a likelihood function for a set of empirical observations. The observations will consist of measures of the stimulus, x_i, and a note of whether or not the event, E_i, occurs. Thus E_i is a binary variable taking the values of 0 (the event does not occur) and 1 (the event does occur).

Consider now an experiment on a number of recipients of the stimulus, in which each recipient is subjected to a specific level of the stimulus, λ_j. Define the probability that a recipient responds to the stimulus as p_j and the probability that a recipient does not respond as q_j, where q_j is equal to $(1-p_j)$. It is assumed the responses of any two individuals are independent of each other. In other words, the response of one recipient is not affected by any other recipient's response to the stimulus. Suppose the experiment is made initially with two recipients. The probability that either recipient responds is p_j and the probability that either one does not respond is q_j.

Therefore the probability that both recipients respond is p_j^2. The probability that neither recipient responds is, similarly, q_j^2. The probability that the first recipient responds and not the second is $p_j q_j$ and the probability that the second recipient responds and the first does not is $q_j p_j$. If the experimenter is indifferent about which of the two responds and which does not, but needs only to know that one did and one did not, the probability of this occurrence is $2p_j q_j$.

Suppose now the experiment is conducted on three individuals. By a similar reasoning, and assuming that the identity of the recipients who respond is of no concern, the probabilities of the possible outcomes are: three responses will occur with a probability of p_j^3, two responses will occur with a probability of $3p_j^2q_j$, one response with a probability of $3p_jq_j^2$, and no responses with a probability of q_j^3.

It is notable that the probabilities of the various outcomes are the terms of the expansion of a function $(p+q)^2$ for two recipients and $(p+q)^3$ for three recipients. If there were n recipients, the probabilities would be the terms of the expansion of $(p+q)^n$. The probability of the occurrence of a particular outcome of the experiment, say r_j responses, is the likelihood of that outcome. Therefore the likelihood can be expressed as the general expansion term of $(p+q)^n$.

$$L(r_j \mid n) = \frac{n!}{r_j!(n-r_j)!} \, p_j^{r_j} q_j^{(n-r_j)} \qquad (14.11)$$

Consider now the repetition of the experiment on k independent samples, each of n individuals. The likelihood of obtaining r_j responses in all k experiments is obtained by multiplying together the likelihoods for each outcome.

$$L(r_j, r_j, \ldots, r_j \mid n) = \prod_{j=1}^{k} \left[\frac{n!}{r_j!(n-r_j)!} \, p_j^{r_j} q_j^{(n-r_j)} \right] \qquad (14.12)$$

Since the samples are independent samples from the same population and the stimulus level is the same in each experiment, the values of p_j and q_j will not vary across the experiments. This description of the experimental situation is not fully appropriate for many applications in the social sciences, although it may be appropriate for some engineering applications. Without loss of generality, it is acceptable to assume that, for any experiment, the value of n is 1 and r_j must be 1 or 0. Bearing in mind that such an assumption would be acceptable, it is appropriate to determine the maximum-likelihood solution for the values of α and β from equation 14.10.

The normal process for maximizing the likelihood, shown in equation 14.12, would be to take partial differentials of the likelihood with respect to the two unknowns, α and β. However, equation 14.12 is a rather complex equation and differentiation will be a cumbersome business. It is clear that the equation could be simplified by taking the natural logarithm of the likelihood. Since the log of the likelihood is monotonically related to the likelihood, maximizing the log of the likelihood will also maximize the likelihood. The log of the likelihood is

$$\ln L = \sum_j \ln\left[\frac{n!}{r_j!(n-r_j)!}\right] + \sum_j r_j \ln p_j + \sum_j (n-r_j)\ln q_j \qquad (14.13)$$

For a given experimental outcome, $r_j = r_0$ for all j, the first term of equation 14.13 is constant for all values of α and β. Therefore, this term contributes nothing to maximizing the log of the likelihood and may be ignored. Also, for constant $r_j = r_0$, the values of r_j and $(n-r_j)$ will be invariant over the summations. Using these factors, the value to be maximized, L', is

$$L' = r_0 \sum_j \ln p_j + (n-r_0) \sum_j \ln q_j \qquad (14.14)$$

To maximize L', it must be differentiated with respect to each of α and β, bearing in mind that p_j and $q_j (= 1-p_j)$ are both functions of α and β. These differentials are set to zero to solve for the maximum. Differentiating equation 14.14 with respect to α and β and setting these equal to zero produces

$$\frac{\partial L'}{\partial \alpha} = r_0 \sum_j \frac{1}{p_j} \frac{\partial p_j}{\partial \alpha} + (n-r_0) \sum_j \frac{1}{q_j} \frac{\partial q_j}{\partial \alpha} = 0 \qquad (14.15)$$

$$\frac{\partial L'}{\partial \beta} = r_0 \sum_j \frac{1}{p_j} \frac{\partial p_j}{\partial \beta} + (n-r_0) \sum_j \frac{1}{q_j} \frac{\partial q_j}{\partial \beta} = 0 \qquad (14.16)$$

Now since q_j is equal to $(1-p_j)$, the above equations can be simplified.

$$\frac{\partial q_j}{\partial \alpha} = - \frac{\partial p_j}{\partial \alpha} \qquad (14.17)$$

Equation 14.17 will also apply analogously to the partial differentials with respect to β. Using this, equations 14.15 and 14.16 become

$$r_0 \sum_j \frac{1}{p_j} \frac{\partial p_j}{\partial \alpha} - (n-r_0) \sum_j \frac{1}{(1-p_j)} \frac{\partial p_j}{\partial \alpha} = 0 \qquad (14.18)$$

$$r_0 \sum_j \frac{1}{p_j} \frac{\partial p_j}{\partial \beta} - (n-r_0) \sum_j \frac{1}{(1-p_j)} \frac{\partial p_j}{\partial \beta} = 0 \qquad (14.19)$$

Since the two equations are of identical form, they could be generalized to partial differentials with respect to a general unknown parameter, θ_t.

$$r_0 \sum_j \frac{1}{p_j} \frac{\partial p_j}{\partial \theta_t} - (n-r_0) \sum_j \frac{1}{(1-p_j)} \frac{\partial p_j}{\partial \theta_t} = 0 \qquad (14.20)$$

This equation can be simplified to make the solution easier. First, both terms are multiples of $\partial p_j/\partial \theta_t$. Therefore equation 14.20 can be written

$$\sum_j \left[\left(\frac{r_0}{p_j} - \frac{(n-r_0)}{(1-p_j)} \right) \frac{\partial p_j}{\partial \theta_t} \right] = 0 \qquad (14.21)$$

This may be simplified to equation 14.22 by gathering terms over a common denominator.

$$\sum_j \left[\left(\frac{r_0(1-p_j) - (n-r_0)p_j}{p_j(1-p_j)} \right) \frac{\partial p_j}{\partial \theta_t} \right] = 0 \qquad (14.22)$$

A further simplification results in

$$\sum_j \left[\frac{n(r_0/n - p_j)}{p_j(1-p_j)} \frac{\partial p_j}{\partial \theta_t} \right] = 0 \qquad (14.23)$$

The value r_0/n is an estimate of the probability of a response from each experiment and represents the estimate of a prior probability. Writing this as P, equation 14.23 can be expressed

$$\sum_j \left[\frac{n(P-p_j)}{p_j(1-p_j)} \frac{\partial p_j}{\partial \theta_t} \right] = 0 \qquad (14.24)$$

It must be remembered that there are two parameters, θ_t, in the simple bivariate case considered here, where those values were originally designated as α and β. Since p_j is expressed as a function of both those parameters, and since further simplification of equation 14.24 will not result in an equation form from which p_j can be eliminated in solving the two simultaneous equations, it follows that a unique algebraic solution to equation 14.24 cannot be obtained. In fact, p_j can be written as equation 14.25 (adapting from equations 14.8 and 14.10);

$$p_j = \frac{1}{\sqrt{2\pi}} \int_{-\infty}^{\theta_1+\theta_2 x_j} \exp[-\tfrac{1}{2}u^2]du \qquad (14.25)$$

where $\theta_1 = \alpha$
and $\quad \theta_2 = \beta$

It should now be clear that a direct algebraic solution of equation 14.24 is not possible. Therefore, it is necessary to devise an iterative procedure of solution, starting from some initial estimates, θ_1' and θ_2', and determining adjustments, $\delta\theta_1'$ and $\delta\theta_2'$, to move the calibrated model to closer fit to observed data.

At the first iteration, it is assumed initial estimates, θ_1' and θ_2', have been obtained, so the desired estimates from the first iteration are $(\theta_1'+\delta\theta_1')$ and $(\theta_2'+\delta\theta_2')$. Using these as estimates, the partial differentials of L' with respect to these new estimates are to be set equal to zero.

$$\frac{\partial L'}{\partial(\theta_1'+\delta\theta_1')} = 0 \qquad (14.26)$$

$$\frac{\partial L'}{\partial(\theta_2'+\delta\theta_2')} = 0 \qquad (14.27)$$

By solving these two equations for $\delta\theta'_1$ and $\delta\theta'_2$, the improved parameter estimates can be obtained and a check made to see if these new estimates are sufficiently close to maximizing the likelihood. If not, a new iteration must be undertaken. In such a case, if the estimates from the previous iteration are written as $\theta''_1(=\theta'_1+\delta\theta'_1)$ and $\theta''_2(=\theta'_2+\delta\theta'_2)$ it is now necessary to find further adjustments, $\delta\theta''_1$ and $\delta\theta''_2$, that will move the estimates closer to the maximum likelihood.

The values of $\delta\theta'_1$ and $\delta\theta'_2$ can be found by using a Taylor-MacLaurin expansion of equations 14.26 and 14.27 and ignoring second-order small quantities.

$$\frac{\partial L'}{\partial\theta'_1} + \delta\theta'_1\frac{\partial^2 L'}{\partial\theta'^2_1} + \delta\theta'_2\frac{\partial^2 L'}{\partial\theta'_1\partial\theta'_2} = 0 \tag{14.28}$$

$$\frac{\partial L'}{\delta\theta'_2} + \delta\theta'_2\frac{\partial^2 L'}{\delta\theta'^2_2} + \delta\theta'_1\frac{\partial^2 L'}{\partial\theta'_1\partial\theta'_2} = 0 \tag{14.29}$$

To determine the values of $\delta\theta'_1$ and $\delta\theta'_2$, it is now necessary to substitute into equations 14.28 and 14.29 expressions for the differentials, as determined in terms of differentials of p_j with respect to θ'_1 and θ'_2.

Equations 14.15 and 14.24 can be used to provide an estimate of each of $\partial L'/\partial\theta'_1$ and $\partial L'/\partial\theta'_2$.

$$\frac{\partial L'}{\partial\theta'_1} = \sum_j\left[\frac{n(P-p_j)}{p_j(1-p_j)}\frac{\partial p_j}{\partial\theta'_1}\right] \tag{14.30}$$

$$\frac{\partial L'}{\partial\theta'_2} = \sum_j\left[\frac{n(P-p_j)}{p_j(1-p_j)}\frac{\partial p_j}{\partial\theta'_2}\right] \tag{14.31}$$

Taking the second differentials of the above two equations is rather complex and is much simplified by returning to the original first differential, equation 14.15, written in terms of a general parameter θ'_t.

$$\frac{\partial L'}{\partial\theta'_t} = r_0\sum_j\frac{1}{p_j}\frac{\partial p_j}{\partial\theta'_t} + (n-r_0)\sum_j\frac{1}{q_j}\frac{\partial q_j}{\partial\theta'_t} \tag{14.32}$$

The second partial differential with respect to θ'_t is

$$\frac{\partial^2 L'}{\partial\theta'^2_t} = r_0\sum_j\left[-\frac{1}{p^2_j}\left(\frac{\partial p_j}{\partial\theta'_t}\right)^2\right] + (n-r_0)\sum_j\left[-\frac{1}{q^2_j}\left(\frac{\partial q_j}{\partial\theta'_t}\right)^2\right] \tag{14.33}$$

Replacing q_j with $(1-p_j)$, gathering terms, and simplifying, equation 14.33 can be rewritten as equation 14.34, again using $P=r_0/n$.

$$\frac{\partial^2 L'}{\partial\theta'^2_t} = \sum_j -\left[\frac{n(P-2Pp_j+p^2_j)}{p^2_j(1-p_j)^2}\right]\left(\frac{\partial p_j}{\partial\theta_t}\right)^2 \tag{14.34}$$

Similarly, the second partial differentials of the form $\partial^2 L'/\partial\theta_t'\partial\theta_s'$ are

$$\frac{\partial^2 L'}{\partial\theta_t'\partial\theta_s'} = \sum_j -\left[\frac{n(P-2Pp_j+p_j^2)}{p_j^2(1-p_j)^2}\right]\frac{\partial p_j}{\partial\theta_t'}\frac{\partial p_j}{\partial\theta_s'} \qquad (14.35)$$

Substituting equations 14.30, 14.31, 14.34, and 14.35 into equations 14.28 and 14.29, the solutions for $\delta\theta_1'$ and $\delta\theta_2'$ can be determined by solving the simultaneous equations 14.36 and 14.37.

$$\delta\theta_1'\sum_j\left[\frac{n(P-2Pp_j+p_j^2)}{p_j^2(1-p_j)^2}\right]\left(\frac{\partial p_j}{\partial\theta_1'}\right)^2$$

$$+ \delta\theta_2'\sum_j\left[\frac{n(P-2Pp_j+p_j^2)}{p_j^2(1-p_j)^2}\right]\left(\frac{\partial p_j}{\partial\theta_1'}\frac{\partial p_j}{\partial\theta_2'}\right)$$

$$= \sum_j\left[\frac{n(P-p_j)}{p_j(1-p_j)}\right]\left(\frac{\partial p_j}{\partial\theta_1'}\right) \quad (14.36)$$

$$\delta\theta_1'\sum_j\left[\frac{n(P-2Pp_j+p_j^2)}{p_j^2(1-p_j)^2}\right]\left(\frac{\partial p_j}{\partial\theta_2'}\frac{\partial p_j}{\partial\theta_1'}\right)$$

$$+ \delta\theta_2'\sum_j\left[\frac{n(P-2Pp_j+p_j^2)}{p_j^2(1-p_j)^2}\right]\left(\frac{\partial p_j}{\partial\theta_2'}\right)^2$$

$$= \sum_j\left[\frac{n(P-p_j)}{p_j(1-p_j)}\right]\left(\frac{\partial p_j}{\partial\theta_2'}\right) \quad (14.37)$$

The process may then be reiterated for estimates of $\delta\theta_1''$ and $\delta\theta_2''$ and subsequent estimates until a satisfactory fit is obtained. To determine the fit, it is first necessary to evaluate the second partial differentials, equation 14.34, to determine if the values are negative, positive, or zero, after finding the actual values of the first differentials. When the first differentials are close to zero and the second differentials are negative, a maximum has been reached.

Equations 14.36 and 14.37 can be simplified by returning to the original probit models, equations 14.8 and 14.10. From equation 14.8, it is possible to determine $\partial P_j/\partial Y$.

$$\frac{\partial P_j}{\partial Y} = \frac{1}{\sqrt{2\pi}}\exp[-\tfrac{1}{2}Y_j^2] \qquad (14.38)$$

In turn, $\partial Y_j/\partial\theta_1$ and $\partial Y_j/\partial\theta_2$ are given by equations 14.39 and 14.40, where $\theta_1 = \alpha$ and $\theta_2 = \beta$.

$$\frac{\partial Y_j}{\partial\theta_1} = 1 \qquad (14.39)$$

$$\frac{\partial Y_j}{\partial\theta_2} = x_j \qquad (14.40)$$

Hence $\partial P_j/\partial\theta_1$ and $\partial P_j/\partial\theta_2$ can be determined. To simplify the notation, these may be expressed

$$\frac{\partial P_j}{\partial\theta_1} = Z_j \tag{14.41}$$

$$\frac{\partial P_j}{\partial\theta_2} = Z_j x_j \tag{14.42}$$

where $Z_j = \dfrac{1}{\sqrt{2\pi}}\exp[-\frac{1}{2}Y_j^2]$

Equations 14.36 and 14.37 can now be rewritten in terms of Z_j, x_j, and p_j.

$$\delta\theta_1'\sum_j\left[\frac{n(P-p_j)^2}{p_j^2(1-p_j)^2}\right]Z_j^2 + \delta\theta_2'\sum_j\left[\frac{n(P-p_j)^2}{p_j^2(1-p_j)^2}\right]Z_j^2 x_j$$

$$= \sum_j\left[\frac{n(P-p_j)}{p_j(1-p_j)}\right]Z_j \tag{14.43}$$

$$\delta\theta_1'\sum_j\left[\frac{n(P-p_j)^2}{p_j^2(1-p_j)^2}\right]Z_j^2 x_j + \delta\theta_2'\sum_j\left[\frac{n(P-p_j)^2}{p_j^2(1-p_j)^2}\right]Z_j^2$$

$$= \sum_j\left[\frac{n(P-p_j)}{p_j(1-p_j)}\right]Z_j x_j \tag{14.44}$$

These equations can be simplified further by writing $(P-p_j)/p_j(1-p_j)$ as ω_j.

$$\delta\theta_1'\sum_j n\omega_j^2 Z_j^2 + \delta\theta_2'\sum_j n\omega_j^2 Z_j^2 x_j = \sum_j n\omega_j Z_j \tag{14.45}$$

$$\delta\theta_1'\sum_j n\omega_j^2 Z_j^2 x_j + \delta\theta_2'\sum_j n\omega_j^2 = \sum_j n\omega_j Z_j x_j \tag{14.46}$$

These simultaneous equations are solved by estimating p_j, P (and hence ω_j), and Z_j from the initial trial values of θ_1' and θ_2' and given observations on x_j.

As noted in chapter 13, the variance of a maximum-likelihood estimator (MLE) is the inverse of the second partial differential of the log likelihood with respect to the parameter estimate. Thus on completing the iterative solutions for θ_1 and θ_2, $\hat{\theta}_1$ and $\hat{\theta}_2$, the variances are

$$V(\hat{\theta}_t) = \left(\frac{\partial^2 L'}{\partial\hat{\theta}_t^2}\right)^{-1} \tag{14.47}$$

Given the asymptotic normality property of the MLE,[8] a t statistic can be formed for testing against a zero value for $\hat{\theta}_t$,

$$t = \frac{\hat{\theta}_t}{\sigma_{\hat{\theta}_t}} \tag{14.48}$$

where $\sigma_{\hat{\theta}_t}$ = the standard deviation of $\hat{\theta}_t$, that is, $[V(\hat{\theta}_t)]^{1/2}$

The value t is compared with the table value of t for a two-tailed test at the $(100-\alpha)\%$ confidence level and with $(n-2)$ degrees of freedom, where n is the number of observations used to calibrate the model. Recalling the notions of likelihood developed in chapter 13, it is possible to develop a test of the calibrated model, in addition to testing each of the parameters, $\hat{\theta}_1$ and $\hat{\theta}_2$. Consider the null hypothesis that the true values, θ_1 and θ_2, are both zero. This is tantamount to assuming that there is a 0.5 probability of any individual responding to any level of the stimulus, x_j. The likelihood for this situation can be written as $L(0,0)$. A likelihood ratio, λ, can be formed from the calibrated model and this null likelihood.

$$\lambda = \frac{L(0,0)}{L(\hat{\theta}_1,\hat{\theta}_2)} \tag{14.49}$$

It has been shown[9] that $-2 \ln \lambda$ is distributed like χ^2 with degrees of freedom equal to the number of parameters that are given different values in the two likelihoods. Since log likelihoods are estimated in the calibration of the probit model, the χ^2 statistic, $-2 \ln \lambda$, can be written

$$-2 \ln \lambda = L'(0,0) - L'(\hat{\theta}_1,\hat{\theta}_2) \tag{14.50}$$

where L' is defined by equation 14.14

In the case being considered here, the χ^2 statistic will have two degrees of freedom. If the calculated χ^2 is larger than the table value for $\alpha\%$, it may be concluded that the calibrated probit model estimates probabilities better than an equal-shares model.

A second more stringent test is obtained by choosing for the null hypothesis the situation in which a constant is estimated that will replicate the observed proportion that responds to the stimulus. Denoting the likelihood for this situation as $L(\theta_0,0)$, a χ^2 statistic can again be formed from the likelihood ratio.

$$-2\ln\lambda = L'(\theta_0,0) - L'(\hat{\theta}_1,\hat{\theta}_2) \tag{14.51}$$

This test is often called a "market-share" test. In this case, the degrees of freedom for χ^2 are 1 and the test is done as before.

Multivariate Probit Analysis

It may be clear already to the reader that the calibration procedure developed in the preceding section can be extended quite readily to a multivariate probit model. However, since the multivariate model is of more interest for applications in the social sciences and engineering, the extension is shown in some detail in this section. This derivation is originally due to Tobin.[10]

Suppose the model to be developed is one to forecast whether or not a household will buy a refrigerator. Define an index I as comprising some linear combination of various independent variables, $X_1, X_2, ..., X_m$, and the value of I as determining whether the dependent variable W has the value 1 or 0 for the household, that is, whether or not the household will buy a new refrigerator. For the ith household, the relationship is

$$I_i = \alpha_0 + \alpha_1 X_{1i} + \alpha_2 X_{2i} + ... + \alpha_m X_{mi} \qquad (14.52)$$

Let $\bar{I}_i$ be the critical value of I_i for the ith household, that is, the value at which they will decide to buy a new refrigerator. If I_i is equal to or greater than $\bar{I}_i$, then W_i will be 1; and if it is less than $\bar{I}_i$, then W_i will be 0. This is shown mathematically by

$$W_i = 1 \qquad I_i \geqslant \bar{I}_i \qquad (14.53)$$

$$W_i = 0 \qquad I_i < \bar{I}_i \qquad (14.54)$$

Conventionally in probit analysis, $\bar{I}_i$ is assumed to be normally distributed with mean 5 and variance 1. (The mean value of 5 was chosen so that negative values would be very unlikely.) This distribution reflects random differences in response by each individual or household not accounted for by the variables in the index.

For a given value of the index I, W will be equal to 1 for households whose $\bar{I}_i \geqslant I$ and will be 0 for households whose $\bar{I}_i < I$. The probability that for given I, W will be 1 is

$$P(I) = \frac{1}{\sqrt{(2\pi)}} \int_{-\infty}^{I} \exp(-\tfrac{1}{2}u^2)du \qquad (14.55)$$

Similarly, the probability that $W = 0$ for given I is

$$Q(I) = \frac{1}{\sqrt{(2\pi)}} \int_{I}^{\infty} \exp(-\tfrac{1}{2}u^2)du \qquad (14.56)$$

Suppose there is a sample of observations of the values of $X_1, X_2, X_3, ..., X_m$ at each of N values of the Xs. These are represented as $X_{1j}, X_{2j}, X_{3j}, ..., X_{mj}$ where $j = 1,2,3, ..., N$. Suppose that n_j is the number of observations at the jth point. Now let r_j be the number of those observations for which W was observed to be 1, and $n_j - r_j$ is then the number of observations for which W was 0. The likelihood of the sample is a function of the values $(\hat{a}_0, \hat{a}_1, \hat{a}_2, ..., \hat{a}_m)$ where these are the estimated values for the population parameters $(\alpha_0, \alpha_1, \alpha_2, ..., \alpha_m)$. This likelihood is

$$L(\hat{a}_0, \hat{a}_1, \hat{a}_2, ..., \hat{a}_m) = \prod_{j=1}^{N} [P(a_0 + a_1 X_{1j} + ...$$

$$+ a_m X_{mj})]^{r_j} [Q(a_0 + a_1 X_{1j} + ... + a_m X_{mj})]^{n_j - r_j} \qquad (14.57)$$

To find the maximum-likelihood estimates, the likelihood function of equation 14.57 must be maximized. It is in fact easier to maximize $\ln L$ instead of L, as before. Hence the function to be maximized can be written

$$\ln L'(\hat{a}_0,\hat{a}_1,...,\hat{a}_m) = \sum_{j=1}^{s} [r_j \ln P_j + (n_j - r_j) \ln Q_j] \qquad (14.58)$$

where $P_j = P(a_0 + a_1 X_{1j} + a_2 X_{2j} + ... + a_m X_{mj})$
and $Q_j = Q(a_0 + a_1 X_{1j} + a_2 X_{2j} + ... + a_m X_{mj})$

It is more convenient to denote $\ln L'(\hat{a}_0,\hat{a}_1,...,\hat{a}_m)$ as $L^*(\hat{a}_0,\hat{a}_1,...,\hat{a}_m)$. The condition for a maximum is obtained by equating all the partial derivatives of the likelihood function, with respect to the parameters a_0, $a_1,...,a_m$, to zero. The ith equation for determining the maximum likelihood is

$$L_i^*(\hat{a}_0,\hat{a}_1,\hat{a}_2,...,\hat{a}_m)$$

$$= \sum_{j=1}^{N} \left[r_j \frac{X_{ij}Z_j}{P_j} - (n_j - r_j) \frac{X_{ij}Z_j}{Q_j} \right] = 0 \qquad \text{(for } i=1,2,...,m) \quad (14.59)$$

where $L_i(\hat{a}_0,\hat{a}_1,\hat{a}_2,...,\hat{a}_m) = \dfrac{\partial L^*(\hat{a}_0,\hat{a}_1,\hat{a}_2,...,\hat{a}_m)}{\partial \hat{a}_i}$

and $\qquad\qquad\qquad Z_j = \dfrac{\partial P_j}{\partial Y_j} = -\dfrac{\partial Q_j}{\partial Y_j}$

These nonlinear equations can be solved by an iterative process in which a set of trial values of the coefficients $(\hat{a}_0,\hat{a}_1,\hat{a}_2,...,\hat{a}_m)$, say $(\tilde{a}_{00},\tilde{a}_{10},\tilde{a}_{20},...,\tilde{a}_{m0})$, are used.

New estimates of these coefficients can be obtained by solving a set of $(m+1)$ linear equations, as was shown for the bivariate case. Let the new estimates be $(\tilde{a}_{00} + \Delta a_0, \tilde{a}_{10} + \Delta a_1,...,\tilde{a}_{m0} + \Delta a_m)$. It is assumed that all the L_i^* are linear between the trial solution and the real solution, so equation 14.60 holds, again using the Taylor-MacLaurin expansion.

$$L_i^*(\tilde{a}_{00} + \Delta a_0, \tilde{a}_{10} + \Delta a_1,..., \tilde{a}_{m0} + \Delta a_m) = L_i^*(\tilde{a}_{00}, \tilde{a}_{10}, \tilde{a}_{20},...,\tilde{a}_{m0})$$

$$+ \sum_{k=0}^{m} a_k L_{ik}^*(\tilde{a}_{00}, \tilde{a}_{10}, \tilde{a}_{20},...,\tilde{a}_{m0}) = 0 \quad (14.60)$$

L_{ik}^* are the second-order derivatives and may be expressed

$$L_{ik}^*(\tilde{a}_{00}, \tilde{a}_{10}, \tilde{a}_{20},...,\tilde{a}_{m0})$$

$$= \sum_{j=1}^{N} \left[\frac{r_j X_{ij} X_{kj}(-P_j Y_j Z_j - Z_j^2)}{P_j^2} \right.$$

$$\left. - \frac{(n_j - r_j) X_{ij} X_{kj}(-Q_j Y_j Z_j + Z_j^2)}{Q_j^2} \right] \quad (14.61)$$

The solution for the $\hat{a}_k$s now proceeds exactly as for the bivariate case, using repeated iterations to adjust the estimates of the parameters until both the adjustment is infinitesimal and the second partial differentials show that a maximum has been reached.

As for the bivariate case, the variances of the parameter estimates can be determined from the inverse of the second partial differentials and these variances used to obtain t statistics for the parameters. Also, a chi-square test can be carried out for the entire model, using either of the two null hypotheses discussed in the preceding section, that is, equal shares or market shares. Some other statistical tests, described in chapter 16, are also applicable to probit-analysis models and may assist in determining the merits of various alternative models.

Comments on Probit Analysis

The computational procedure for bivariate and multivariate binary probit models, described in the preceding sections, represents one possible method of calibrating probit models. Other techniques exist, including graphical methods for bivariate cases, as described by Finney,[11] and a maximum-likelihood procedure using table lookup values to iterate the parameter estimates.[12] At the time of writing, a new procedure has been developed for satisfactorily fitting multinomial (multiple-choice) probit models.[13] Thus applications of probit analysis are no longer restricted to situations in which only two outcomes are possible, for example, purchase of a commodity, failure of a concrete slab, failure of an electrical component, and so forth.

It is important to note that there is no equivalent of stepwise procedures for the calibration of probit models. The iterative nature of the calibration process requires that the model be estimated in its entirety. Therefore, it becomes extremely important to construct well-thought-out hypotheses of the relevant variables before commencing the calibration process. If any variables are found to be nonsignificant, it becomes necessary to try all possible combinations of the nonsignificant variables, eliminating one, two, and so on until all eventualities have been explored, allowing for the effects of interdependencies among the variables. This can clearly be a lengthy and expensive process if several variables are found to be nonsignificant. Hence there is a need to try through prior hypotheses to structure the initial model very carefully.

As discussed in more detail in chapter 16, it is not possible to develop a correlation coefficient that is useful for probit analysis. This is so because the observed values of the dependent variable are not designed to be the same as the model estimates of the dependent variable. Observations cannot be made of the probabilities of occurrences, but only whether or not an event occurs. There is therefore an apparent "loss of variance" between the observed dependent variable values and the ones desired from the model. This loss cannot be calculated but renders a correlation coefficient useless.

In most computer programs for probit analysis, it is not necessary for the user to provide first estimates of the coefficients. It has been found that the use of zero values as the initial values generally provides an efficient starting point for the fitting process. In some cases, a nonzero value may be assigned to the constant term, the value being the one which will produce an estimate over the whole data set of the observed market shares. That is, the value of the constant is set by

$$\frac{r}{n} = \frac{1}{\sqrt{2\pi}} \int_{-\infty}^{a_0} \exp(-\tfrac{1}{2}u^2)du \qquad (14.62)$$

Either of these starting positions produces acceptable maximum-likelihood estimates in rather few iterations (5–10 usually). With modern high-speed computers, this is quite efficient. In addition, the starting position automatically provides an estimate of the likelihood for one of the two alternative null hypotheses used in the chi-square test of the model. Clearly, this is a useful advantage.

Finally, a cautionary comment is in order concerning the need to include a constant term in the specified linear probit. First, it should be clear from the original bivariate model that a constant term is essential in the fitted model. In addition, it has been shown[14] that when sampling or measurement errors are present in the data, inclusion of a constant in the model will generally permit the parameter estimates for all but the constant to be unbiased. Failure to include a constant will lead to biased parameter estimates. It is also important to note that the likelihood-ratio test (chi-square) with the market-share null hypothesis is invalid if no constant is specified in the calibrated model.

An Example

For this example, the same data are used as for discriminant analysis in chapter 12. Table 14–1 shows the correlation matrix for these data, from which it can be seen that strong correlations between choice (a 0,1 variable) and the independent variables do not appear to exist. As discussed in chapter 16, this is not surprising and does *not* indicate that a good probit model *cannot* be expected to result.

In the program used, a Newton-Raphson hill-climbing technique is used to maximize the likelihood. The initial estimates used are zero for all

Table 14–1
Correlation Matrix for Example Data from Chapter 12

Variable	Choice	Time	Cost	CA1	CA2	Income	Sex	Age	Const.
Choice	1.00								
Time	−.19	1.00							
Cost	.24	−.17	1.00						
CA1	.03	.06	−.21	1.00					
CA2	.23	.07	.03	−.22	1.00				
Income	.26	.12	−.03	.33	−.12	1.00			
Sex	.14	.11	−.17	.27	.36	−.02	1.00		
Age	.19	.07	−.02	.35	−.08	.31	.09	1.00	
Const.	0	0	0	0	0	0	0	0	1.00

coefficients except the constant (the problem is a binary one and the program is for binary problems only), which is set to a value to produce market shares. Since there are 98 transit users and 61 auto users, the market shares are 0.6164 and 0.3836. The initial estimate of the constant can be determined as being 0.3961 from the normal distribution and produces a log likelihood of −105.87.

Given this starting configuration, the program takes three iterations to find a maximum solution for these data at the given closure criterion. The coefficient estimates for this solution are given in table 14–2. Transit use was coded as 1 and auto use as zero. Therefore, the model of table 14–2 gives coefficients for estimating transit use probabilities. Since time and cost are differences between transit and auto, the negative signs of these coefficients are intuitively correct, showing that increasing transit time or cost will reduce transit choice probabilities and increasing automobile time or cost will increase transit choice probabilities.

From table 14–2, it can be seen that each variable has a significant coefficient, except the car-owning dummy, CA1, and sex. All the situational variables except CA1 show that increases in their values lead to a greater propensity to ride transit, ceteris paribus. Since CA1 is 1 for the

Table 14–2
Probit Analysis Results for Data of Chapter 12

			Variable Coefficients (t Scores)						
Time	Cost	CA1	CA2	Income	Sex	Age	Const.	Log Likeli- hood	χ^2 (d.f.)
−0.0831	−0.0112	−0.213	1.273	2.062	0.700	0.0673	−2.492	−79.2	53.4
(3.3)	(3.0)	(0.6)	(2.9)	(3.8)	(1.5)	(2.0)	(3.0)		(7)

case of the worker being the only licensed driver or a single-adult house-hold, while CA2 is 1 when there are two workers and one car, or one worker and no car, these results seem reasonable, with the possible exception of income. It seems more likely that increasing income would result in lower transit ridership. This does not seem to be the case, however.

The likelihood-ratio statistic is determined to have a value of 53.4 with seven degrees of freedom.

$$-2\ln(L_0 - L_1) = -2[-105.87 - (-79.2)] = 53.4 \qquad (14.63)$$

The table value of χ^2 with seven degrees of freedom and at 99.5% confidence is 20.3. Therefore, one may have better than 99.5% confidence that the model did not arise by chance, representing a random process. Rather, the model seems very likely to be based on a real nonrandom improvement in understanding.

It is useful to see how the probit model is used to predict probabilities. To do this, it must be remembered that the probit model is

$$p_i^f = \frac{1}{\sqrt{2\pi}} \int_{-\infty}^{Y_i} \exp[-u^2/2]du \qquad (14.64)$$

where Y_i is

$$Y_i = a_0 + a_1\Delta T_i + a_2\Delta C_i + a_3CA1 + a_4CA2$$
$$+ a_5\text{income} + a_6\text{sex} + a_7\text{age} \qquad (14.65)$$

Consider now the three individuals given in table 12–3 for discriminant analysis. The values of Y_i and p_i are given in table 14–3. The probabilities are those of being a transit user. It can be seen that individual 1 is more likely to be a car user, while individuals 2 and 3 are highly likely to be transit users. It is also useful to see what the effect is of a travel time change on each of these three individuals. Suppose automobile time increases by 10 minutes for all three individuals, so the time differences become 0, −30, and −25, respectively. The probabilities of transit use now become 0.688, 0.9999, and 0.986, respectively. On the other hand, if transit time increased by 10 minutes, the time differences would become

Table 14–3
Probit Predictions for Data of Table 12–3

Individual	Y_i	p_i^f
1	−0.341	0.367
2	2.856	0.998
3	1.377	0.916

20, −10, and −5, respectively, and the probabilities of transit use become 0.121, 0.979, and 0.707, respectively. Note that as a result of the non-linearity of the probit curve, the changes are not symmetrical for a 10-minute change in each direction for travel time. It should also be noted that the probabilities are obtained by looking up the probabilities for a unit normal curve corresponding to the value of the probit.

Notes

1. D.J. Finney, *Probit Analysis*, 3d ed., Cambridge, England: Cambridge University Press, 1971, pp. 8–19.

2. Finney, *Probit Analysis*, pp. 9–11.

3. Ibid, p. 23.

4. Ibid, pp. 20–49, Appendix I, 2d ed., 1964.

5. See, for example, the 1947 and 1952 editions of Finney, *Probit Analysis*.

6. Finney, *Probit Analysis*, pp. 50–57.

7. See chapter 13, p. 294.

8. See chapter 13, p. 295.

9. See chapter 13, p. 293.

10. J. Tobin, "The Application of Multivariate Probit Analysis to Economic Survey Data," Cowles Foundation Discussion Paper No. 1, 1955.

11. Finney, *Probit Analysis*, pp. 25–34.

12. J. Cornfield and N. Mantel, "Some New Aspects of the Application of Maximum Likelihood to the Calculation of the Dosage Response Curve," *Journal of the American Statistical Association*, 1950, vol. 45, no. 250, pp. 181–210; F. Garwood, "The Application of Maximum Likelihood to Dosage Mortality Curves," *Biometrika*, 1941, vol. 32, pp. 46–58.

13. Carlos F. Daganzo, Fernando Bouthelier, and Yosef Sheffi, "Multinomial Probit and Qualitative Choice: A Computationally Efficient Algorithm," *Transportation Science*, 1977, vol. 11, no. 4, pp. 338–358; Charles F. Manski, Steven R. Lerman, and Richard Albright, *An Estimator for the Generalized Multinomial Probit Model*, Series of Technical Memos for Ongoing Projects, Cambridge Systematics, Inc., 1977.

14. Charles F. Manski and Steven R. Lerman, "The Estimation of Choice Probabilities from Choice-Based Samples," *Econometrica*, 1977, Vol. 45, No. 8, pp. 1977–1988.

15

Logit Analysis

Introduction

The general type of application of the logit model is described in chapter 14, since it has similar statistical applicability to the probit model. The logit model appears to have been put forward originally by Berkson[1] as a simplified version, in essence, of the probit model. At the time of Berkson's work, probit models could be calibrated only by the tedious graphical methods mentioned in chapter 14 and described at length by Finney.[2] No further justification of the model form was given at the time, it being noted only that the logit and probit models produced very similar symmetrical ogives.

In recent years, there has been a sudden upsurge of interest in the logit model, particularly in consumer economics and transportation planning. In this new work, theoretical developments of the logit model have been put forward, showing the model to be derivable from certain microeconomic assumptions of behavior and from specific distributional assumptions on an error term. Since these derivations are well-documented, they are not repeated in detail here.[3] Suffice it to say that the model is developed in a utility context, in which it is assumed that one maximizes one's utility in making a choice from a set of possible purchases or courses of action, and the utility is expressed by

$$U(X_j,S_i) = U'(X'_j,S_i) + \epsilon_{ji} \qquad (15.1)$$

where $U(X_j,S_i)$ = the utility of alternative j to individual i as a function of characteristics, X_j, of the alternative and characteristics, S_i, of the individual

$U'(X'_j,S_i)$ = the observable utility for alternative j to individual i,

and ϵ_{ji} = the unobservable utility (or error)

The error term, ϵ_{ji}, is assumed to be identically and independently distributed as a Weibull distribution.[4] Based on this assumption, the general form of the multinomial logit model can be derived,

$$p^i_j = \frac{\exp[U'(X'_j,S_i)]}{\sum\limits_{k}\exp[U'(X'_k,S_i)]} \qquad (15.2)$$

where p^i_j = the probability that individual i chooses alternative j
and k = subscript denoting any alternative from the set K

So far, the derivation does not specify the functional form of the observable utility, $U'(X'_j,S_i)$. In fact, the analyst may choose any functional form that seems useful or applicable. In the remainder of the chapter, however, the simplest functional form will be used, that is, a linear form.

$$U'(X'_j,S_i) = \alpha_{0j} + \sum_s \alpha_s X'_{sj} + \sum_t \alpha_{tj} S_i \qquad (15.3)$$

Thus the observable utility is assumed to consist of an alternative-specific dummy variable (coefficient α_{0j}), a set of alternative-specific characteristics with generic coefficients ($\alpha_s X'_{sj}$), and a set of individual specific characteristics with alternative-specific coefficients ($\alpha_{tj} S_i$).[5]

The model developed above is known as the multinomial logit (MNL) model. Much of the early work in these models used a simpler form of the multinomial logit model. The MNL model is a general model that holds for any number (finite) of alternatives, K. A simple form that is also useful for exploring many of the properties of the model is the binary logit model, shown for choice of alternative 1.

$$p_1^i = \frac{\exp[U'(X'_1,S_i)]}{\exp[U'(X'_1,S_i)] + \exp[U'(X'_2,S_i)]} \qquad (15.4)$$

Similarly, the binary logit model for the choice of alternative 2 is

$$p_2^i = \frac{\exp[U'(X'_2,S_i)]}{\exp[U'(X'_1,S_i)] + \exp[U'(X'_2,S_i)]} \qquad (15.5)$$

This model form is used in much of the discussion in the remainder of the chapter because of its simple form.

Some Properties of the Logit Model

Before proceeding further on the statistical properties of the model, there are certain other properties that should be explored since they affect the conceptual value of the logit model.

First, both the multinomial and binary forms of the model (see equations 15.2 and 15.4) are symmetrical. That is, the probability of an alternative being chosen is directly proportional to a function of the utility of the alternative and is inversely proportional to a sum of functions of all the utilities. This latter term will be the same for all alternatives. Hence the logit model is symmetrical.

However, the model is frequently met in an asymmetrical form. Suppose, in the binary case, one were to divide the numerator and denominator of equation 15.4 by $\exp[U'(X'_2,S_i)]$. This would produce equation 15.6, which is still identical in fact to equation 15.4.

$$p_1^i = \frac{\exp[U'(X',S_i)-U'(X'_2,S_i)]}{\exp[U'(X'_1 S_i)-U'(X'_2,S_i)]+1} \qquad (15.6)$$

A similar manipulation of equation 15.5 yields

$$p_2^i = \frac{1}{\exp[U'(X_1',S_i)-U'(X_2',S_i)]+1} \quad (15.7)$$

While equations 15.6 and 15.7 are in fact identical to equations 15.4 and 15.5, respectively, the appearance of symmetry has been lost. Given the equivalence of the models, it should also be noted that one could equally well have chosen to divide numerator and denominator by $\exp[U'(X_1',S_i)]$. For the multinomial case, any alternative may be selected as the base (i.e., the one used to divide numerator and denominator) and a general form produced as equations 15.8 and 15.9 where t is the arbitrary base alternative.

$$p_j^i = \frac{\exp[U'(X_j',S_i)-U'(X_t',S_i)]}{1+\sum_{k \neq t}\exp[U'(X_k',S_i)-U'(X_t',S_i)]} \quad j \neq t \quad (15.8)$$

$$p_t^i = \frac{1}{1+\sum_{k \neq t}\exp[U'(X_k',S_i)-U'(X_t',S_i)]} \quad (15.9)$$

Several important concepts emerge from this process of manipulation. First, it reveals that the logit model is structured around the concept that choice (or probability of occurrence of an event) is determined by the *difference* between the respective utilities. Thus the logit model is based on an implicit decision rule that is revealed by this manipulation. Second, given a linear utility function, it should now be clear why the coefficients of the S_i characteristics in equation 15.3 were designated as being alternative specific. If they were in fact generic, the S_i characteristics would have no effect on the probabilities, since all would cancel out in the difference formulation. The third concept derives from the model form used for calibration. As discussed later in the chapter, the model form for calibration is always that of equations 15.8 and 15.9. From this, it can be seen that all the alternative-specific dummy variables will be in difference form, that is, $(\alpha_{0k}-\alpha_{0t})$. Hence it will only be possible to fit a maximum of $K-1$ such dummies, where there are K alternatives in the data set.

The second property of interest for the logit model concerns the shape of the curve that it yields. For the binary case, this curve is shown in figure 15-1, where the model is

$$p_j^i = \frac{\exp(U_j^i)}{\exp(U_j^i)+\exp(U_k^i)} \quad (15.10)$$

If a probit curve is drawn for equation 15.11, this is found to produce a very similar curve, as shown in figure 15-1.

$$p_j^i = \frac{1}{\sqrt{2\pi}} \int_{-\infty}^{U_j} \exp(-\tfrac{1}{2}u^2)du \quad (15.11)$$

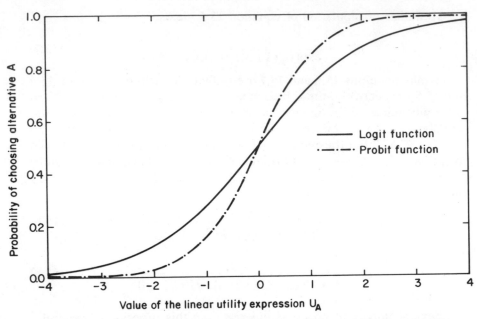

Figure 15–1. Logit and Probit Curves for Equations 15.10 and 15.11

In fact, calibrating probit and logit curves on the same data produces curves that are generally statistically indistinguishable from each other.[6] It must be noted, however, that the assumptions and concepts behind these two models are significantly different and these differences may have far-reaching effects upon the performance and use of the models.

A third property of the logit model is that it is a member of a set of models called "general-share" models.[7] These models have the general form that the dependent variable for case k is proportional to a function of characteristics of case k divided by a summation of a function of the characteristics of all cases. There are many other models that fall into this group, such as those of the gravity type,[8] as transformed for many uses, such as shopping attraction,[9] and trip distribution.[10]

This general-share structure means that new alternatives can be added into the set fairly easily, provided alternative-specific coefficients do not occur in the model. The addition of a new alternative is shown by

$$p_j^i = \frac{\exp(U_j^i)}{\sum_{k=1}^{K} \exp(U_k^i)} \tag{15.12}$$

$$p'^i_j = \frac{\exp(U_j^i)}{\sum_{k=1}^{K+1} \exp(U_j^i)} \tag{15.13}$$

Thus each prior share, p_j^i, is reduced by the addition to the denominator of $\exp(U_{K+1}^i)$ and the share of the new alternative is

$$p_{K+1}^i = \frac{\exp(U_{K+1}^i)}{\sum\limits_{k=1}^{K+1}\exp(U_k^i)} \qquad (15.14)$$

Because the denominator is the same for all alternatives before the addition of the new alternative, the probabilities are proportional to the numerators for each alternative. The effect of the addition of a new alternative is simply to increase this constant denominator. Thus each of the probabilities for the original alternatives will be reduced *in proportion to the original shares* of those alternatives.

Consider now the ratio of the probabilities for two alternatives, m and n.

$$\frac{p_m^i}{p_n^i} = \frac{\exp(U_m^i)}{\exp(U_n^i)} \qquad (15.15)$$

It is clear from equation 15.15 that the ratio of the probabilities of two alternatives is independent of the other alternatives in the choice set. This is another property of the logit model and of all general-share models. The property was formalized by Luce and Raiffa[11] as the "Axiom of the Independence of Irrelevant Alternatives" or the IIA property. The effects of this property in the MNL model have been studied at some length in transportation-planning applications.[12] In general, it may be concluded that this property brings certain advantages of simple structure and ability to handle additions to the set of alternatives. However, problems arise when there is difficulty in defining what represents a *distinct* alternative or when the MNL model is underspecified. It is not the intention here to explore this property in detail, but simply to inform the reader that it is a property of the MNL model and of all general-share models.

A further property of importance concerns the set of alternatives in the multinomial logit model. In equation 15.12, K denoted the set of alternatives appropriate for a particular choice situation. However, there is no requirement in the MNL model that all individuals have all alternatives available to them. It is necessary, rather, that the subset of alternatives available to any individual be completely included in the set specified in the MNL model. For calibrating a MNL model, the restrictions are slightly more stringent. Each individual must not only have a subset of alternatives available but also have at least two alternatives in his or her subset, one of which must be chosen.

To illustrate, consider the decision of mode of travel to work. Suppose the set is defined as car driver, car passenger, bus, rapid transit, commuter rail, and walk all the way. An individual who has only bus, rapid transit, and car passenger available is valid for inclusion in the model for both calibration and prediction. (For prediction, it will be necessary to set the system performance measures for the unavailable alternatives to very large—*not* zero—values, so the probabilities of use of these alternatives are effectively zero. These values will be large positive for measures such as time and cost, but zero or large negative for measures like frequency, reliability, and safety.) An individual who has only car driver available is not valid for calibration, but could be used for prediction. An individual with the alternatives of bus, rapid transit, and taxi is not valid for this model, since this subset of alternatives is not included completely in the model set.

Calibrating the Multinomial Logit Model

Having considered the structure and properties of the MNL model, it is appropriate to turn attention to methods of calibrating the model from sample data. As for the probit model, it is clear that one cannot observe values of the dependent variable, the probability, unless one considers the frequency of outcomes as providing an estimate of the probabilities. Since most applications result in one observation of each individual, such prior probabilities could not be inferred, except on a group basis. This immediately raises a problem for calibration. It suggests the possibility, however, of using a maximum-likelihood fitting procedure, analogous to that used in probit analysis, and this indeed is currently the most widely used process. Again, consider the binary model.

$$p_j^i = \frac{\exp(U_j^i)}{\exp(U_j^i) + \exp(U_k^i)} \tag{15.16}$$

Suppose two individuals are observed in an identical choice situation. Three observations are possible: both may choose alternative j, one may choose j and the other k, or both may choose k. As for the probit model, the likelihood of these three observations is given by p_j^2, $2p_j p_k$, and p_k^2, respectively. Again, these likelihoods represent the terms in the expansion of $(p_j + p_k)^2$. If the experiment were extended to three individuals, four outcomes are possible, given by the expansion of $(p_j + p_k)^3$. It is assumed here that each individual responds independently of the others. Thus it is not more likely that one individual will respond similarly to other individuals, because of the decisions or choices they make. In general, the probability of a particular set of binary choices by n individuals is

$$p(r \mid n) = \frac{n!}{r!(n-r)!} p_j^r p_k^{(n-r)} \tag{15.17}$$

As before, this leads to the likelihood of r responses.

$$L(r_l \mid n) = \prod_{l=1}^{L} \left| \frac{n!}{r_l!(n-r_l)!} \, p_j^{r_l} p_k^{(n-r_l)} \right. \tag{15.18}$$

To determine estimates of the coefficients of the logit model, the likelihood of equation 15.18 is maximized by choosing values of these coefficients that will yield the maximum value for the expression. As is so often the case, it is mathematically easier to maximize the log of the likelihood.

$$\ln L(r_l \mid n) = \sum_l \ln \frac{n!}{r_l!(n-r_l)!} + \sum_l r_l \ln p_j + \sum_l (n-r_l) \ln p_k \tag{15.19}$$

The first term on the right side of equation 15.19 will not change for given n and r_1, as one chooses different sets of coefficients to maximize the likelihood. Hence this term may be omitted from the expression to be maximized. Thus the maximand is

$$L' = \sum_l r_l \ln p_j + \sum_l (n-r_l) \ln p_k \tag{15.20}$$

Before proceeding further, consider the meaning of the summations in equation 15.20. The likelihood expressed in equation 15.17 is the likelihood of observing r choices of alternative j from a group of n individuals and $(n-r)$ choices of alternative k. To estimate the coefficients of the logit model that underlie the probabilities, p_j and p_k, one must have a number of observations of n individuals for each of which r_l choices of alternative j are observed and $(n-r_l)$ choices of alternative k. Without loss of generality, the analyst may assume that n is one and that r_l is zero or one. Then the summation over l is a summation over the set of observations of L individuals who are all faced with a choice between alternatives j and k. Bearing this in mind, one may proceed with the solution as for probit analysis (chapter 14).

To find the values of the coefficients in the logit model that maximize the likelihood, it is necessary to differentiate the maximand (equation 15.20) with respect to each unknown coefficient, set the differentials to zero, and solve the resulting equations. Following the notions of equation 15.3, suppose the utility of the logit model to be fitted is

$$U_j^i = a_0 + a_1 X_{1j} + a_2 X_{2j} + a_3 X_{3j} \tag{15.21}$$

For the general coefficient, a_t, the partial differential of the maximand is

$$\frac{\partial L'}{\partial a_t} = \sum_l r_l \left(\frac{1}{p_j} \frac{\partial p_j}{\partial a_t} \right) + \sum_l (n-r_l) \frac{1}{p_k} \frac{\partial p_k}{\partial a_t} = 0 \tag{15.22}$$

To solve this equation, the partial differentials, $\partial p_j / \partial a_t$ and $\partial p_k / \partial a_t$, must each be determined. This is done most easily by using the binary difference form of the logit model, as developed earlier in the chapter and shown here by

$$p_j^i = \frac{\exp[U_j^i - U_k^i]}{1 + \exp[U_j^i - U_k^i]} \tag{15.23}$$

Writing the expression $[U_j^i - U_k^i]$ as ΔU^i, equation 15.23 can be simplified to

$$p_j^i = \frac{\exp\Delta U^i}{1 + \exp\Delta U^i} \tag{15.24}$$

To obtain the partial differentials, $\partial p_j/\partial a_t$ and $\partial p_k/\partial a_t$, two steps must now be undertaken. First, one must evaluate each of the partial differentials, $\partial p_j/\partial \Delta U^i$ and $\partial p_k/\partial \Delta U^i$.

$$\frac{\partial p_k}{\partial \Delta U^i} = \frac{\exp[\Delta U^i]}{(1+\exp[\Delta U^i])^2} \tag{15.25}$$

$$\frac{\partial p_j}{\partial \Delta U^i} = \frac{\exp[\Delta U^i]}{(1+\exp[\Delta U^i])^2} \tag{15.26}$$

The second step is to evaluate the partial differentials, $\partial \Delta U^i/\partial a_t$, by recollecting that ΔU^i is given in terms of the a_t.

$$\Delta U^i = a_0^i + \sum_{t=1}^{T} a_t^i(X_{tj} - X_{tk}) \tag{15.27}$$

The partial differentials are

$$\frac{\partial \Delta U^i}{\partial a_t} = (X_{tj} - X_{tk}) = \Delta X_t \tag{15.28}$$

Hence the partial differentials required in equation 15.22 are

$$\frac{\partial p_j}{\partial a_t} = \frac{(\exp[\Delta U^i])\Delta X_t}{(1+\exp[\Delta U^i])^2} \tag{15.29}$$

$$\frac{\partial p_k}{\partial a_t} = \frac{-(\exp[\Delta U^i])\Delta X_t}{(1+\exp[\Delta U^i])^2} \tag{15.30}$$

These equations can be simplified further by observing that the two partial differentials of equations 15.25 and 15.26 can be rewritten as equations 15.31 and 15.32, respectively.

$$\frac{\partial p_j}{\partial \Delta U^i} = p_j p_k \tag{15.31}$$

$$\frac{\partial p_k}{\partial \Delta U^i} = -p_j p_k \tag{15.32}$$

Hence equations 15.29 and 15.30 can be simplified to

$$\frac{\partial p_j}{\partial a_t} = p_j p_k \Delta X_t \tag{15.33}$$

$$\frac{\partial p_k}{\partial a_t} = -p_j p_k \Delta X_t \tag{15.34}$$

Returning to equation 15.22, this can be expressed as shown in equation 15.35 by substituting equations 15.33 and 15.34 into 15.22.

$$\frac{\partial L'}{\partial a_t} = \sum_l r_l p_k \Delta X_t - \sum_l (n - r_l) p_j \Delta X_t = 0 \tag{15.35}$$

As in probit analysis, a direct algebraic solution cannot be obtained to the set of equations given by equation 15.35. Instead, an iterative solution is required, in which trial values of the a_ts are proposed and the set of partial differentials evaluated. Initial estimates, a_t', are proposed and the first iteration is used to find amended estimates, a_t''.

$$a_t'' = a_t' + \delta a_t' \tag{15.36}$$

New partial differentials with respect to a_t'' are then found and evaluated, where these differentials are

$$\frac{\partial L'}{\partial (a_t' + \delta a_t')} = 0 \tag{15.37}$$

These equations are solved for the $\delta a_t'$, where the differentials are expanded by a truncated Taylor-MacLaurin expansion of the form shown in equation 15.38, where a_s' is a specific coefficient.

$$\frac{\partial L'}{\partial (a_s' + \delta a_s')} = \frac{\partial L'}{\partial a_s'} + \sum_{t=1}^{T} \delta a_s' \frac{\partial^2 L'}{\partial a_s' \partial a_t'} \tag{15.38}$$

The solution for the first partial differentials, $\partial L'/\partial a_s'$, has been shown in equation 15.35. The second partial differentials may be found by differentiating equation 15.22 again. In general, this is

$$\frac{\partial^2 L'}{\partial a_t \partial a_s} = \sum_l r_l \frac{-1}{p_j^2} \left(\frac{\partial p_j}{\partial a_t}\right) \left(\frac{\partial p_j}{\partial a_s}\right) + \sum_l (n - r_l) \frac{-1}{p_k^2} \left(\frac{\partial p_k}{\partial a_t}\right) \left(\frac{\partial p_k}{\partial a_s}\right) \tag{15.39}$$

By substituting the appropriate expressions for the partial differentials of equation 15.39, equation 15.40 results.

$$\frac{\partial L'}{\partial a_t \partial a_s} = \sum_l \left[-r_l p_k^2 \Delta X_t \Delta X_s - (n - r_l) p_j^2 \Delta X_t \Delta X_s \right] \tag{15.40}$$

From this, solutions can be obtained for the $\delta a_t'$ and the process is repeated until the $\delta a_t'$ become very small, the first partial differentials,

$\partial L'/\partial a_t$, approach zero, and the second partial differentials, $\partial^2 L'/\partial a_t^2$ are negative, showing that a maximum has been reached.

Having achieved a satisfactory maximization of the likelihood function, it is again possible to conduct tests of significance of the fitted coefficients by asymptotic t tests. As for probit analysis, the t value is

$$t = \frac{\hat{a}_t}{\sigma_{\hat{a}_t}} \tag{15.41}$$

where

$$\sigma_{\hat{a}_t} = \left(\frac{\partial^2 L'}{\partial a_t^2}\right)^{-1/2}$$

It is also possible to conduct tests of the entire model using the estimated maximum of the likelihood function, as described in the next section.

Statistical Properties of the MLE Logit Model

As with any maximum-likelihood estimate, certain statistical tests can be conducted on the logit model fitted by the process described in the preceding section. The t test on the coefficients has already been described. The tests described here are on the whole model.

Likelihood-Ratio Tests of the Fitted Model

First, one may put forward the null hypothesis that the phenomenon being modeled can be described by apportioning equal probabilities to all outcomes. Thus if there are n possible alternatives or outcomes, each one would be assigned a probability of $1/n$ under the null hypothesis. This is equivalent to hypothesizing that none of the fitted coefficients, $\hat{a}_t$, is significantly different from zero. For this situation, the abbreviated log likelihood of equation 15.20 can be determined for the null hypothesis by substituting $1/n$ for each p_j and p_k.

Referring to chapter 13, it will be remembered that a likelihood-ratio test can be performed whereby two hypotheses, H_0 and H_1, can be compared and an assessment made of the statistical significance of H_1 relative to H_0. In this case, H_0, the null hypothesis, is that one can predict the relevant probabilities on the basis of equal shares. H_1 is the hypothesis that one can predict the relevant probabilities using the fitted model. Writing the log likelihood of the fitted model as L_1' and that for equal shares as L_0' and the likelihoods as L_1 and L_0, respectively, the desired likelihood ratio is

$$\lambda = \frac{L_0}{L_1} \tag{15.42}$$

Taking logs of equation 15.42 produces

$$\ln\lambda = \ln L_0 - \ln L_1 = L_0' - L_1' \tag{15.43}$$

It must be noted that L_0 should be expected to be smaller than L_1, so λ would be less than one and $\ln\lambda$ would be negative.

The log of the inverse of λ^2 is distributed like the χ^2 statistic. Thus one may say that $-2\ln\lambda$ is the desired statistic to calculate, where a value of $-2\ln\lambda$ greater than the table value of χ^2 at the $\alpha\%$ confidence level indicates that the model can be considered to be significantly better than the equal-shares model with $\alpha\%$ confidence. The degrees of freedom of the χ^2 statistic, in this case, are the number of parameters in the fitted model (including constants).

A more stringent test can be applied by considering an alternative null hypothesis. Clearly, having collected data for fitting a logit model, additional information is available for determining probabilities beyond that of assuming equal shares. For example, the sample data provide estimates of shares for each alternative, without the use of a model. Suppose the sample contains n observations. Suppose further there are three alternative outcomes and n_1 individuals choose the first alternative, n_2 the second, and n_3 the third (where $n = n_1 + n_2 + n_3$). Then prior probabilities can be assigned to every observation, such that $p_1 = n_1/n, p_2 = n_2/n$, and $p_3 = n_3/n$ (and $p_1 + p_2 + p_3 = 1$). These prior probabilities if converted to shares of the sample for each alternative may be termed the "market shares." It would seem logical to use market shares as an alternative null hypothesis. Since the market shares contain more information than the equal shares, the test is necessarily more stringent.

To apply the test for the market shares, L_0' is evaluated for the prior probabilities instead of equal probabilities. In effect, this amounts to fitting a model with constants only and no coefficients, as shown in equations 15.44 through 15.46 for the three-alternative case just described.

$$p_1 = \frac{n_1}{n} = \frac{\exp(a_0^1)}{1 + \exp(a_0^1) + \exp(a_0^3)} \tag{15.44}$$

$$p_2 = \frac{n_2}{n} = \frac{1}{1 + \exp(a_0^1) + \exp(a_0^3)} \tag{15.45}$$

$$p_3 = \frac{n_3}{n} = \frac{\exp(a_0^3)}{1 + \exp(a_0^1) + \exp(a_0^3)} \tag{15.46}$$

Using these values, L'_0 can be evaluated and the same χ^2 test undertaken as before. It must be noted now that the degrees of freedom for this test are the number of distinct coefficients in the fitted model, not counting the alternative-specific constants. Also, the test should only be performed for a model in which a full set of alternative-specific constants is used.

These two alternative tests are basic for a fitted logit model (or any other similar maximum-likelihood model). In general, however, it must be noted that the χ^2 test applied in this way is a weak statistical test, that is, it is easily satisfied. Second, the computed χ^2 value from the ratio of the likelihoods is found to be sensitive to the sample size, increasing as the sample size increases. The tabulated χ^2 is not, however, a function of sample size. Thus models developed from large samples ($n > 150$) will rarely fail either χ^2 test.

Likelihood-Ratio Tests between Fitted Models

Frequently, several different models may be constructed in a search for the best model. In this case, one may construct likelihood-ratio tests to test the models. In addition to the assessment of each individual model against either an equal-share or market-share null hypothesis, the models may more usefully be tested against each other.[13]

Consider the following situation. A data set has been obtained from which M variables are available for use in a logit model. Suppose there are T alternatives available to the sample. A model is built using all M variables (with generic coefficients) and the full set of $(T-1)$ alternative-specific constants. A second model is fitted using a subset of M' variables $(M' - M)$ and the $(T-1)$ alternative-specific constants. For each model specified, a log likelihood can be estimated. These are denoted as L'_M and $L'_{M'}$. A hypothesis can be constructed that the model with the M variables is better than the model with the M' variables. A statistic can be calculated for the likelihood ratio.

$$-2\ln\lambda = -2[L'_{M'} - L'_M] \qquad (15.47)$$

The statistic, $-2\ln\lambda$, is distributed like χ^2, with $M - M'$ degrees of freedom. As before, if the table value of χ^2 is exceeded by $-2\ln\lambda$ at $\alpha\%$, one may state with $\alpha\%$ confidence that the model with M variables is significantly better than the one with M' variables.

Another interesting application of this test is to determine the effect on the goodness-of-fit of the alternative-specific constants. This may be done by fitting one model with the full set of $(T-1)$ constants and a second model with none. The test is carried out by taking the difference in the log likelihoods, as shown in equation 15.47. The degrees of freedom of the χ^2 statistic are $(T-1)$.

The computation of these various likelihood-ratio tests can be facilitated by noting a useful feature. Most computer programs produce the equal-shares χ^2 statistic.

$$-2\ln\lambda = -2[L_1' - L_0'] \tag{15.48}$$

Consider now the test between two models containing M and M' variables, respectively. For each model, a likelihood-ratio test would be performed.

$$-2\ln\lambda_M = -2[L_M' - L_0'] \tag{15.49}$$

$$-2\ln\lambda_{M'} = -2[L_{M'}' - L_0'] \tag{15.50}$$

By inspection, it can be seen that the χ^2 statistic for the test between the models is given by the difference in the two χ^2 statistics.

$$-2\ln\lambda_{M'-M} = -2\ln\lambda_{M'} - (-2\ln\lambda_M) \tag{15.51}$$

This follows because in subtracting the right side of equation 15.49 from equation 15.50, the L_0' terms cancel and the expression becomes $-2[L_{M'}' - L_M']$ which is identical to equation 15.47.

A difficulty arises with the test discussed above. It is not clear what a "better" model constitutes.[14] Adding another variable must increase the likelihood, so the notion of better is rather difficult to define. An alternative way to view this test is to reverse the hypothesis and consider a test of a more restricted model. In other words, the issue becomes whether the model can be simplified without significant loss of explanatory power. The null hypothesis is that the $M - M'$ variables can be dropped without loss of explanatory power. If the χ^2 value obtained from equation 15.51 is larger than the table value of χ^2 for the $\alpha\%$ level, the hypothesis can be rejected with $\alpha\%$ confidence. If the value is less than the table value, one would say with $\alpha\%$ confidence that there is no evidence to reject the hypothesis.

It is also important to note that this test is only valid for the situation in which M' is a proper subset of M. This is so since the test represents an example of nested hypotheses and is not valid for the nonnested case. Thus if any of the variables in M' are not in M, the test is invalid. Similarly, the set of observations for M and M' must be the same. One point should be noted here. It is frequently found that the model with M variables is based upon fewer observations than that with M' variables. This can occur where some observations are rejected because of missing values for one or more of the $M - M'$ variables. In such a case, L_0' for the more complex model (with M variables) will be a larger value (smaller negative) than that of the simpler model. This situation should be corrected before making the comparative likelihood-ratio test. There are two ways to correct this. One should restrict the data set for estimating M' to the same observations as used for M variables. Alternatively, the likelihood for the model with M' variables can be factored by the ratio of the two equal-shares likelihoods. This is done by

$$\text{adj.}(L'_{M'}) = \frac{L'_{0M}}{L'_{0M'}}L'_{M'} = \frac{N_M}{N_{M'}}L'_{M'} \tag{15.52}$$

where

$\quad\quad \text{adj.}(L'_{M'})$ = adjusted log likelihood for the model with M' variables

$\quad\quad\quad\quad L'_{0M}$ = equal-shares log likelihood for the model with M variables

$\quad\quad\quad\quad L'_{0M'}$ = equal-shares log likelihood for the model with M' variables

$\quad\quad\quad\quad N_M$ = number of cases in model with M variables

and $\quad\quad\quad N_{M'}$ = number of cases in model with M' variables

Equation 15.52 is strictly correct only for data in which there are no replicates, that is, the dependent variable is a 0,1 variable and not a measure of frequency. If frequencies are used, the number of cases must be replaced by the number of replicates. Alternatively, the ratio of the equal-shares log likelihoods can be used.

A similar test may also be performed between models developed on different subsets of the same data set.[15] The test is performed in an identical manner to the test described above. Suppose a data set is divided into K subsets, with n_k observations in each. If an identically specified model is fitted to each data subset, a likelihood-ratio test can be performed between each model;

$$-2\ln\lambda_{l-k} = -2[L'_{Ml} - L'_{Mk}] \tag{15.53}$$

where L'_{M_l} = log likelihood for stratum l, using M variables

and $\quad\, L'_{M_k}$ = log likelihood for stratum k, using M variables

More usefully, a test can be made between the segmented models and a pooled model, such as

$$-2\ln\lambda_{p-G} = -2[L'_{M_p} - \sum_i L'_{M_i}] \tag{15.54}$$

where L'_{M_p} = log likelihood for the unsegmented model using M variables

$\quad\quad L'_{M_i}$ = log likelihood for the model of segment i, using M variables

and $\quad p-G$ = test designation for pooled minus segmented (group) models

The degrees of freedom for this test are

$$\text{d.f.} = M(N_G - 1) \qquad (15.55)$$

where M = number of variables
and N_G = number of segments

This test is only valid when all models (pooled and each segment) are identically specified. If the specification is changed in any model, intuitive judgment is the only test basis available.

Frequently, because of missing information on the segmentation variable, the sum of the observations does not equal (is less than) the number of observations used to build the pooled model. In this case, an approximate adjustment is suggested;

$$\text{adj.}(\sum_i L'_{Mi}) = \frac{N_p}{\sum_i N_i} L'_{Mi} \qquad (15.56)$$

where N_p = number of cases for the pooled model
and N_i = number of cases for the model of the ith segment

If a model is built using frequencies instead of 0,1 data, the appropriate numbers to be used are those of replicates, not of cases.

Other Assessments of MLE Logit Models

An index has been proposed[16] that behaves somewhat like the correlation coefficient used in regression and other procedures. The likelihood-ratio index is

$$\rho^2 = 1 - L'_1/L'_0 \qquad (15.57)$$

where L'_1 = log likelihood for fitted model
and L'_0 = log likelihood for equal-shares or market-shares hypothesis

Because the likelihood is always a small positive number ($\ll 1$), the log likelihood will be a large negative number. The null likelihood (L_0) will be equal to or smaller than the fitted likelihood (L_1), that is, the log likelihood, L'_0, will be a larger negative number than L'_1. If the fitted model adds nothing to the null hypothesis, then L'_1 and L'_0 will be approximately equal, so L'_1/L'_0 will approach 1 and ρ^2 will approach zero. As the fitted model diverges further from the null hypothesis, providing a better explanation of the data, the log likelihood of the model, L'_1, will become smaller negative and the ratio L'_1/L'_0 will decline from 1. The absolute maximum likelihood is unity, for which the log likelihood is zero. As this value is approached, the ratio will approach zero and hence ρ^2 will approach unity. Thus ρ^2 behaves in a similar fashion to R^2. However, large values of ρ^2 are much less likely to occur than large values of R^2.[17]

Another test of models of this type can be accomplished by examining the predictive capabilities of the model. In some computer programs, this measure is output as the "percent correctly predicted." In such cases, the measure would be calculated as described below.

For each observation, the predicted probabilities are determined for each alternative from the fitted model. It is assumed that an individual is most likely to choose the alternative that has the highest probability. If this alternative coincides with the observed choice, a one is scored for this observation; if not, a zero is scored. Scores for all observations are summed and expressed as a percentage of all observations. This value is the percent correctly predicted.

McFadden[18] suggested an alternative process that yields a "success index." Suppose the number of observations predicted to choose alternative k who actually choose alternative k is denoted N_{kk}, so that

$$N_{kk} = \sum_{n=1}^{N} S_{kn} P_{kn} \tag{15.58}$$

where $S_{kn} = 1$ if k is chosen by individual n, and zero otherwise
$\qquad P_{kn} =$ calculated probability of individual n choosing alternative k
and $\qquad N_{kk} =$ as previously defined

The proportion successfully predicted, θ, is

$$\theta = \frac{\sum\limits_{k} N_{kk}}{N} \tag{15.59}$$

The proportion of success for a particular alternative, k, can also be calculated;

$$\theta_k = \frac{N_{kk}}{N_k} \tag{15.60}$$

where $N_k =$ observed number of observations choosing alternative k

A success index, σ_k, can then be calculated that corrects for the aggregate share for the alternative.

$$\sigma_k = \frac{N_{kk}}{N_k} - \frac{N_k}{N} \tag{15.61}$$

In effect, the N_k/N fraction represents the market-share hypothesis. Therefore σ_k is corrected for the success that could be achieved with a market-shares assumption. The index will be nonnegative (provided that a full set of alternative-specific constants are used) and will have a maximum value of $(1-N_k/N)$. If one desires the index to lie between zero and one, then σ_k should be normalized to σ_k'.

$$\sigma_k' = \left[\frac{N_{kk}}{N_k} - \frac{N_k}{N}\right] \Big/ \left[1 - \frac{N_k}{N}\right] \qquad (15.62)$$

An overall success index, σ, can be obtained by summing σ_k over all alternatives.

$$\sigma = \sum_{k=1}^{K}\left[\frac{N_{kk}}{N_k} - \frac{N_k}{N}\right] \qquad (15.63)$$

For convenience, it may be more useful to express the overall success index in terms more in line with the proportion successfully predicted.

$$\sigma = \sum_{k=1}^{K}\left[\frac{N_{kk}}{N} - \left[\frac{N_k}{N}\right]^2\right] \qquad (15.64)$$

Again, this index may be normalized to take values between zero and one by dividing through by the maximum value of $[1-\sum_{k}(N_k/N)^2]$ if desired.

The above statistics may also be calculated againt a classification criterion. Instead of using predicted probabilities in calculating N_{kk}, these values are determined by assuming that each individual will choose the alternative for which his or her probability is largest; thus N_{kk} is

$$N_{kk} = \sum_{n}S_{kn} \qquad (15.65)$$

All the other statistics are then calculated in the same way as before.

Some Alternative Fitting Procedures for the MNL Model

This chapter has provided a detailed review of the maximum-likelihood process for fitting the multinominal logit (MNL) model. However, a number of other fitting procedures have been proposed and used from time to time. In this section, several of these methods are reviewed and comments are made on their usefulness and appropriateness.

Linear Least Squares

Before the relatively wide availability of MLE programs for fitting logit models, the use of linear least squares in several forms was proposed. The most direct use was proposed by Theil[19] and is described here briefly.

Consider the binary logit model expressed as

$$p_j = \frac{\exp[U_j - U_k]}{1 + \exp[U_j - U_k]} \tag{15.66}$$

$$p_k = 1 - p_j = \frac{1}{1 + \exp[U_j - U_k]} \tag{15.67}$$

Dividing equation 15.66 by equation 15.67 produces

$$\frac{p_j}{1 - p_j} = \exp[U_j - U_k] \tag{15.68}$$

This equation can be linearized by taking logs of both sides of the equation.

$$\ln\left[\frac{pj}{1 - pj}\right] = [U_j - U_k] \tag{15.69}$$

Since it was proposed in equation 15.3 that the utility function be linear in characteristics, equation 15.69 may be rewritten

$$\ln\left|\frac{p_j}{1 - p_j}\right| = \alpha_{0j} + \sum_s \alpha_s X'_{sj} + \sum_t \alpha_{tj} S_i \tag{15.70}$$

This is now a linear equation for which, given values of p_j, the X'_{sj} and S_i, a linear least-squares solution could be obtained.

Unfortunately, three problems arise for this solution, none of which can be surmounted in a fully satisfactory manner. The first problem is that a solution for individual observations is not possible. This arises because the probabilities are not observable. Observations will be of the form of zero or one. For these values, the left side of equation 15.70 is indeterminate (that is, $\ln[0]$ or $\ln[\infty]$, where the convention is used that $0/1 = 0$ and $1/0 = \infty$). This problem can be surmounted by grouping data in some fashion to produce aggregate proportions of choice. These aggregate proportions will then be the observed values of the dependent variable. Such aggregation, however, introduces both error and arbitrariness into the process, unless it is achieved by multiple observations of each individual.

It is important to consider the form of the error term in this case. If an observation is in error, it can only be that a choice was incorrectly observed. Thus the error term can take only one of two values, according to whether a choice of alternative j was observed incorrectly or correctly. It can be asserted then that the distribution of the error term is bipolar and not normal. Hence the estimates obtained from the linear regression are not maximum-likelihood estimators, although they are still consistent estimators. This constitutes the second reason for not using a linear transform of the logit model.

The third problem relates to the statistical assessment of the fitted model. The model of interest is the untransformed model of equations 15.66 and 15.67. The fitted model, however, would be that of equation 15.70. As pointed out in chapter 8, the goodness-of-fit statistics that are based on normality assumptions or even just central limit theorems for the error term will not hold for the original model, if developed from the transformed model. Thus the usual statistics, such as t and F scores, apply only to the linearized model and not to the model that would be used for prediction.

Direct Linear Fit

Another suggested fitting procedure is based on the fact that the middle portion of the logit curve is approximately linear.[20] Again, this process was put forward before the current emergence of maximum-likelihood fitting procedures. The process is similar to that of linear least squares but avoids the problem of linearization of the logit model. For this method, it is necessary either to compute prior probabilities by grouping data, thus incurring the penalty of arbitrariness once again, or to regress directly onto the 1s and 0s, incurring distributional problems on the error term.

Having identified the prior probabilities between 0.3 and 0.7, a linear model is fitted over this range using standard least-squares procedures;

$$p' = mX + c \tag{15.71}$$

where X = an array of characteristics for each alternative
$\quad\quad m$ = a vector of generic coefficients
and $\quad c$ = a vector of alternative-specific constants

Consider a simple case in which there are two alternatives and one independent variable. By examining the midpoint of the logit model, where $X = 0$, it may be noted that equations 15.72 and 15.73 both hold and are identical.

$$p' = c \tag{15.72}$$

$$p = \frac{\exp(a_0)}{1+\exp(a_0)} \tag{15.73}$$

By solving these two equations (equating p' and p), the constant is found to be

$$a_0 = \ln\left(\frac{c}{1-c}\right) \tag{15.74}$$

In a similar manner, the curve may be examined at the point where $p = p' = 0.5$. It can be shown that at this point the slope of the logit curve is 0.250, where the logit model is

$$p = \frac{\exp(a_0 + a_1 \Delta X)}{1 + \exp(a_0 + a_1 \Delta X)} \tag{15.75}$$

Thus, at this point, equation 15.76 should hold.

$$m = 0.25a_1 \tag{15.76}$$

This method can be extended to a multivariate case by applying the model in such a form as

$$p' = c + m \left[\frac{a_1}{m} X_1 + \frac{a_2}{m} X_2 + \cdots + \frac{a_k}{m} X_k \right] \tag{15.77}$$

The method has not been used for multinomial cases and it seems unlikely that it would work in such situations. While this method circumvents some of the problems of the linear least-squares method, it still does not allow the use of standard goodness-of-fit statistics. It is also a process with a degree of inherent arbitrariness, stemming from the assumptions about the prior probabilities. The method is therefore not recommended.

Nonlinear Least Squares

More recently, McFadden[21] and others noted that maximum-likelihood methods have certain undesirable properties. In particular, investigations have suggested that maximum-likelihood estimates are unduly sensitive to observations with very low calculated probabilities (i.e., observations in the tails of the logit curve). Hence the estimates are likely to be relatively nonrobust if there are errors in such observations. Noting this, more attention has been given to trying to find a fitting procedure that may be more robust.

Manski and McFadden[22] investigated the use of a number of such possible, more robust estimators. One on which they reported is nonlinear least squares, formed by minimizing the function given in equation (15.78).

$$M = \sum_n [S_{kn} - P_{kn}]^2 \tag{15.78}$$

where $S_{kn} = 1$ if k is chosen and zero otherwise
and P_{kn} = calculated probability from a model

Manski and McFadden point out that this procedure gives coefficient estimates that are consistent, but less efficient than maximum-likelihood estimators. Investigations of the robustness of the models fitted by this procedure revealed that the models did appear to be somewhat more robust than MLE models. However, significant differences in the estimated coefficients were not found.

Choice-based Samples

The method of choice-based sampling is described in chapter 6. As indicated there, choice-based sampling has special application to logit models. In much of the early work on applying logit models in transportation planning, choice-based samples were used because of their particular suitability for choice models and their relatively low costs.[23] Because of the extensive use of such samples in this area of work, some seminal works on choice-based samples have been written in the area of transportation applications,[24] and specific properties of logit models calibrated with such samples have been researched.

Of particular interest in this context is the finding by McFadden, reported by Manski and Lerman,[25] that a logit model that is calibrated with a full set of alternative-specific constants on a choice-based sample will provide unbiased estimators for all coefficients *except* the alternative-specific constants. Furthermore, the biases in the alternative-specific constants can be removed by the use of supplementary data. The bias correction is a relatively simple calculation that can be performed after the model is calibrated; the correction is

$$a_{k0}^* = \hat{a}_{k0} - \ln(H_k/Q_k) \qquad (15.79)$$

where a_{k0}^* = unbiased estimate of the alternative-specific constant for alternative k

$\hat{a}_{k0}$ = biased estimate of the alternative-specific constant for alternative k

H_k = market share of alternative k in the choice-based sample

and Q_k = market share of alternative k in the population

As can be seen from equation 15.79, the only supplementary information required is the market shares of the total population for the choice process of concern. Thus, for the particular case of a logit model with a full set of alternative-specific constants, a biased data set from choice-based sampling can be used to produce an unbiased model.

One further property of logit models that is useful and was developed out of a study of choice-based sample properties relates to the updating of estimates from subsequent samples. In most cases, it is expected that the data from two different samples must be pooled first, and a new model estimated from the pooled data. In the case of logit analysis, reestimation is not necessary from pooled data. Suppose a one-variable model has been estimated from a sample of data, producing an estimated coefficient of α_1. A second sample of data is collected and it is desired to update the original model with these data. To do this, a model is estimated, using the same independent variable, to produce an estimate of the coefficient of α_1'. Then the updated coefficient is given by equation 15.80;[26]

$$\alpha_1'' = \frac{\dfrac{\alpha_1}{V(\alpha_1)} + \dfrac{\alpha_1'}{V(\alpha_1')}}{1/V(\alpha_1) + 1/V(\alpha_1')} \tag{15.80}$$

where α_1'' = updated coefficient estimate
and $V(\ )$ = variance of

It is important to note that equation 15.80 holds regardless of the sampling method or situation for the two samples. It is not even necessary for the two samples to have been selected by the same method. The two samples may also derive from different locations or from different points in time.

This process of updating may be extremely time-saving, since it is clearly not necessary to go to the time and cost of pooling data sets and reestimating on the larger data set. This is particularly important because the cost of calibrating a logit model generally rises quadratically with sample size. Hence, if pooling doubles the size of the data set, calibration costs may quadruple.

Equation 15.80 applies to a one-variable model only. The procedure may, however, be extended to a multivariate model, as explained by Lerman, Manski, and Atherton.[27] Briefly, the updating procedure is

$$\theta_v = (\Sigma_1^{-1} + \Sigma_2^{-1})(\Sigma_1^{-1}\theta_1 + \Sigma_2^{-1}\theta_2) \tag{15.81}$$

where θ_v = vector of updated coefficients
θ_1, θ_2 = vectors of coefficients from the two samples
and Σ_1, Σ_2 = matrices of variances and covariances of the coefficients from the two samples

Clearly, the sample that gives the lowest coefficient variances will dominate the estimation of updated coefficients, as is desired. The multivariate equation 15.81 is subject to no restrictions apart from those indicated for equation 15.80. Thus there is still no restriction on how the two samples are drawn and the equation is correct for both geographic and temporal updating.

An Example

For the example of the use of logit analysis, the same data are used as for the examples on discriminant analysis and probit analysis (chapters 12 and 14). The program used to construct the logit model is the same one used to calibrate the probit model of chapter 14, and again uses the Newton-Raphson hill-climbing technique. The estimation proceeds as for probit analysis, using initial estimates of zero for all coefficients except the constant, which is set again as the value to produce the market shares. That value is found, in this case, to be 0.4743.

$$\frac{\exp(0.4743)}{1 + \exp(0.4743)} = 0.6164 \qquad (15.82)$$

This value gives a market-share log likelihood of -105.87.

$$L'(a_0,0,0,0,\ldots0) = 98 \ln(0.6164) + 61 \ln(0.3836) \qquad (15.83)$$

One can also calculate the equal-share log likelihood.

$$L'(0,0,0,\ldots,0) = 159\ln(0.5) = -110.21 \qquad (15.84)$$

In this case, two iterations were sufficient to find a maximum of the likelihood, the log of which was found to be -81.52. The coefficient estimates are shown in table 15–1. As for the probit model, the t scores are all significant at 95%, except for the coefficients of CA1 and sex. A comparison of table 15–1 with table 14–2 shows that the t scores are very similar for both models. However, the coefficient values are fairly different, as would be expected, since this is a different model. The ratio of the cost and time coefficients is very similar in the probit and logit models, as would be expected from figure 15–1.

The likelihood-ratio statistic for the market-shares hypothesis is equal to 48.6, as given by equation 15.85 and has seven degrees of freedom.

$$-2\ln(L_0-L_1) = -2[-105.87-(-81.5)] = 48.6 \qquad (15.85)$$

Table 15–1
Logit Analysis Results for Data of Chapter 12

		Variable Coefficients (t Scores)						Statistics	
Time	Cost	CA1	CA2	Income	Sex	Age	Const.	Log Like-lihood	χ^2 d.f.
−0.134	−0.018	−0.276	2.253	3.369	1.096	0.093	−3.958	−81.5	48.6
(3.4)	(3.1)	(0.5)	(3.3)	(4.1)	(1.5)	(2.0)	(3.1)		(7)

Again, this value is much larger than the table value of χ^2 at 99.5% confidence, so one may have confidence that the model has added significantly to knowledge about the choice process.

One may also conduct an equal-shares test which is less stringent than market-shares. To do this, L_0 is replaced with the likelihood obtained by allocating a probability of 0.5 of transit use to everyone; this yields

$$-2\ln(L_0' - L_1) = -2[-110.21 - (-81.5)] = 57.4 \qquad (15.86)$$

As expected, the value from equation 15.85 is larger than that of equation 15.86 indicating again that the model contributes information that is significantly better than an equal-shares hypothesis. The fact that the value of the likelihood-ratio statistic in equation 15.86 is larger than that of equation 15.85 leads to the notion that the equal-shares test is a "weaker" test, that is, it is more easily satisfied.

For these data, a success table can be constructed using either probabilities or a classification process, as described earlier in the chapter.[28] The classification table, obtained by using equation 15.65, is given in table 15–2 and that using probabilities—equation 15.58—is shown in table 15–3. Based on a classification procedure, 73.6% of the sample is correctly predicted and the overall success index is 0.442 (after standardizing to a maximum value of 1). Using the calculated probabilities, these figures decline to 64.8% and 0.256, respectively. One can also see that there is a greater success in predicting transit ridership (alternative 2) than car use—82.7% to 59% and 71.4% to 54.1%. However, correcting for the prior probabilities, where transit is chosen by 61.6% of the population, the success indices become much more similar for the two modes.

In addition, one may calculate ρ^2 for this model, using equation 15.57.

$$\rho^2 = 1 - L_1'/L_0' = 1 - \frac{81.5}{105.87} \qquad (15.87)$$

Table 15–2
Success Table Using Classification Process

| | Predicted | | Total |
	1	2	Observed
Observed			
1	36	25	61
2	17	81	98
Total predicted	53	106	159
Successes	0.590	0.827	0.736
Success index	0.206	0.211	0.209
Standardized success index	0.334	0.550	0.442

Table 15–3
Success Table Using Calculated Probabilities

	Predicted		Total
	1	2	Observed
Observed			
1	33	28	61
2	28	70	98
Total predicted	61	98	159
Successes	0.541	0.714	0.648
Success index	0.157	0.098	0.121
Standardized success index	0.255	0.255	0.256

This yields a value of 0.230 for ρ^2 for the market-shares hypothesis. A value of ρ^2 of 0.261 is obtained against the equal-shares hypothesis. As indicated by McFadden,[29] these values indicate excellent fit. From all these tests, one may conclude that the model is useful and performs well against the various measures available.

It is also useful to examine two other aspects of this model. The data are choice-based data collected from workers in downtown Chicago who used either car or rapid transit to work from a specific suburb. Suppose information exists that shows the true modal split in the population is 45% transit riders and 55% car users. Since the model of table 15–1 used an alternative-specific constant for transit, where only two alternatives exist, it can be assumed that the coefficients are unbiased estimates, but the constant is biased and should be corrected by using equation 15.79; this is shown by

$$a_0^* = -3.958 - \ln(61.64/45) \tag{15.88}$$

This yields a corrected value of -4.273, which is as expected since the model, as calibrated, is forced to allocate a larger portion of the population to transit than would be found for a random sample. The negative constant represents a bias away from transit and therefore should increase to a larger negative value.

Second, it is useful to see what happens when one updates a model. In this case, a simpler model was built from the same data, but only travel time and an alternative-specific constant were used. This model produced a coefficient for travel time of -0.2688 with a variance of 0.01184. A second data set with 733 observations produced a coefficient of -0.300 with a variance of 0.001587. Applying equation 15.80 to these figures results in

$$\theta_v = \frac{(-0.2688/0.01184) + (-0.300/0.001587)}{(1/0.01184) + (1/0.001587)} \qquad (15.89)$$

This produces an estimate of -0.2963 for the pooled coefficient of travel time, with an estimated variance of 0.001399 and an estimated t score of 7.92. As expected, the estimate from the second data set, which is more reliable since its variance is only about one-tenth of that of the first data set, dominates the estimate of the updated coefficient.

To update the model of table 15–1 would require the use of the entire variance-covariance matrix for the coefficients from both samples. For simplicity, this updating is not dealt with in this text.

Notes

1. J. Berkson, "Application of the Logistic Function to Bio-Assay," *Journal American Statistical Association,* 1944, vol. 39, pp. 357–365.

2. D.J. Finney, *Probit Analysis*, Cambridge, England: Cambridge University Press, 1965.

3. See, for example, Daniel McFadden and Thomas Domencich, *Urban Travel Demand,* Amsterdam: North Holland Press, 1975, chapter 4; Peter R. Stopher and Arnim H. Meyburg, *Urban Transportation Modeling and Planning*, Lexington, Mass.: Lexington Books, D.C. Heath and Co., 1975, chapter 16.

4. W. Weibull, "A Statistical Distribution Function of Wide Applicability," *Journal of Applied Mechanics,* 1951, vol. 18, pp. 293, 297.

5. For further discussion of these terms and their behavioral implications, see Stopher and Meyburg, *Urban Transportation Modeling,* chapter 16.

6. Peter R. Stopher and John O. Lavender, "Disaggregate Behavioral Demand Models: Empirical Tests of Three Hypotheses," *Transportation Research Forum Proceedings,* 1972, vol. 13, no. 1, pp. 321–336.

7. Marvin L. Manheim, "Practical Implications of Some Fundamental Properties of Travel Demand Models," *Highway Research Record No. 422,* 1973, pp. 21–38.

8. Isaac Newton, *Philosophiae Naturalis Principia Mathematica,* London 2d ed., 1713, Book III, Prop. VII and Corr. II, p. 414.

9. Torsten R. Astrom, "Laws of Traffic and Their Applications to Traffic Forecasts with Special Reference to the Sound Bridge Project," Stockholm: Royal Institute of Technology, 1973.

10. Stopher and Meyburg, *Urban Transportation Modeling,* chapter 8.

11. R.D. Luce and H. Raiffa, *Games and Decisions*, New York: John Wiley & Sons, 1957.

12. For an extensive treatment of the IIA property, see Charles River Associates, "Disaggregate Travel Demand Models, Volume II," Final Report on Project 8-13, *National Cooperative Highway Research Program*, Washington D.C., 1977.

13. Peter L. Watson and Richard B. Westin, "Reported and Revealed Preferences as Determinants of Mode Choice Behavior," *Journal of Marketing Research*, 1975, vol. 12, pp. 282–289.

14. This point was suggested in correspondence with Professor Richard B. Westin.

15. Watson and Westin, "Mode Choice Behavior."

16. Daniel McFadden, "Quantitative Methods for Analyzing Travel Behavior of Individuals: Some Recent Developments," in David A. Hensher and Peter R. Stopher, *Behavioral Travel Modeling* (to be published, 1978).

17. McFadden notes that values of ρ^2 of 0.2 to 0.4 represent excellent fit, while such values for R^2 would tend to represent indifferent to poor fit.

18. McFadden, "Quantitative Methods."

19. Henry Theil, "A Multinomial Extension of the Linear Logit Models," *International Economic Review*, 1969, vol. 10, no. 3, pp. 251–259.

20. Peter R. Stopher, "A Probabilistic Model of Travel-Mode Choice," *Highway Research Record No. 238*, 1969, pp. 57–65.

21. McFadden, "Quantitative Models."

22. Charles F. Manski and Daniel McFadden (1977), see Hensher and Stopher, *Behavioral Travel Modeling*, chapter 13.

23. Stopher and Meyburg, *Urban Transportation Modeling*, chapter 15.

24. For example, Steven R. Lerman, Charles F. Manski, and Terence J. Atherton, *Non-Random Sampling in the Calibration of Disaggregate Choice Models*, Final Report to U.S. Department of Transportation, No. PO-6-3-0021, December 1975; Charles F. Manski and Steven R. Lerman, "The Estimation of Choice Probabilities from Choice-Based Samples," *Econometrica*, 1977, Vol. 45, No. 8, pp. 1977–1988.

25. Manski and Lerman, "Estimation of Choice Probabilities."

26. Lerman, Manski, and Atherton, *Non-Random Sampling*.

27. Ibid.

28. See page 335.

29. McFadden, "Quantitative Methods."

16 The Correlation Ratio

The Need for an Alternative Goodness-of-Fit Measure[1]

Much of the latter part of this book is concerned with a variety of modeling methods, a few of which are linear but three are nonlinear. In any experimental work, it may be desirable to test several alternative hypotheses, among which may be the alternatives of a linear or a non-linear relationship among variables. In the various descriptions given of the model-fitting procedures, there is no apparent consistency of goodness-of-fit measures that would permit such hypotheses to be tested. Linear-regression analysis and the allied techniques are assessed by means of a correlation coefficient, R, and an associated F statistic. Discriminant analysis can be tested with a Mahalanobis' D^2 and an F statistic that applies to a different notion than that of the F statistic for regression. Finally, probit and logit analyses can be assessed by means of a likelihood-ratio test that produces a statistic distributed like chi-square. Clearly, these various measures provide no comparability.

A second problem arises with all these methods of curve fitting. Observations are impossible to obtain on the actual dependent variable, since it is a choice probability. The observations are generally of the form of a 0,1 binary variable. Obviously, it is not expected or desired that the curve fit exactly to the observed points. Indeed, if one considers a particular set of values of the independent variables, one would expect to find a scatter of 0 and 1 values of the dependent variable, where the relative frequency of each value would give an approximate measure of the probability of choice associated with that set of values of the independent variables. The curve should not therefore pass through any of the observed points, but should rather be as close as possible to the point represented by the mean of the dependent-variable observations. The correlation coefficient measures the extent of the match between the observed values of the dependent variable and the estimated values obtained from the fitted curve. Thus the correlation coefficient is an inappropriate measure for the models under consideration here.[2]

This point can be demonstrated rather easily by considering the value of R^2, the coefficient of determination, for the situation of a 0,1 dependent variable. Suppose a sample of n observations has been drawn, in which np observations take the value of 1 for the dependent variable and nq observations take the value of 0. In general, the coefficient of determination is

$$R^2 = \frac{[\text{cov}(y,x)]^2}{V(x)V(y)} \tag{16.1}$$

347

The elements of equation 16.1 can be evaluated for the case in question, as shown in the next few steps. The general expression for the covariance (numerator of equation 16.1) is

$$\text{cov}(y,x) = \frac{1}{n}\Sigma (x-\bar{x})(y-\bar{y}) \tag{16.2}$$

When y is a 0,1 variable, then $\bar{y} = p$. Thus equation 16.2 can be rewritten

$$\text{cov}(y,x) = \frac{1}{n}\Sigma (x-\bar{x})(y-p) \tag{16.3}$$

This expression may be partitioned by considering values of x for which $y = 1$ and $y = 0$. Using this partitioning, equation 16.3 can be rewritten

$$\text{cov}(y,x) = \frac{1}{n}\Sigma \{[(x\,|\,y=1)-\bar{x}](1-p) + [(x\,|\,y=0)-\bar{x}](-p)\} \tag{16.4}$$

As noted earlier, there are np values for which $y = 1$ and nq for which $y = 0$. Denoting $E(x\,|\,y=1)$ as $\mu_{x.1}$, and $E(x\,|\,y=0)$ as $\mu_{x.0}$, equation 16.4 may be simplified to

$$\text{cov}(y,x) = pq\mu_{x.1} - pq\mu_{x.0} \tag{16.5}$$

Writing $(\mu_{x.1}-\mu_{x.0})$ as Δ, equation 16.5 becomes

$$\text{cov}(y,x) = pq\Delta \tag{16.6}$$

The variance of the dependent variable can be manipulated in a similar manner to obtain an expression in terms of p and q. The general expression for the variance of y is

$$V(y) = \frac{1}{n}\Sigma y^2 - \bar{y}^2 \tag{16.7}$$

In terms of p and q, equation 16.7 may be rewritten

$$V(y) = \frac{1}{n}[np] - p^2 \tag{16.8}$$

Simplifying, this becomes

$$V(y) = p(1-p) = pq \tag{16.9}$$

Finally, the variance of x may be rewritten by partitioning values according to whether they yield a 0 or 1 for y. Defining $V(x\,|\,y=1)$ as $\sigma^2_{x.1}$ and $V(x\,|\,y=0)$ as $\sigma^2_{x.0}$, the variance of x may be written

$$V(x) = p\sigma^2_{x.1} + q\sigma^2_{x.0} + pq\Delta^2 \tag{16.10}$$

Thus equation 16.10 may be rewritten

$$R^2 = \frac{(pq\Delta)^2}{pq[p\sigma^2_{x.1}+q\sigma^2_{x.0}+pq\Delta^2]} \tag{16.11}$$

Dividing throughout by pq results in

$$R^2 = \frac{pq\Delta^2}{p\sigma_{x.1}^2 + q\sigma_{x.0}^2 + pq\Delta^2} \tag{16.12}$$

Perfect fit is considered to be achieved only when the value of R^2 is unity. It is therefore worth considering under what conditions R^2, as given in equation 16.12, will be unity. Each term in equation 16.12 must be positive, since p and q must lie between 0 and 1, and all other terms are squared values. A value of unity for R^2 can therefore be achieved only under the conditions given by

$$pq\Delta^2 = p\sigma_{x.1}^2 + q\sigma_{x.0}^2 + pq\Delta^2 \tag{16.13}$$

Given the nonnegativity of all the terms in equation 16.13, the value of unity can occur only under the conditions given by

$$p\sigma_{x.1}^2 = q\sigma_{x.0}^2 = 0 \tag{16.14}$$

Clearly, this equation implies that both $\sigma_{x.1}$ and $\sigma_{x.0}$ are zero, which can occur only if all the values of x for which $y = 0$ are also identical to each other. In other words, x must also be a binary variable which is perfectly partitioned with y.

That an indication of perfect fit can only be provided by R^2 under such extremely restrictive conditions attests to the inadequacy of R^2 under these circumstances. Hence, it is clear that an alternative measure for R^2 is required for this general category of models.

An Alternative Measure of Goodness of Fit

It was suggested earlier in the chapter that a more appropriate measure of goodness-of-fit would be one that examined the closeness of the mean of the values of the dependent variables for a specific set of values of the independent variables, and the value estimated from the fitted curve. This notion suggests the possible applicability of a measure called the *correlation ratio*.[3] This measure has been used in statistics in general for many years, but its applicability to this situation has been recognized only very recently.[4]

Consider a set of x and y values where y is a continuous dependent variable. Suppose the xs are integer values that represent the means of a set of class intervals of the independent variable. Thus there will be an array of y values for each x value. Denote the ith array mean by x_i, and a value within the array as y_{ij}. Within each array, a mean and variance of the y values in the array may be defined;

$$V(y_i) = \frac{1}{n_i} \sum_j (y_{ij} - \bar{y}_i)^2 \tag{16.15}$$

where y_{ij} = a set of values of y in the ith array
$\bar{y}_i$ = the mean of the y values in the ith array
and n_i = number of observations in the ith array

The sum of squares of the deviations of the ys from their array means over the entire data set may then be written

$$S_y^2 = \frac{1}{N}\sum_i \sum_j (y_{ij} - \bar{y}_i)^2 \tag{16.16}$$

where $N = \sum_i n_i$

If each array of ys were represented by its mean, $\bar{y}_i$, then S_y^2 represents the variance of the ys which would be lost by such a procedure. The total variance of the y_{ij}s is σ_y^2.

$$\sigma_y^2 = \frac{1}{N}\sum_i \sum_j (y_{ij} - \bar{y})^2 \tag{16.17}$$

The variance, σ_y^2, represents the variance of the points y_{ij} about the global mean, $\bar{y}$. A third variance can also be estimated, namely, the variance of the array means about the global mean.

$$\sigma_{\bar{y}}^2 = \frac{1}{N}\sum_i n_i(\bar{y}_i - \bar{y})^2 \tag{16.18}$$

The two variances, S_y^2 and $\sigma_{\bar{y}}^2$, must sum to the total variance, σ_y^2.

$$\sigma_y^2 = \sigma_{\bar{y}}^2 + S_y^2 \tag{16.19}$$

The variance, S_y^2, is the within-array variance (or within-group variance) and $\sigma_{\bar{y}}^2$ is the between-array variance. If all the observations in each array were at the mean of the array, then S_y^2 would be zero and $\sigma_{\bar{y}}^2$ would be the same as σ_y^2. If all the array means are identical and there is an identical scatter of points within each array, then $\sigma_{\bar{y}}^2$ is zero and S_y^2 is identical to σ_y^2. Consider a variance ratio, as given by

$$\eta_{yx}^2 = 1 - S_y^2/\sigma_y^2 \tag{16.20}$$

This ratio gives a measure of the ability of the means, $\bar{y}_i$, to represent the individual observations, the y_{ij}s. Clearly, if all the values y_{ij} are at the means $\bar{y}_i$, then S_y^2 is 0 and η_{yx}^2 is 1.

Similarly, S_y^2 cannot exceed σ_y^2, so η_{yx}^2 behaves like R^2 in that its values lie between 0 and 1, and it is a measure of association. (Conventionally, η_{yx} is always considered to be the positive root of η_{yx}^2, unlike the sign considerations of R.)

The relationship between η_{yx} and R may also be examined. The sum of squares of deviations is always least when measured from the mean of the arrays. Consider a curve, fitted to the values y_{ij}, where the estimated values from the curve within each array are given by $\hat{y}_i$. Consider further the sums of squares within the arrays for the entire data set. The sum of squares about the array estimates is

$$SR_y^2 = \sum_i \sum_j (y_{ij} - \hat{y}_i)^2 \tag{16.21}$$

As already stated, the minimum sum of squares will be that of the y_{ij}s about the array means, $\bar{y}_i$. Hence equation 16.22 follows.

$$NS_y^2 \leqslant SR_y^2 \tag{16.22}$$

However, equation 16.21 represents the residual sum of squares from a regression. The coefficient of determination can be expressed in terms of this residual sum of squares and the total variance.

$$R_{yx}^2 = 1 - \frac{\frac{1}{N} SR_y^2}{\sigma_y^2} \tag{16.23}$$

Given equations 16.20 and 16.22, it follows that η_{yx}^2 must always be equal to or greater than R_{yx}^2. Further, neither R_{yx}^2 nor η_{yx}^2 can be greater than 1 or less than 0. Hence the relationships are

$$1 \geqslant \eta_{yx}^2 \geqslant R_{yx}^2 \geqslant 0 \tag{16.24}$$

A further association can be demonstrated between R^2 and η_{yx}^2. If there is a linear relationship between y and x, the means of the arrays of y, $\bar{y}_i$, all lie on the straight line. Thefore $\hat{y}_i$ is equal to $\bar{y}_i$ and S_y^2 is equal to $SR_{y/N}^2$. In this case, R^2 is equal to η_{yx}^2. Thus the difference $(\eta_{yx}^2 - R^2)$ is an indication of a departure from linearity in the relationship between y and x.

The correlation ratio is clearly not dependent upon the shape of the relationship between x and y. As S_y^2 approaches zero, the closer are the individual values, y_{ij}, to the means of the arrays, $\bar{y}_i$. If S_y^2 is zero, all the values of y_{ij} are at the means and lie on a curve joining the means across values of x_i. Thus there is a functional relation of unspecified geometric form between x and y. Hence η_{yx}^2 may be considered to be a measure of association between y and x, irrespective of the precise functional form of the relationship.

The mathematical expression for η_{yx}^2, given in equation 16.20, is not very convenient for computation. Using similar properties to those used for the coefficient of determination, η_{yx}^2 may also be expressed as

$$\eta_{yx}^2 = \frac{\sigma_{\bar{y}}^2}{\sigma_y^2} \tag{16.25}$$

Tests of Significance

An F statistic can be formed by taking the ratio of two independent variances, as discussed in chapter 7. To test the significance of a given correlation ratio, the F statistic of equation 16.26 can be used, so that

$$F = \frac{\sigma_{\bar{y}}^2/(h-1)}{S_y^2/(N-h)} \tag{16.26}$$

where h = the number of arrays.

The degrees of freedom of this F statistic are $(h-1)$ and $(N-h)$, respectively. By calculating the value of equation 16.26 and comparing it with the table value of F at $\alpha\%$, one may have $(100-\alpha)\%$ confidence that the value of η_{yx}^2 did not arise by chance, if the calculated value exceeds the table value.

It can be seen that the F statistic can also be expressed in terms of the correlation ratio, since equations 16.25 and 16.20 can be rearranged to give S_y^2 and $\sigma_{\bar{y}}^2$ as functions of η_{yx}^2.

$$S_y^2 = (1-\eta_{yx}^2)/\sigma_y^2 \tag{16.27}$$

$$\sigma_{\bar{y}}^2 = \eta_{yx}^2/\sigma_y^2 \tag{16.28}$$

Hence equation 16.26 can be rewritten as equation 16.29 which is analogous to the F statistic derived for the correlation coefficient.

$$F = \frac{\eta_{yx}^2(N-h)}{(1-\eta_{yx}^2)(h-1)} \tag{16.29}$$

Application to a 0,1 Dependent Variable

One of the major reasons given at the beginning of the chapter for rejecting the correlation coefficient was its unsuitability to the 0,1 dependent-variable case for nonlinear models such as logit, probit, and discriminant analysis. It is necessary therefore to consider whether or not the correlation ratio is applicable to this case.

Consider the values of the two mean squares, $\sigma_{\bar{y}}^2$ and σ_y^2, for the 0,1 case. In determining the value of R^2 for the 0,1 case, it was established that σ_y^2 was given by equation 16.30 (see equation 16.9).

$$\sigma_y^2 = p(1-p) \tag{16.30}$$

The variance of the array means about the global mean, $\sigma_{\bar{y}}^2$, is

$$\sigma_{\bar{y}}^2 = \frac{1}{N}\sum_i n_i \, (\bar{y}_i - \bar{y})^2 \tag{16.31}$$

where N = total number of observations
and n_i = number of observations in the ith array

If the mean of the ith array is written as P_i and W_i is the proportion of observations in the ith array, then $\sigma_{\bar{y}}^2$ can be written

$$\sigma_{\bar{y}}^2 = \frac{1}{N}\sum_i NW_i\,(P_i-P)^2 \tag{16.32}$$

Thus the correlation ratio can be written as equation 16.33 for the 0,1 dependent-variable case.

$$\eta_{yx}^2 = \frac{\sum W_i\,(P_i-P)^2}{P(1-P)} \tag{16.33}$$

In this case, it is not possible to demonstrate whether or not the correlation ratio can achieve its maximum value of 1.0 under less restrictive conditions than those applying to the correlation coefficient. It is necessary, instead, to work a few examples to see how the correlation ratio behaves, both in absolute terms and relative to the correlation coefficient.

An Example

In table 16–1 data are presented that indicate a perfect fit to a dichotomous situation. The data consist of 10 observations and inspection shows that given a value of x_i, y_i can be predicted with absolute certainty—hence the assertion that the data represent perfect fit. Consider, first, the correlation coefficient for these data. Using equation 16.12, the values shown in table 16–2 are obtained, yielding a value of R^2 of 0.73. Similarly, the correlation ratio can be calculated by splitting the data into four arrays. The values obtained from this are shown in table 16–3. From these values, using equation 16.33, η^2 is found to be 1.0.

Hence the correlation ratio correctly reflects the data situation of perfect fit, while the correlation coefficient and coefficient of determination yield values that suggest less than perfect fit.

Table 16–1
Data for a Nonlinear Binary Dependent Variable

x	y	x	y
1	0	3	1
1	0	3	1
2	0	3	1
2	0	4	1
2	0	4	1

Table 16–2
Values for Calculating R^2 from Data of Table 16–1

Measure	Value
$\mu_{x.0}$	1.6
$\mu_{x.1}$	3.4
Δ	1.8
$\sigma^2_{x.0}$	0.3
$\sigma^2_{x.1}$	0.3

Table 16–3
Values for Calculating η^2 from Data of Table 16–1

Measure	Value
$p(1-p)$	0.25
$W_1(p_1-p)^2$	0.05
$W_2(p_2-p)^2$	0.075
$W_3(p_3-p)^2$	0.075
$W_4(p_4-p)^2$	0.05

Uses of the Correlation Ratio

The principal use of the correlation ratio is to assist in the assessment of the goodness-of-fit of a model, or in comparing the goodness-of-fit of several alternative models. So far, the correlation ratio has been demonstrated as a measure between two variables, y and x. To assess the goodness-of-fit of a model requires the rather simple notion of obtaining the correlation ratio between the observed values of a dependent variable, y_j, and the estimated values, $\hat{y}_j$.

This use of the correlation ratio raises a problem. As formulated, the correlation ratio assumes there is a natural basis for classifying the data into discrete arrays, based on the x variable. The actual values of the x variable then become irrelevant. What is important is only the behavior of the y values within the arrays. If one now applies the correlation ratio to the assessment of goodness-of-fit of a model with an observed 0,1 dependent variable, a question arises as to how to group the data into arrays. One variable is a binary variable which would thus restrict the correlation ratio to two arrays only. As shown later in the chapter, two arrays are too few for the assessment of goodness-of-fit. The other variable—the estimated values of y, $\hat{y}_j$—is a continuous variable, which is unlikely to exhibit a natural grouping into arrays. In a simple case, it may be possible to consider each separate value of $\hat{y}_j$ to represent an array, provided there

Table 16–4
Logit Model Estimates for Table 16–1

x	p	x	p
1	.378	3	.622
1	.378	3	.622
2	.5	3	.622
2	.5	4	.731
2	.5	4	.731

are sufficient values in each array to permit a reasonably reliable estimate to be made of the array mean and within-array variance. This problem is discussed further in the next section.

Consider again the data used in the preceding example. A logit model might be fitted to this, using the values

$$p = \frac{\exp[0.5x-1]}{1 + \exp[0.5x-1]} \tag{16.34}$$

This model yields the values shown in table 16–4. Now consider table 16–5 for the correlation ratio. If the data are grouped by the distinct values of $\hat{y}_j$, it can be seen that the situation becomes identical to that of table 16–1, so the correlation ratio will take the value of 1.0. Note that the actual values of $\hat{y}_j$ do not affect the value of the correlation ratio, but only the grouping against y_j.

This example reveals an interesting property of the correlation ratio with respect to 0,1 variables. If the data are perfectly dichotomized on the independent variable(s), that is, a perfect split of values can be obtained for the independent variables in relation to the 0 values of y and the 1 values of y, the correlation ratio will indicate perfect fit between the dependent (0,1) variable and the independent variable(s). This also means that any monotonic relationship will fit the data, exhibiting perfect fit on the basis of the correlation ratio. For the example above, the reader can

Table 16–5
Observed and Estimated Dependent-Variable Values for the Model of Equation 16.34

$\hat{y}_j$	y_i	$\hat{y}_j$	y_j
.378	0	.622	1
.378	0	.622	1
.5	0	.622	1
.5	0	.731	1
.5	0	.731	1

Table 16–6
Imperfect-fitting Data for a 0,1 Dependent Variable

x_j	y_i	x_j	y_i
1	0	3	0
1	0	3	0
1	0	3	1
1	0	3	1
1	0	3	1
2	0	4	1
2	0	4	1
2	0	4	1
2	1	4	1
2	1	4	1

quickly confirm that any monotonically increasing relationship between x and y will yield perfect fit.

Considering this property of the correlation ratio, it would seem to be of interest to investigate its behavior when the data on x and y do not exhibit perfect fit. A different set of data is shown in table 16–6, where a perfect dichotomy does not exist.

For these data, table 16–7 shows the appropriate statistics for calculating each of R^2 and η^2_{yx}. From these figures, R^2 is found to be 0.48, while η^2_{yx} is 0.83. In this case, the contrast between the two statistics is even more pronounced than in the first example. It is notable on this occasion, however, that η^2_{yx} is less than unity. In fact, this example and an inspection of the form of the correlation ratio show that perfect fit for a 0,1 variable is defined only by the situation of complete dichotomy of the

Table 16–7
Statistics for Data of Table 16–6

Statistic	Value
$\mu_{x.0}$	1.7
$\mu_{x.1}$	3.3
Δ	1.6
$\sigma^2_{x.0}$	0.689
$\sigma^2_{x.1}$	0.689
p	0.5
q	0.5
$W_1(p_1-p)^2$	0.1
$W_2(p_2-p)^2$	0.004
$W_3(p_3-p)^2$	0.004
$W_4(p_4-p)^2$	0.1
$p(1-p) = pq$	0.25

values of x (or $\hat{y}$) against the 0 and 1 values of y. This shows that with respect to 0,1 variables the correlation ratio represents only a marginal improvement on the correlation coefficient in terms of the conditions under which perfect fit will occur. To recap, the correlation coefficient indicates perfect fit only when both the variables concerned are binary and are perfectly partitioned. The correlation ratio indicates perfect fit only when the dependent variable can be perfectly partitioned with respect to the independent variable(s).

Some Problems

The major problem relating to the correlation ratio concerns the assumption that the independent variable is a naturally partitioned variable. In many desired applications of the correlation ratio to binary data and models developed from such data, the independent variable(s) and the estimated dependent variable, $\hat{y}_j$, will be continuous. Grouping of one of the variables or sets of variables must now be done arbitrarily. Unfortunately, the values of η^2_{yx} and F are dependent on the number of class intervals used. This affects η^2_{yx} principally through the estimation of within-array and between-array variances and means. It affects F through both the value of η^2_{yx} and the numbers of degrees of freedom of each of the two sums of squares. Work by Sucher[5] has demonstrated the problem using two data sets, one with 211 observations and one with 1308 observations. For each data set, a model was estimated and the correlation ratio and associated F statistic were calculated with each of 5, 10, and 20 class intervals. It is important to note that in line with the original notions of a naturally classified independent variable, the arrays were made of equal size, relative to the independent variable. The results are shown in table 16–8.[6]

Table 16–8
Effects on Correlation Ratio and F Statistic of Varying the Number of Class Intervals

Data Set	Number of Intervals	η^2	F	Degrees of Freedom D_1	D_2	Table F at 0.1%
1	5	0.109	6.27	4	206	4.80
	10	0.125	3.19	9	201	3.30
	20	0.150	1.78	19	191	2.30
2	5	0.265	117.24	4	1303	4.66
	10	0.288	58.28	9	1298	3.14
	20	0.293	28.15	19	1288	2.42

It can be seen that the correlation ratio increases in value as more class intervals are used, while the F value falls. A further problem was also noted, although it is not apparent from the figures in table 16–8. This was that for the smaller data set, 1, the number of observations in some of the intervals when 20 class intervals were used decreased to one or none, thus not permitting means or variances to be calculated. Even with 10 class intervals, some class intervals contained very few observations, casting some doubts on the validity of the within-array statistics.

In this context, it is useful to note some measures proposed by Elderton.[7] Elderton suggested that a minimum significant value of η^2 is

$$\min(\eta^2) = E(\eta^2) + 2 \text{ s.e. } (\eta^2) \tag{16.35}$$

where

$$E(\eta^2) = \frac{h-1}{N-1}$$

$$\text{s.e. } (\eta^2) = \frac{1}{N-1}\sqrt{\frac{2(h-1)(N-h)}{(N+1)}}$$

and N, h are as defined before.

This value can be calculated for any value of h, thus permitting a further assessment of the number of class intervals proposed. Alternatively, from empirical work, one might be able to propose an acceptable value for the minimum of η^2 and use this to determine the number of class intervals to use.

For the results of table 16–8, the minimum value of η^2 can be calculated and is shown in table 16–9.[8] Figure 16–1 summarizes many of these figures. It can be seen that for the smaller data set the minimum value of η^2 is almost the same as the calculated value for 20 class intervals. This clearly shows the problems of using this many class intervals with such a

Table 16–9
Minimum Values of η^2 from Elderton's Formula

Data Set	Class Intervals	η^2	η^2_{min}
1	5	0.109	0.045
	10	0.125	0.083
	20	0.150	0.146
2	5	0.265	0.007
	10	0.288	0.013
	20	0.293	0.024

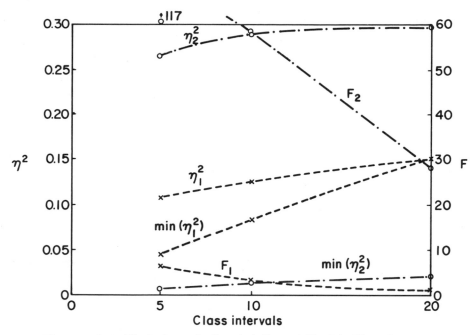

Figure 16–1. Variation of η^2, min(η^2), and F with Class Intervals

small data set. For the larger data set, the minimum value is far less than the calculated value for all class intervals. Also, figure 16–1 shows that over this range of class intervals the calculated value of η^2 is growing faster than the minimum η^2.

Summary

In the beginning of the chapter, the correlation coefficient was shown to be inappropriate for limited dependent variables and thus not to provide a means for comparing models developed from procedures such as discriminant, probit, and logit analyses. The correlation ratio has been demonstrated to be a more appropriate measure than the correlation coefficient, since it can produce a perfect-fit indication for such data under less restricted conditions and is more readily calculable for limited dependent variables. Unfortunately, when the independent variable is not naturally classified into intervals, it has been shown that the arbitrary choice of intervals may have significant effects upon the values of the correlation ratio and its associated F statistic.

In the light of the above considerations, it seems best to recommend some caution on the use of the correlation ratio. As Stopher noted:

> . . . provided that a consistent number of class intervals is used in a set of comparisons of models, the correlation ratio remains an appropriate and useful comparative goodness-of-fit measure for nonlinear models. Until the arbitrariness of determining class intervals is adequately researched, the correlation ratio must be regarded as a comparative, not an absolute measure.[9]

It should be apparent from this and earlier chapters that there still are insufficient means to establish the adequacy and goodness-of-fit of nonlinear models constructed from limited dependent variables. Hence much further research should be undertaken in this topic, particularly given the increasing interest in and use of such models.

A Practical Example

To illustrate the use of the correlation ratio, two situations are examined. First, the reader will recall that the same data were used to calibrate a discriminant-analysis model in chapter 12, a probit model in chapter 14, and a logit model in chapter 15. As these models were reported, it is difficult to make comparisons between them for goodness-of-fit. The discriminant analysis model provided values of D^2 and F of 1.564 and 8.08, respectively. The probit model gave a χ^2 value of 53.4 and ρ^2 can be computed as 0.252 for the market-shares case. Finally, the logit model gave a value of χ^2 of 48.6 and ρ^2 of 0.230, also for the market-shares case. The issue then is to see if it is possible to obtain some comparative measure of these three models. All three models use 159 observations and 7 variables, plus a constant.

The correlation ratio and its associated F statistic was computed for all three models, using 10 equal-sized class intervals for the predicted choice. The results of these calculations are shown in table 16–10. In all cases, F has 9 and 149 degrees of freedom. The table value of F with these

Table 16–10
Correlation Ratios for the Models from Chapters 12, 14, and 15

Model	η	F
Discriminant	0.246	5.40
Probit	0.367	9.59
Logit	0.351	8.95

degrees of freedom at 0.1% is 3.36. One would conclude that all three models are therefore highly significant. However, it appears that the logit and probit models are superior to the discriminant model, although these two models are virtually indistinguishable from one another.

Notes

1. This chapter is based heavily on Peter R. Stopher, "Goodness-of-Fit Measures for Probabilistic Travel Demand Models," *Transportation,* 1975, vol. 4, pp. 67–83.

2. J. Neter and E.S. Maynes, "On the Appropriateness of the Correlation Coefficient with a 0,1 Dependent Variable," *Journal of American Statistical Association,* 1970, vol. 65, no. 330, pp. 501–509.

3. Neter and Maynes, "On the Correlation Coefficient"; N.L. Johnson and Frederick C. Leone, *Statistics and Experimental Design in Engineering and the Physical Sciences,* New York: John Wiley & Sons, 1964, vol. 2, pp. 78–84; C.E. Weatherburn, *A First Course in Mathematical Statistics,* Cambridge, England: Cambridge University Press, 1962, pp. 88–95, 221–222.

4. Neter and Maynes, "On the Correlation Coefficient"; Stopher, "Goodness-of-Fit Measures."

5. Peter O. Sucher, *An Intra-City Study of A Proxy Variable for Convenience,* unpublished M.S. thesis, Department of Environmental Engineering, Cornell University, Ithaca, N.Y., 1973.

6. Table 16–8 is reproduced from Stopher, "Goodness-of-Fit Measures."

7. W.P. Elderton, *Frequency Curves and Correlation,* New York: Cambridge University Press, 1938.

8. Table 16–9 is reproduced from Stopher, "Goodness-of-Fit Measures."

9. Stopher, "Goodness-of-Fit Measures," p. 80.

Appendix
Statistical Tables and Tables of Random Numbers

Table A-1
Random Numbers

08210	32973	08003	54512	64863	78634	36344	73293	77660	90199
56822	81546	04735	15228	37475	79057	48749	89898	95554	39734
89182	97110	37211	11688	69132	97318	83419	82573	72506	55631
72674	35938	58483	08612	23129	95675	47142	74290	97894	17531
90431	35286	24567	05594	75288	52097	07687	25548	36274	59709
26899	07988	28526	97601	03588	46466	55143	65558	25847	59878
18345	22317	08035	94062	39380	80045	80231	29109	50908	15304
59231	50655	21753	70621	91045	11318	73098	42457	45905	58889
25446	92080	34212	70395	64493	86192	96633	11311	62422	32794
64825	37691	90465	51579	83918	22217	55262	47065	36168	65350
56902	23614	90057	12880	89522	81453	41294	14603	39279	54730
94781	64586	97431	01310	04978	25833	61253	48959	53841	12864
55140	32776	03346	61125	91434	84200	73664	45914	90551	83073
71284	19521	04812	76028	20044	78244	58015	05867	53847	04736
12938	37575	74532	60769	26928	29666	84318	21283	17537	31834
74888	92697	43016	56171	27506	72769	60939	50508	37332	77234
92592	82165	90227	88560	86908	32283	83958	92590	97910	40983
13741	37073	04809	23138	34852	45986	73823	08534	45867	18800
84984	54147	59332	70137	24535	17957	21012	51279	67206	63087
60528	16871	20195	14420	27684	35554	90414	80832	86089	61328
74340	27455	43144	85543	85639	91044	58951	85251	69761	57118
22071	81837	14952	50079	08851	56337	76769	83059	23254	73901
46002	37283	12700	60802	93248	12585	88835	97516	69585	78330
08389	69818	88126	63220	09217	89809	41827	63143	94194	69001
80520	76260	83873	30245	69153	36222	71188	50184	69731	00382
86956	16358	66811	24846	57071	40649	92444	63235	89322	99933
53136	21244	18794	69981	67197	20656	01157	43212	71717	36254
36225	08162	77121	36757	48280	01627	42182	90286	33607	39507
94504	39905	37479	59621	90903	25258	02601	40884	18925	19143
62903	81216	82151	05623	29557	17355	75959	26697	92117	50471
49684	94165	26723	59644	97156	95754	14613	62639	19567	24652
74368	66450	66013	32765	29673	05637	18535	59269	40732	90216
65947	65189	25064	99907	41696	83361	05711	44600	31021	34714
48562	64985	78945	43273	93601	41398	47898	87513	71023	77141
56879	44941	45205	84341	92222	50193	34172	63957	65014	93068
56352	13144	30800	60403	84654	28738	48297	30614	71784	23440
02272	83016	25229	43692	36151	96505	73151	82126	30327	15817
33209	67546	92428	91522	75736	59065	00668	43097	75282	81904
04042	27494	05204	86656	04755	03880	91082	33651	63806	19334
06255	03873	64275	81962	00650	39269	73457	07644	21562	67255
02637	88632	65996	41115	33863	95116	66522	55347	30163	50594
66788	56790	14606	63123	05831	14117	04288	63610	48924	01827
75185	54242	23276	86228	68663	31051	10832	72330	70330	91769
97678	41209	34692	03790	01489	45480	18809	16099	52339	41146
98264	76970	35868	18540	25973	28110	52129	77706	18505	89583
85483	81260	62113	64994	60821	40518	94273	64432	48779	05310
11364	56369	72114	94681	74135	49670	21072	78489	39630	89270
52663	97730	76199	52766	34479	03055	92954	43490	39790	46437
86276	71122	87556	29504	32565	48447	08719	83014	23615	14317
92295	34604	61004	19707	97387	90740	48389	57337	76910	82215

Reproduced from "A Million Random Digits with 100,000 Normal Deviates," by the RAND Corp.; reprinted with permission of the Free Press.

Table A-1 (continued)

91225	47297	05208	09509	83287	98993	04792	82551	59606	88054
48832	04241	71986	08556	40419	69537	86871	54707	41149	16991
83516	35332	54964	28304	46934	61746	09772	20208	36456	51403
55814	15346	17425	41510	13329	09591	71725	31094	34654	45090
85716	12864	61976	24101	23601	62813	47996	57362	30232	35867
77799	89902	53499	34027	44773	91246	93487	85827	35988	31423
89346	94359	64580	88245	21215	78937	18180	62989	17247	96211
22821	26700	43247	48748	35591	77935	97016	92278	91298	23566
19651	46588	74048	25245	88242	89392	74849	23163	74727	89559
21738	10422	44197	57245	23564	05076	18267	27692	18681	49264
14439	16349	58690	24767	66401	63240	44038	15142	81338	70308
25482	05354	72238	80246	75754	88446	87496	92774	28165	06299
14606	94425	14315	64213	96364	29901	94156	13008	34784	34997
47291	66501	04111	98604	76249	16047	95252	69177	23764	57974
00097	39513	26145	50286	37804	95165	97489	83770	80511	71298
44474	18685	83439	63916	76277	87092	43999	65474	45455	17684
80188	55310	74084	41674	80282	46222	74965	69025	10428	30224
99909	70398	88267	96784	22232	74548	18681	71053	49820	54954
58968	12199	67836	95022	67725	67527	86541	97150	74569	90047
19893	22171	37003	03270	40464	39309	71950	31827	28303	62957
31180	66582	07814	48192	79581	82781	59678	20881	03922	96690
55358	46206	28790	27657	47210	39684	69566	95109	17541	67975
45265	25613	50103	93017	49489	63137	42899	46824	55305	68436
78752	50062	52099	49755	47455	85377	13404	12583	42142	94438
77026	65887	30936	69948	52651	44038	14192	65084	94240	30663
39276	97558	34925	86347	06528	94788	98409	12127	61672	09999
47532	77074	39717	09655	69029	12061	62872	18773	11799	42629
99298	62008	14744	81394	50813	60959	17941	99294	68438	54384
23713	29543	20617	02525	49301	62333	84918	38377	45095	89424
70125	93654	46311	61173	48844	38937	03812	05838	34285	08267
74948	69730	38268	45877	74220	17727	68357	92038	16486	72612
01975	51053	74679	33939	04308	29308	00031	52498	46210	21401
19636	08802	65859	83454	29762	95675	80618	46154	81250	49413
37063	11564	68775	32383	78364	35447	70729	31821	41957	96850
06570	48472	76950	25543	37661	13124	05752	28250	06892	32216
67187	70029	32276	51020	16715	26725	00374	24518	85007	95592
74318	16668	14616	51147	63823	28920	63506	67422	21521	62018
84658	32328	48257	69420	57437	18892	88152	43925	07585	13485
43578	54413	29390	82628	06420	48451	80697	68097	22577	12231
65336	91369	07765	92143	34215	96303	03353	71515	55424	68205
12297	99455	36506	53575	42859	03056	54436	72004	90550	24695
07592	19189	36976	54389	52519	88593	12640	63742	52863	57294
72348	55701	98604	75531	73266	45496	74386	51293	20682	99981
70909	48599	36829	27150	21839	05236	20499	47538	84775	44543
16013	75265	65054	51584	65837	44116	49457	46055	92802	10073
39954	51272	93372	19705	20047	81087	62993	40227	95610	75971
61131	59612	43759	27369	68613	88117	88168	62985	01794	51874
01608	31737	72572	47112	73336	86842	54882	81541	97497	42052
59312	10832	96622	32093	71354	71923	25833	55831	35692	71534
90697	91454	99243	74995	80926	93834	49471	55910	09853	12529

Table A–2
Random Normal Deviates

2.689	.408-	1.144-	.234-	1.212	.600	.846-	.485	1.775	2.397-
.287	1.442-	.747	.223	2.070-	.624-	.828	1.147	.141-	.841
1.136	.706	.221-	1.333	.575	.195-	1.386-	.497	1.338-	.986
1.482-	.447-	1.145-	.675-	1.619	.677	2.443	.588-	.497	1.297
1.107-	.503	1.831	1.497	1.533	1.367-	.731-	1.259	1.534	.276
.110-	.356-	.298	.105	.069-	.683	1.485	.368	2.496	2.025
.614	1.007-	.198-	1.510-	.794-	1.119-	.279-	.380-	.262	.191
1.385-	.869	.483	1.207-	.150	.596-	.818	.671-	.192	1.265-
.636-	.610	.853-	2.817	.331-	.959-	.603	.823-	1.900	1.559
.492	1.721-	.561-	1.172	.334-	.721-	1.729-	1.328-	.829	.298-
.698-	.985	1.780	2.934-	.298-	.446-	2.239	.585	.602-	.410-
.055	1.740-	1.338	.741-	.131-	1.071-	2.042	.326	1.409	2.189-
.487	.024	.755-	.158	.066	.480-	1.134	1.081	.549-	.083-
.087-	.163-	.240-	1.032-	.071-	1.197	1.609-	.520-	.427-	1.236
.599	.299-	1.346	1.099-	.696	1.550-	.295-	.544-	1.129-	.532
.157	.019	.112-	1.665	2.587-	2.037	.610-	.335	.433	1.021-
.405	.918	.483-	1.311	.051-	.709	.802	.545-	3.066	.677
.717	.445-	.212	1.463	.040-	.522	2.481-	1.069	1.124	.763-
.459-	.074	.355	.393-	1.198	.761	1.343-	.730-	.915-	.279-
.955	2.253	1.188	2.261-	.147	1.615-	.074	.196	.981-	1.087-
.092-	.669-	2.217-	1.110-	1.085-	.386	.843	.364-	.011-	.775-
.453	.223	.365	.523	.039-	.262-	1.517	.419-	.563	.153-
.777-	.441-	.913-	.314	.276	1.424-	.804	.824	.210	.627-
.174-	.024-	.102-	.629	.576	1.196-	.198	.149	.737	1.224
.122	.192-	.826	1.677	.669	.906-	1.366	.614	.813-	.815
1.027	.511-	.032-	1.117	.702-	.773-	1.259-	.333	1.920-	.667
.684-	.714	.715-	.754-	.715-	1.385-	.626-	1.440	.860	1.114
.185-	1.151-	.648-	.506	1.982-	1.030	.642	.671	1.196-	2.169-
.002	.726	.143	.703-	.117-	.053-	1.074	.508	.447	1.314-
.227-	.261	2.677	.852-	.551-	2.044	1.130	.860	.488	.218-
.736-	1.046	.060-	1.737-	.188	.515	.637	1.945	1.504-	.148-
.093-	.093	.321-	.810	.535-	1.897-	1.031	.138-	.644	.214
.418-	.022-	.581-	.192	.093	.801-	.520	.056	.875-	.774-
.343-	.743	1.041	1.299	.828	.340-	.003	1.084-	.248-	.563-
.698	.998-	.909-	.568-	2.846-	1.132-	1.210	1.521-	.869	.154
.268	.769	1.563	.143-	.276-	.052-	.462	.916	1.556-	.789
.209	.215	.960-	1.161	.277-	.861-	.745-	.362-	.086-	.032
2.246	.514-	.407	.851-	.139-	.017	1.831-	.339	.282-	1.161
.275-	.181	.208	1.691-	.504-	1.134-	1.317-	.288	1.749-	.789
1.395-	.358-	1.356	.268	.733-	.525-	.030	1.952	.053	1.143-
.435	.172-	.593	.486-	.372	1.887	1.233	.118	1.103-	.422-
1.386-	.941-	.947	1.598	.345-	.367	1.802	.543-	.905	.696-
1.127-	.030	.488-	.984	.567	1.632	.590	.914-	.615-	2.176
.147-	1.567	.344	1.530	.506-	.919-	.313	.663-	.163	.011-
.353	.782-	.915-	1.210-	1.279-	.718	.024-	1.220-	.559-	1.371-
.274-	1.364	.249	.736	1.351	.750	.467-	1.220	.918	.912-
.254-	1.156-	.809	1.196	2.020	2.101	1.732	1.001	.623-	1.012-
.925-	.726	1.160	.520-	1.475-	1.352	1.735	.020-	1.244	.482
1.719-	1.263-	.357	1.114	1.437-	.866-	.268	.805	.219-	1.488
1.820-	.351	2.012	.297-	.321-	.450-	.689-	1.942-	.348-	1.060

Table A–2 (continued)

.322-	.402-	1.130-	.076-	1.297-	.093	1.178-	.196	.966-	2.066-
.895	1.071-	2.171-	.973	1.193	2.017-	1.677	.282	2.271	.218
.543	.052-	1.538-	.401	.705-	.971	2.169	2.169-	.665-	.587
1.459-	.375	.002	.415-	.219-	1.209-	1.211	1.068-	1.066-	1.226
.681-	1.101	.408-	.038-	.137-	.474-	.441-	1.084	1.489	.229
1.238-	.878-	.838	.472-	1.236-	.678	.189-	1.439	1.472-	1.440
.551-	1.199	.404-	.503-	2.099	.654-	.248	1.076	.798-	.972
.255-	.874	.655	.558-	.952	.085	.262-	.404	1.095-	.338
.141-	1.019	1.054-	1.567	1.885-	1.955-	1.553	.276	.942	.120-
.264	.458	.494	.482-	.717-	.530-	1.425-	1.237	.217-	.863
.156-	.058-	1.778-	.100-	1.157	.748	.057-	.343-	.330-	1.584
1.245-	.194-	.054-	.005	.855-	1.546-	2.032-	.693-	.236	.498
.775	.379-	.725-	.603-	.665	.169-	.788	.254	.172-	1.592
1.205-	1.360	.734-	.583-	.033	2.249-	2.248-	.865-	.956	.532
.871-	.491	1.323	.182-	.157-	.083	1.983	.527-	.790	.848
.176	1.874-	.090	.446	.130	.471-	.205	1.010	.515-	1.080-
.915	.420	2.309	1.716-	.992	.727	.291	1.585	.459-	.527-
2.128	.985	.457	1.356-	.405	.394-	1.084	.544	.724-	1.062
.329	1.720-	.644	1.407	.158-	.206-	.773-	.216	1.119-	.903-
.177-	1.412	.526	1.139-	.403	.965-	1.010-	.207	.437	.541-
.876	.237-	.262-	1.762-	.020	.846-	1.136-	.476	1.238	1.608-
1.748-	1.015	.203-	.315-	.854	.502	.562	2.089	.728-	.788-
.976-	1.613	.551-	.844	1.783	.298-	1.075-	1.566-	.018-	.544-
1.196	.967	.432-	1.825-	.639	1.254	.915-	.967-	.509	.905-
.581	.418	2.592-	.636-	.849-	.333-	1.320	1.076-	.451-	1.825-
1.431	1.211-	1.140	.091	.287-	.519	.675-	.057	.771-	1.257
.005-	.557-	.722-	.459-	1.361-	.961-	.648	.321	.433-	1.184
.261	.969-	1.680	2.453-	1.062	1.185-	.697-	.166-	.023-	.771
.250	.847-	.125-	.309-	1.569-	.309	.231	.631-	1.242	.443-
.392-	.471-	1.005-	.866-	.255-	1.211	.264-	1.157	.174-	1.155
1.008	.232	.775	1.048-	.237-	.159	1.200-	.876-	1.202-	.505-
.779	.023-	.362-	.715	.455-	.084-	1.293-	.477	.470	.499
.367-	.964-	1.495	.070	.178-	.756-	.499-	.335	.395	.065-
2.139-	.698	1.596	.326	.474	1.570-	.126	.615-	.177-	.870
.995-	.019-	1.012	.006-	.329	.473-	.577	.781	1.354-	.668
.628	1.548-	1.400-	.434	.116-	.669-	1.286	1.263	2.352	.198
.937-	.110-	.129	.911-	1.866-	1.831	.643-	.208-	1.459	.557-
.254	.821	2.036	.755-	.683	.966	.410	.175-	1.534-	1.056-
.415	2.194-	.013	.611	.542	.535-	1.285	.843-	.539-	.235-
.084	.267	.981	1.583-	.225-	.905-	.159-	.899	1.661	1.429
1.228	.535	.485	.540-	.143	1.025	.486-	1.534	1.540-	.746-
.283-	.455-	.250	.339	1.215-	.769-	.500	1.702-	.134	1.728-
1.356-	.354-	.330-	.396-	1.292	.236	.933-	.204-	.161-	.510-
1.331	.578	.581	1.214-	1.074	.357-	.739	1.054-	.604-	.697-
.352	.681	1.413-	2.343-	1.745	.157	.593-	1.126	.036	1.815
1.147-	.261	.094-	.258	1.055	1.302	.671-	.824-	.503-	.057
.583-	.761-	1.230-	1.183	.951-	.295	.043-	.872	.199	1.060
.390-	.779-	1.640-	.997-	.408	1.697	.893-	.021	1.140	.508
1.477-	.056	1.626	.463	2.800	1.267-	2.698-	.581	.713	.193-
2.192	.753-	.674	.426	1.448-	.142-	.910	.885	1.837	.537-

Table A–3
Percentage Points of the χ^2 Distribution

ν \ Q	0·990	0·950	0·900	0·750	0·500	0·250
1	157088.10^{-9}	393214.10^{-8}	0·0157908	0·1015308	0·454936	1·32330
2	0·0201007	0·102587	0·210721	0·575364	1·38629	2·77259
3	0·114832	0·351846	0·584374	1·212534	2·36597	4·10834
4	0·297100	0·710723	1·063623	1·92256	3·35669	5·38527
5	0·554298	1·145476	1·61031	2·67460	4·35146	6·62568
6	0·872090	1·63538	2·20413	3·45460	5·34812	7·84080
7	1·239043	2·16735	2·83311	4·25485	6·34581	9·03715
8	1·64650	2·73264	3·48954	5·07064	7·34412	10·2189
9	2·08790	3·32511	4·16816	5·89883	8·34283	11·3888
10	2·55821	3·94030	4·86518	6·73720	9·34182	12·5489
11	3·05348	4·57481	5·57778	7·58414	10·3410	13·7007
12	3·57057	5·22603	6·30380	8·43842	11·3403	14·8454
13	4·10692	5·89186	7·04150	9·29907	12·3398	15·9839
14	4·66043	6·57063	7·78953	10·1653	13·3393	17·1169
15	5·22935	7·26094	8·54676	11·0365	14·3389	18·2451
16	5·81221	7·96165	9·31224	11·9122	15·3385	19·3689
17	6·40776	8·67176	10·0852	12·7919	16·3382	20·4887
18	7·01491	9·39046	10·8649	13·6753	17·3379	21·6049
19	7·63273	10·1170	11·6509	14·5620	18·3377	22·7178
20	8·26040	10·8508	12·4426	15·4518	19·3374	23·8277
21	8·89720	11·5913	13·2396	16·3444	20·3372	24·9348
22	9·54249	12·3380	14·0415	17·2396	21·3370	26·0393
23	10·19567	13·0905	14·8480	18·1373	22·3369	27·1413
24	10·8564	13·8484	15·6587	19·0373	23·3367	28·2412
25	11·5240	14·6114	16·4734	19·9393	24·3366	29·3389
26	12·1981	15·3792	17·2919	20·8434	25·3365	30·4346
27	12·8785	16·1514	18·1139	21·7494	26·3363	31·5284
28	13·5647	16·9279	18·9392	22·6572	27·3362	32·6205
29	14·2565	17·7084	19·7677	23·5666	28·3361	33·7109
30	14·9535	18·4927	20·5992	24·4776	29·3360	34·7997
40	22·1643	26·5093	29·0505	33·6603	39·3353	45·6160
50	29·7067	34·7643	37·6886	42·9421	49·3349	56·3336
60	37·4849	43·1880	46·4589	52·2938	59·3347	66·9815
70	45·4417	51·7393	55·3289	61·6983	69·3345	77·5767
80	53·5401	60·3915	64·2778	71·1445	79·3343	88·1303
90	61·7541	69·1260	73·2911	80·6247	89·3342	98·6499
100	70·0649	77·9295	82·3581	90·1332	99·3341	109·141
X	−2·3263	−1·6449	−1·2816	−0·6745	0·0000	+0·6745

$Q = Q(\chi^2|\nu) = 1 - P(\chi^2|\nu) = 2^{-\frac{1}{2}\nu}\{\Gamma(\frac{1}{2}\nu)\}^{-1}\int_{\chi3}^{\infty}e^{-\frac{1}{2}x}x^{\frac{1}{2}\nu-1}dx$. For $\nu>100$ take $\chi^2 = \nu\left\{1 - \dfrac{2}{9\nu} + X\sqrt{\dfrac{2}{9\nu}}\right\}^3$ or $\chi^2 = \frac{1}{2}\{X + \sqrt{(2\nu - 1)}\}^2$, according to the degree of accuracy required. X is the standardized normal deviate corresponding to $P = 1 - Q$, and is shown in the bottom line of the table.

Table A–3 (continued)

0·100	0·050	0·025	0·010	0·005	0·001
2·70554	3·84146	5·02389	6·63490	7·87944	10·828
4·60517	5·99146	7·37776	9·21034	10·5966	13·816
6·25139	7·81473	9·34840	11·3449	12·8382	16·266
7·77944	9·48773	11·1433	13·2767	14·8603	18·467
9·23636	11·0705	12·8325	15·0863	16·7496	20·515
10·6446	12·5916	14·4494	16·8119	18·5476	22·458
12·0170	14·0671	16·0128	18·4753	20·2777	24·322
13·3616	15·5073	17·5345	20·0902	21·9550	26·125
14·6837	16·9190	19·0228	21·6660	23·5894	27·877
15·9872	18·3070	20·4832	23·2093	25·1882	29·588
17·2750	19·6751	21·9200	24·7250	26·7568	31·264
18·5493	21·0261	23·3367	26·2170	28·2995	32·909
19·8119	22·3620	24·7356	27·6882	29·8195	34·528
21·0641	23·6848	26·1189	29·1412	31·3194	36·123
22·3071	24·9958	27·4884	30·5779	32·8013	37·697
23·5418	26·2962	28·8454	31·9999	34·2672	39·252
24·7690	27·5871	30·1910	33·4087	35·7185	40·790
25·9894	28·8693	31·5264	34·8053	37·1565	42·312
27·2036	30·1435	32·8523	36·1909	38·5823	43·820
28·4120	31·4104	34·1696	37·5662	39·9968	45·315
29·6151	32·6706	35·4789	38·9322	41·4011	46·797
30·8133	33·9244	36·7807	40·2894	42·7957	48·268
32·0069	35·1725	38·0756	41·6384	44·1813	49·728
33·1962	36·4150	39·3641	42·9798	45·5585	51·179
34·3816	37·6525	40·6465	44·3141	46·9279	52·618
35·5632	38·8851	41·9232	45·6417	48·2899	54·052
36·7412	40·1133	43·1945	46·9629	49·6449	55·476
37·9159	41·3371	44·4608	48·2782	50·9934	56·892
39·0875	42·5570	45·7223	49·5879	52·3356	58·301
40·2560	43·7730	46·9792	50·8922	53·6720	59·703
51·8051	55·7585	59·3417	63·6907	66·7660	73·402
63·1671	67·5048	71·4202	76·1539	79·4900	86·661
74·3970	79·0819	83·2977	88·3794	91·9517	99·607
85·5270	90·5312	95·0232	100·425	104·215	112·317
96·5782	101·879	106·629	112·329	116·321	124·839
107·565	113·145	118·136	124·116	128·299	137·208
118·498	124·342	129·561	135·807	140·169	149·449
+1·2816	+1·6449	+1·9600	+2·3263	+2·5758	+3·0902

Table A–4
Percentage Points of the t Distribution

ν	Q=0·4 2Q=0·8	0·25 0·5	0·1 0·2	0·05 0·1	0·025 0·05	0·01 0·02	0·005 0·01	0·0025 0·005	0·001 0·002	0·0005 0·001
1	0·325	1·000	3·078	6·314	12·706	31·821	63·657	127·32	318·31	636·62
2	·289	0·816	1·886	2·920	4·303	6·965	9·925	14·089	22·327	31·598
3	·277	·765	1·638	2·353	3·182	4·541	5·841	7·453	10·214	12·924
4	·271	·741	1·533	2·132	2·776	3·747	4·604	5·598	7·173	8·610
5	0·267	0·727	1·476	2·015	2·571	3·365	4·032	4·773	5·893	6·869
6	·265	·718	1·440	1·943	2·447	3·143	3·707	4·317	5·208	5·959
7	·263	·711	1·415	1·895	2·365	2·998	3·499	4·029	4·785	5·408
8	·262	·706	1·397	1·860	2·306	2·896	3·355	3·833	4·501	5·041
9	·261	·703	1·383	1·833	2·262	2·821	3·250	3·690	4·297	4·781
10	0·260	0·700	1·372	1·812	2·228	2·764	3·169	3·581	4·144	4·587
11	·260	·697	1·363	1·796	2·201	2·718	3·106	3·497	4·025	4·437
12	·259	·695	1·356	1·782	2·179	2·681	3·055	3·428	3·930	4·318
13	·259	·694	1·350	1·771	2·160	2·650	3·012	3·372	3·852	4·221
14	·258	·692	1·345	1·761	2·145	2·624	2·977	3·326	3·787	4·140
15	0·258	0·691	1·341	1·753	2·131	2·602	2·947	3·286	3·733	4·073
16	·258	·690	1·337	1·746	2·120	2·583	2·921	3·252	3·686	4·015
17	·257	·689	1·333	1·740	2·110	2·567	2·898	3·222	3·646	3·965
18	·257	·688	1·330	1·734	2·101	2·552	2·878	3·197	3·610	3·922
19	·257	·688	1·328	1·729	2·093	2·539	2·861	3·174	3·579	3·883
20	0·257	0·687	1·325	1·725	2·086	2·528	2·845	3·153	3·552	3·850
21	·257	·686	1·323	1·721	2·080	2·518	2·831	3·135	3·527	3·819
22	·256	·686	1·321	1·717	2·074	2·508	2·819	3·119	3·505	3·792
23	·256	·685	1·319	1·714	2·069	2·500	2·807	3·104	3·485	3·767
24	·256	·685	1·318	1·711	2·064	2·492	2·797	3·091	3·467	3·745
25	0·256	0·684	1·316	1·708	2·060	2·485	2·787	3·078	3·450	3·725
26	·256	·684	1·315	1·706	2·056	2·479	2·779	3·067	3·435	3·707
27	·256	·684	1·314	1·703	2·052	2·473	2·771	3·057	3·421	3·690
28	·256	·683	1·313	1·701	2·048	2·467	2·763	3·047	3·408	3·674
29	·256	·683	1·311	1·699	2·045	2·462	2·756	3·038	3·396	3·659
30	0·256	0·683	1·310	1·697	2·042	2·457	2·750	3·030	3·385	3·646
40	·255	·681	1·303	1·684	2·021	2·423	2·704	2·971	3·307	3·551
60	·254	·679	1·296	1·671	2·000	2·390	2·660	2·915	3·232	3·460
120	·254	·677	1·289	1·658	1·980	2·358	2·617	2·860	3·160	3·373
∞	·253	·674	1·282	1·645	1·960	2·326	2·576	2·807	3·090	3·291

$Q = 1 - P(t|\nu)$ is the upper-tail area of the distribution for ν degrees of freedom, appropriate for use in a single-tail test. For a two-tail test, $2Q$ must be used.

Table A–5
Percentage Points of the F Distribution

Upper 10% points

ν_2 \ ν_1	1	2	3	4	5	6	7	8	9	10	12	15	20	24	30	40	60	120	∞
1	39·86	49·50	53·59	55·83	57·24	58·20	58·91	59·44	59·86	60·19	60·71	61·22	61·74	62·00	62·26	62·53	62·79	63·06	63·33
2	8·53	9·00	9·16	9·24	9·29	9·33	9·35	9·37	9·38	9·39	9·41	9·42	9·44	9·45	9·46	9·47	9·47	9·48	9·49
3	5·54	5·46	5·39	5·34	5·31	5·28	5·27	5·25	5·24	5·23	5·22	5·20	5·18	5·18	5·17	5·16	5·15	5·14	5·13
4	4·54	4·32	4·19	4·11	4·05	4·01	3·98	3·95	3·94	3·92	3·90	3·87	3·84	3·83	3·82	3·80	3·79	3·78	3·76
5	4·06	3·78	3·62	3·52	3·45	3·40	3·37	3·34	3·32	3·30	3·27	3·24	3·21	3·19	3·17	3·16	3·14	3·12	3·10
6	3·78	3·46	3·29	3·18	3·11	3·05	3·01	2·98	2·96	2·94	2·90	2·87	2·84	2·82	2·80	2·78	2·76	2·74	2·72
7	3·59	3·26	3·07	2·96	2·88	2·83	2·78	2·75	2·72	2·70	2·67	2·63	2·59	2·58	2·56	2·54	2·51	2·49	2·47
8	3·46	3·11	2·92	2·81	2·73	2·67	2·62	2·59	2·56	2·54	2·50	2·46	2·42	2·40	2·38	2·36	2·34	2·32	2·29
9	3·36	3·01	2·81	2·69	2·61	2·55	2·51	2·47	2·44	2·42	2·38	2·34	2·30	2·28	2·25	2·23	2·21	2·18	2·16
10	3·29	2·92	2·73	2·61	2·52	2·46	2·41	2·38	2·35	2·32	2·28	2·24	2·20	2·18	2·16	2·13	2·11	2·08	2·06
11	3·23	2·86	2·66	2·54	2·45	2·39	2·34	2·30	2·27	2·25	2·21	2·17	2·12	2·10	2·08	2·05	2·03	2·00	1·97
12	3·18	2·81	2·61	2·48	2·39	2·33	2·28	2·24	2·21	2·19	2·15	2·10	2·06	2·04	2·01	1·99	1·96	1·93	1·90
13	3·14	2·76	2·56	2·43	2·35	2·28	2·23	2·20	2·16	2·14	2·10	2·05	2·01	1·98	1·96	1·93	1·90	1·88	1·85
14	3·10	2·73	2·52	2·39	2·31	2·24	2·19	2·15	2·12	2·10	2·05	2·01	1·96	1·94	1·91	1·89	1·86	1·83	1·80
15	3·07	2·70	2·49	2·36	2·27	2·21	2·16	2·12	2·09	2·06	2·02	1·97	1·92	1·90	1·87	1·85	1·82	1·79	1·76
16	3·05	2·67	2·46	2·33	2·24	2·18	2·13	2·09	2·06	2·03	1·99	1·94	1·89	1·87	1·84	1·81	1·78	1·75	1·72
17	3·03	2·64	2·44	2·31	2·22	2·15	2·10	2·06	2·03	2·00	1·96	1·91	1·86	1·84	1·81	1·78	1·75	1·72	1·69
18	3·01	2·62	2·42	2·29	2·20	2·13	2·08	2·04	2·00	1·98	1·93	1·89	1·84	1·81	1·78	1·75	1·72	1·69	1·66
19	2·99	2·61	2·40	2·27	2·18	2·11	2·06	2·02	1·98	1·96	1·91	1·86	1·81	1·79	1·76	1·73	1·70	1·67	1·63
20	2·97	2·59	2·38	2·25	2·16	2·09	2·04	2·00	1·96	1·94	1·89	1·84	1·79	1·77	1·74	1·71	1·68	1·64	1·61
21	2·96	2·57	2·36	2·23	2·14	2·08	2·02	1·98	1·95	1·92	1·87	1·83	1·78	1·75	1·72	1·69	1·66	1·62	1·59
22	2·95	2·56	2·35	2·22	2·13	2·06	2·01	1·97	1·93	1·90	1·86	1·81	1·76	1·73	1·70	1·67	1·64	1·60	1·57
23	2·94	2·55	2·34	2·21	2·11	2·05	1·99	1·95	1·92	1·89	1·84	1·80	1·74	1·72	1·69	1·66	1·62	1·59	1·55
24	2·93	2·54	2·33	2·19	2·10	2·04	1·98	1·94	1·91	1·88	1·83	1·78	1·73	1·70	1·67	1·64	1·61	1·57	1·53
25	2·92	2·53	2·32	2·18	2·09	2·02	1·97	1·93	1·89	1·87	1·82	1·77	1·72	1·69	1·66	1·63	1·59	1·56	1·52
26	2·91	2·52	2·31	2·17	2·08	2·01	1·96	1·92	1·88	1·86	1·81	1·76	1·71	1·68	1·65	1·61	1·58	1·54	1·50
27	2·90	2·51	2·30	2·17	2·07	2·00	1·95	1·91	1·87	1·85	1·80	1·75	1·70	1·67	1·64	1·60	1·57	1·53	1·49
28	2·89	2·50	2·29	2·16	2·06	2·00	1·94	1·90	1·87	1·84	1·79	1·74	1·69	1·66	1·63	1·59	1·56	1·52	1·48
29	2·89	2·50	2·28	2·15	2·06	1·99	1·93	1·89	1·86	1·83	1·78	1·73	1·68	1·65	1·62	1·58	1·55	1·51	1·47
30	2·88	2·49	2·28	2·14	2·05	1·98	1·93	1·88	1·85	1·82	1·77	1·72	1·67	1·64	1·61	1·57	1·54	1·50	1·46
40	2·84	2·44	2·23	2·09	2·00	1·93	1·87	1·83	1·79	1·76	1·71	1·66	1·61	1·57	1·54	1·51	1·47	1·42	1·38
60	2·79	2·39	2·18	2·04	1·95	1·87	1·82	1·77	1·74	1·71	1·66	1·60	1·54	1·51	1·48	1·44	1·40	1·35	1·29
120	2·75	2·35	2·13	1·99	1·90	1·82	1·77	1·72	1·68	1·65	1·60	1·55	1·48	1·45	1·41	1·37	1·32	1·26	1·19
∞	2·71	2·30	2·08	1·94	1·85	1·77	1·72	1·67	1·63	1·60	1·55	1·49	1·42	1·38	1·34	1·30	1·24	1·17	1·00

$F = \dfrac{s_1^2}{s_2^2} = \dfrac{S_1/\nu_1}{S_2/\nu_2}$, where $s_1^2 = S_1/\nu_1$ and $s_2^2 = S_2/\nu_2$ are independent mean squares estimating a common variance σ^2 and based on ν_1 and ν_2 degrees of freedom, respectively.

Table A-5 (continued)

Upper 2.5% points

ν_2 \ ν_1	1	2	3	4	5	6	7	8	9	10	12	15	20	24	30	40	60	120	∞
1	647.8	799.5	864.2	899.6	921.8	937.1	948.2	956.7	963.3	968.6	976.7	984.9	993.1	997.2	1001	1006	1010	1014	1018
2	38.51	39.00	39.17	39.25	39.30	39.33	39.36	39.37	39.39	39.40	39.41	39.43	39.45	39.46	39.46	39.47	39.48	39.49	39.50
3	17.44	16.04	15.44	15.10	14.88	14.73	14.62	14.54	14.47	14.42	14.34	14.25	14.17	14.12	14.08	14.04	13.99	13.95	13.90
4	12.22	10.65	9.98	9.60	9.36	9.20	9.07	8.98	8.90	8.84	8.75	8.66	8.56	8.51	8.46	8.41	8.36	8.31	8.26
5	10.01	8.43	7.76	7.39	7.15	6.98	6.85	6.76	6.68	6.62	6.52	6.43	6.33	6.28	6.23	6.18	6.12	6.07	6.02
6	8.81	7.26	6.60	6.23	5.99	5.82	5.70	5.60	5.52	5.46	5.37	5.27	5.17	5.12	5.07	5.01	4.96	4.90	4.85
7	8.07	6.54	5.89	5.52	5.29	5.12	4.99	4.90	4.82	4.76	4.67	4.57	4.47	4.42	4.36	4.31	4.25	4.20	4.14
8	7.57	6.06	5.42	5.05	4.82	4.65	4.53	4.43	4.36	4.30	4.20	4.10	4.00	3.95	3.89	3.84	3.78	3.73	3.67
9	7.21	5.71	5.08	4.72	4.48	4.32	4.20	4.10	4.03	3.96	3.87	3.77	3.67	3.61	3.56	3.51	3.45	3.39	3.33
10	6.94	5.46	4.83	4.47	4.24	4.07	3.95	3.85	3.78	3.72	3.62	3.52	3.42	3.37	3.31	3.26	3.20	3.14	3.08
11	6.72	5.26	4.63	4.28	4.04	3.88	3.76	3.66	3.59	3.53	3.43	3.33	3.23	3.17	3.12	3.06	3.00	2.94	2.88
12	6.55	5.10	4.47	4.12	3.89	3.73	3.61	3.51	3.44	3.37	3.28	3.18	3.07	3.02	2.96	2.91	2.85	2.79	2.72
13	6.41	4.97	4.35	4.00	3.77	3.60	3.48	3.39	3.31	3.25	3.15	3.05	2.95	2.89	2.84	2.78	2.72	2.66	2.60
14	6.30	4.86	4.24	3.89	3.66	3.50	3.38	3.29	3.21	3.15	3.05	2.95	2.84	2.79	2.73	2.67	2.61	2.55	2.49
15	6.20	4.77	4.15	3.80	3.58	3.41	3.29	3.20	3.12	3.06	2.96	2.86	2.76	2.70	2.64	2.59	2.52	2.46	2.40
16	6.12	4.69	4.08	3.73	3.50	3.34	3.22	3.12	3.05	2.99	2.89	2.79	2.68	2.63	2.57	2.51	2.45	2.38	2.32
17	6.04	4.62	4.01	3.66	3.44	3.28	3.16	3.06	2.98	2.92	2.82	2.72	2.62	2.56	2.50	2.44	2.38	2.32	2.25
18	5.98	4.56	3.95	3.61	3.38	3.22	3.10	3.01	2.93	2.87	2.77	2.67	2.56	2.50	2.44	2.38	2.32	2.26	2.19
19	5.92	4.51	3.90	3.56	3.33	3.17	3.05	2.96	2.88	2.82	2.72	2.62	2.51	2.45	2.39	2.33	2.27	2.20	2.13
20	5.87	4.46	3.86	3.51	3.29	3.13	3.01	2.91	2.84	2.77	2.68	2.57	2.46	2.41	2.35	2.29	2.22	2.16	2.09
21	5.83	4.42	3.82	3.48	3.25	3.09	2.97	2.87	2.80	2.73	2.64	2.53	2.42	2.37	2.31	2.25	2.18	2.11	2.04
22	5.79	4.38	3.78	3.44	3.22	3.05	2.93	2.84	2.76	2.70	2.60	2.50	2.39	2.33	2.27	2.21	2.14	2.08	2.00
23	5.75	4.35	3.75	3.41	3.18	3.02	2.90	2.81	2.73	2.67	2.57	2.47	2.36	2.30	2.24	2.18	2.11	2.04	1.97
24	5.72	4.32	3.72	3.38	3.15	2.99	2.87	2.78	2.70	2.64	2.54	2.44	2.33	2.27	2.21	2.15	2.08	2.01	1.94
25	5.69	4.29	3.69	3.35	3.13	2.97	2.85	2.75	2.68	2.61	2.51	2.41	2.30	2.24	2.18	2.12	2.05	1.98	1.91
26	5.66	4.27	3.67	3.33	3.10	2.94	2.82	2.73	2.65	2.59	2.49	2.39	2.28	2.22	2.16	2.09	2.03	1.95	1.88
27	5.63	4.24	3.65	3.31	3.08	2.92	2.80	2.71	2.63	2.57	2.47	2.36	2.25	2.19	2.13	2.07	2.00	1.93	1.85
28	5.61	4.22	3.63	3.29	3.06	2.90	2.78	2.69	2.61	2.55	2.45	2.34	2.23	2.17	2.11	2.05	1.98	1.91	1.83
29	5.59	4.20	3.61	3.27	3.04	2.88	2.76	2.67	2.59	2.53	2.43	2.32	2.21	2.15	2.09	2.03	1.96	1.89	1.81
30	5.57	4.18	3.59	3.25	3.03	2.87	2.75	2.65	2.57	2.51	2.41	2.31	2.20	2.14	2.07	2.01	1.94	1.87	1.79
40	5.42	4.05	3.46	3.13	2.90	2.74	2.62	2.53	2.45	2.39	2.29	2.18	2.07	2.01	1.94	1.88	1.80	1.72	1.64
60	5.29	3.93	3.34	3.01	2.79	2.63	2.51	2.41	2.33	2.27	2.17	2.06	1.94	1.88	1.82	1.74	1.67	1.58	1.48
120	5.15	3.80	3.23	2.89	2.67	2.52	2.39	2.30	2.22	2.16	2.05	1.94	1.82	1.76	1.69	1.61	1.53	1.43	1.31
∞	5.02	3.69	3.12	2.79	2.57	2.41	2.29	2.19	2.11	2.05	1.94	1.83	1.71	1.64	1.57	1.48	1.39	1.27	1.00

$F = \dfrac{s_1^2}{s_2^2} = \dfrac{S_1/\nu_1}{S_2/\nu_2}$, where $s_1^2 = S_1/\nu_1$ and $s_2^2 = S_2/\nu_2$ are independent mean squares estimating a common variance σ^2 and based on ν_1 and ν_2 degrees of freedom, respectively.

Table A-5 (continued)

Upper 1% points

ν_2 \ ν_1	1	2	3	4	5	6	7	8	9	10	12	15	20	24	30	40	60	120	∞
1	4052	4999·5	5403	5625	5764	5859	5928	5981	6022	6056	6106	6157	6209	6235	6261	6287	6313	6339	6366
2	98·50	99·00	99·17	99·25	99·30	99·33	99·36	99·37	99·39	99·40	99·42	99·43	99·45	99·46	99·47	99·47	99·48	99·49	99·50
3	34·12	30·82	29·46	28·71	28·24	27·91	27·67	27·49	27·35	27·23	27·05	26·87	26·69	26·60	26·50	26·41	26·32	26·22	26·13
4	21·20	18·00	16·69	15·98	15·52	15·21	14·98	14·80	14·66	14·55	14·37	14·20	14·02	13·93	13·84	13·75	13·65	13·56	13·46
5	16·26	13·27	12·06	11·39	10·97	10·67	10·46	10·29	10·16	10·05	9·89	9·72	9·55	9·47	9·38	9·29	9·20	9·11	9·02
6	13·75	10·92	9·78	9·15	8·75	8·47	8·26	8·10	7·98	7·87	7·72	7·56	7·40	7·31	7·23	7·14	7·06	6·97	6·88
7	12·25	9·55	8·45	7·85	7·46	7·19	6·99	6·84	6·72	6·62	6·47	6·31	6·16	6·07	5·99	5·91	5·82	5·74	5·65
8	11·26	8·65	7·59	7·01	6·63	6·37	6·18	6·03	5·91	5·81	5·67	5·52	5·36	5·28	5·20	5·12	5·03	4·95	4·86
9	10·56	8·02	6·99	6·42	6·06	5·80	5·61	5·47	5·35	5·26	5·11	4·96	4·81	4·73	4·65	4·57	4·48	4·40	4·31
10	10·04	7·56	6·55	5·99	5·64	5·39	5·20	5·06	4·94	4·85	4·71	4·56	4·41	4·33	4·25	4·17	4·08	4·00	3·91
11	9·65	7·21	6·22	5·67	5·32	5·07	4·89	4·74	4·63	4·54	4·40	4·25	4·10	4·02	3·94	3·86	3·78	3·69	3·60
12	9·33	6·93	5·95	5·41	5·06	4·82	4·64	4·50	4·39	4·30	4·16	4·01	3·86	3·78	3·70	3·62	3·54	3·45	3·36
13	9·07	6·70	5·74	5·21	4·86	4·62	4·44	4·30	4·19	4·10	3·96	3·82	3·66	3·59	3·51	3·43	3·34	3·25	3·17
14	8·86	6·51	5·56	5·04	4·69	4·46	4·28	4·14	4·03	3·94	3·80	3·66	3·51	3·43	3·35	3·27	3·18	3·09	3·00
15	8·68	6·36	5·42	4·89	4·56	4·32	4·14	4·00	3·89	3·80	3·67	3·52	3·37	3·29	3·21	3·13	3·05	2·96	2·87
16	8·53	6·23	5·29	4·77	4·44	4·20	4·03	3·89	3·78	3·69	3·55	3·41	3·26	3·18	3·10	3·02	2·93	2·84	2·75
17	8·40	6·11	5·18	4·67	4·34	4·10	3·93	3·79	3·68	3·59	3·46	3·31	3·16	3·08	3·00	2·92	2·83	2·75	2·65
18	8·29	6·01	5·09	4·58	4·25	4·01	3·84	3·71	3·60	3·51	3·37	3·23	3·08	3·00	2·92	2·84	2·75	2·66	2·57
19	8·18	5·93	5·01	4·50	4·17	3·94	3·77	3·63	3·52	3·43	3·30	3·15	3·00	2·92	2·84	2·76	2·67	2·58	2·49
20	8·10	5·85	4·94	4·43	4·10	3·87	3·70	3·56	3·46	3·37	3·23	3·09	2·94	2·86	2·78	2·69	2·61	2·52	2·42
21	8·02	5·78	4·87	4·37	4·04	3·81	3·64	3·51	3·40	3·31	3·17	3·03	2·88	2·80	2·72	2·64	2·55	2·46	2·36
22	7·95	5·72	4·82	4·31	3·99	3·76	3·59	3·45	3·35	3·26	3·12	2·98	2·83	2·75	2·67	2·58	2·50	2·40	2·31
23	7·88	5·66	4·76	4·26	3·94	3·71	3·54	3·41	3·30	3·21	3·07	2·93	2·78	2·70	2·62	2·54	2·45	2·35	2·26
24	7·82	5·61	4·72	4·22	3·90	3·67	3·50	3·36	3·26	3·17	3·03	2·89	2·74	2·66	2·58	2·49	2·40	2·31	2·21
25	7·77	5·57	4·68	4·18	3·85	3·63	3·46	3·32	3·22	3·13	2·99	2·85	2·70	2·62	2·54	2·45	2·36	2·27	2·17
26	7·72	5·53	4·64	4·14	3·82	3·59	3·42	3·29	3·18	3·09	2·96	2·81	2·66	2·58	2·50	2·42	2·33	2·23	2·13
27	7·68	5·49	4·60	4·11	3·78	3·56	3·39	3·26	3·15	3·06	2·93	2·78	2·63	2·55	2·47	2·38	2·29	2·20	2·10
28	7·64	5·45	4·57	4·07	3·75	3·53	3·36	3·23	3·12	3·03	2·90	2·75	2·60	2·52	2·44	2·35	2·26	2·17	2·06
29	7·60	5·42	4·54	4·04	3·73	3·50	3·33	3·20	3·09	3·00	2·87	2·73	2·57	2·49	2·41	2·33	2·23	2·14	2·03
30	7·56	5·39	4·51	4·02	3·70	3·47	3·30	3·17	3·07	2·98	2·84	2·70	2·55	2·47	2·39	2·30	2·21	2·11	2·01
40	7·31	5·18	4·31	3·83	3·51	3·29	3·12	2·99	2·89	2·80	2·66	2·52	2·37	2·29	2·20	2·11	2·02	1·92	1·80
60	7·08	4·98	4·13	3·65	3·34	3·12	2·95	2·82	2·72	2·63	2·50	2·35	2·20	2·12	2·03	1·94	1·84	1·73	1·60
120	6·85	4·79	3·95	3·48	3·17	2·96	2·79	2·66	2·56	2·47	2·34	2·19	2·03	1·95	1·86	1·76	1·66	1·53	1·38
∞	6·63	4·61	3·78	3·32	3·02	2·80	2·64	2·51	2·41	2·32	2·18	2·04	1·88	1·79	1·70	1·59	1·47	1·32	1·00

$F = \dfrac{s_1^2}{s_2^2} = \dfrac{S_1/\nu_1}{S_2/\nu_2}$, where $s_1^2 = S_1/\nu_1$ and $s_2^2 = S_2/\nu_2$ are independent mean squares estimating a common variance σ^2 and based on ν_1 and ν_2 degrees of freedom, respectively.

Table A-5 (continued)

Upper 0.5 % points

ν_2 \ ν_1	1	2	3	4	5	6	7	8	9	10	12	15	20	24	30	40	60	120	∞
1	16211	20000	21615	22500	23056	23437	23715	23925	24091	24224	24426	24630	24836	24940	25044	25148	25253	25359	25465
2	198.5	190.0	199.2	199.2	199.3	199.3	199.4	199.4	199.4	199.4	199.4	199.4	199.4	199.5	199.5	199.5	199.5	199.5	199.5
3	55.55	49.80	47.47	46.19	45.39	44.84	44.43	44.13	43.88	43.69	43.39	43.08	42.78	42.62	42.47	42.31	42.15	41.99	41.83
4	31.33	26.28	24.26	23.15	22.46	21.97	21.62	21.35	21.14	20.97	20.70	20.44	20.17	20.03	19.89	19.75	19.61	19.47	19.32
5	22.78	18.31	16.53	15.56	14.94	14.51	14.20	13.96	13.77	13.62	13.38	13.15	12.90	12.78	12.66	12.53	12.40	12.27	12.14
6	18.63	14.54	12.92	12.03	11.46	11.07	10.79	10.57	10.39	10.25	10.03	9.81	9.59	9.47	9.36	9.24	9.12	9.00	8.88
7	16.24	12.40	10.88	10.05	9.52	9.16	8.89	8.68	8.51	8.38	8.18	7.97	7.75	7.65	7.53	7.42	7.31	7.19	7.08
8	14.69	11.04	9.60	8.81	8.30	7.95	7.69	7.50	7.34	7.21	7.01	6.81	6.61	6.50	6.40	6.29	6.18	6.06	5.95
9	13.61	10.11	8.72	7.96	7.47	7.13	6.88	6.69	6.54	6.42	6.23	6.03	5.83	5.73	5.62	5.52	5.41	5.30	5.19
10	12.83	9.43	8.08	7.34	6.87	6.54	6.30	6.12	5.97	5.85	5.66	5.47	5.27	5.17	5.07	4.97	4.86	4.75	4.64
11	12.23	8.91	7.60	6.88	6.42	6.10	5.86	5.68	5.54	5.42	5.24	5.05	4.86	4.76	4.65	4.55	4.44	4.34	4.23
12	11.75	8.51	7.23	6.52	6.07	5.76	5.52	5.35	5.20	5.09	4.91	4.72	4.53	4.43	4.33	4.23	4.12	4.01	3.90
13	11.37	8.19	6.93	6.23	5.79	5.48	5.25	5.08	4.94	4.82	4.64	4.46	4.27	4.17	4.07	3.97	3.87	3.76	3.65
14	11.06	7.92	6.68	6.00	5.56	5.26	5.03	4.86	4.72	4.60	4.43	4.25	4.06	3.96	3.86	3.76	3.66	3.55	3.44
15	10.80	7.70	6.48	5.80	5.37	5.07	4.85	4.67	4.54	4.42	4.25	4.07	3.88	3.79	3.69	3.58	3.48	3.37	3.26
16	10.58	7.51	6.30	5.64	5.21	4.91	4.69	4.52	4.38	4.27	4.10	3.92	3.73	3.64	3.54	3.44	3.33	3.22	3.11
17	10.38	7.35	6.16	5.50	5.07	4.78	4.56	4.39	4.25	4.14	3.97	3.79	3.61	3.51	3.41	3.31	3.21	3.10	2.98
18	10.22	7.21	6.03	5.37	4.96	4.66	4.44	4.28	4.14	4.03	3.86	3.68	3.50	3.40	3.30	3.20	3.10	2.99	2.87
19	10.07	7.09	5.92	5.27	4.85	4.56	4.34	4.18	4.04	3.93	3.76	3.59	3.40	3.31	3.21	3.11	3.00	2.89	2.78
20	9.94	6.99	5.82	5.17	4.76	4.47	4.26	4.09	3.96	3.85	3.68	3.50	3.32	3.22	3.12	3.02	2.92	2.81	2.69
21	9.83	6.89	5.73	5.09	4.68	4.39	4.18	4.01	3.88	3.77	3.60	3.43	3.24	3.15	3.05	2.95	2.84	2.73	2.61
22	9.73	6.81	5.65	5.02	4.61	4.32	4.11	3.94	3.81	3.70	3.54	3.36	3.18	3.08	2.98	2.88	2.77	2.66	2.55
23	9.63	6.73	5.58	4.95	4.54	4.26	4.05	3.88	3.75	3.64	3.47	3.30	3.12	3.02	2.92	2.82	2.71	2.60	2.48
24	9.55	6.66	5.52	4.89	4.49	4.20	3.99	3.83	3.69	3.59	3.42	3.25	3.06	2.97	2.87	2.77	2.66	2.55	2.43
25	9.48	6.60	5.46	4.84	4.43	4.15	3.94	3.78	3.64	3.54	3.37	3.20	3.01	2.92	2.82	2.72	2.61	2.50	2.38
26	9.41	6.54	5.41	4.79	4.38	4.10	3.89	3.73	3.60	3.49	3.33	3.15	2.97	2.87	2.77	2.67	2.56	2.45	2.33
27	9.34	6.49	5.36	4.74	4.34	4.06	3.85	3.69	3.56	3.45	3.28	3.11	2.93	2.83	2.73	2.63	2.52	2.41	2.29
28	9.28	6.44	5.32	4.70	4.30	4.02	3.81	3.65	3.52	3.41	3.25	3.07	2.89	2.79	2.69	2.59	2.48	2.37	2.25
29	9.23	6.40	5.28	4.66	4.26	3.98	3.77	3.61	3.48	3.38	3.21	3.04	2.86	2.76	2.66	2.56	2.45	2.33	2.21
30	9.18	6.35	5.24	4.62	4.23	3.95	3.74	3.58	3.45	3.34	3.18	3.01	2.82	2.73	2.63	2.52	2.42	2.30	2.18
40	8.83	6.07	4.98	4.37	3.99	3.71	3.51	3.35	3.22	3.12	2.95	2.78	2.60	2.50	2.40	2.30	2.18	2.06	1.93
60	8.49	5.79	4.73	4.14	3.76	3.49	3.29	3.13	3.01	2.90	2.74	2.57	2.39	2.29	2.19	2.08	1.96	1.83	1.69
120	8.18	5.54	4.50	3.92	3.55	3.28	3.09	2.93	2.81	2.71	2.54	2.37	2.19	2.09	1.98	1.87	1.75	1.61	1.43
∞	7.88	5.30	4.28	3.72	3.35	3.09	2.90	2.74	2.62	2.52	2.36	2.19	2.00	1.90	1.79	1.67	1.53	1.36	1.00

$F = \dfrac{s_1^2}{s_2^2} = \dfrac{S_1/\nu_1}{S_2/\nu_2}$, where $s_1^2 = S_1/\nu_1$ and $s_2^2 = S_2/\nu_2$ are independent mean squares estimating a common variance σ^2 and based on ν_1 and ν_2 degrees of freedom, respectively.

Table A–5 (continued)

Upper 5 % points

ν_1 \ ν_2	1	2	3	4	5	6	7	8	9	10	12	15	20	24	30	40	60	120	∞
1	161·4	199·5	215·7	224·6	230·2	234·0	236·8	238·9	240·5	241·9	243·9	245·9	248·0	249·1	250·1	251·1	252·2	253·3	254·3
2	18·51	19·00	19·16	19·25	19·30	19·33	19·35	19·37	19·38	19·40	19·41	19·43	19·45	19·45	19·46	19·47	19·48	19·49	19·50
3	10·13	9·55	9·28	9·12	9·01	8·94	8·89	8·85	8·81	8·79	8·74	8·70	8·66	8·64	8·62	8·59	8·57	8·55	8·53
4	7·71	6·94	6·59	6·39	6·26	6·16	6·09	6·04	6·00	5·96	5·91	5·86	5·80	5·77	5·75	5·72	5·69	5·66	5·63
5	6·61	5·79	5·41	5·19	5·05	4·95	4·88	4·82	4·77	4·74	4·68	4·62	4·56	4·53	4·50	4·46	4·43	4·40	4·36
6	5·99	5·14	4·76	4·53	4·39	4·28	4·21	4·15	4·10	4·06	4·00	3·94	3·87	3·84	3·81	3·77	3·74	3·70	3·67
7	5·59	4·74	4·35	4·12	3·97	3·87	3·79	3·73	3·68	3·64	3·57	3·51	3·44	3·41	3·38	3·34	3·30	3·27	3·23
8	5·32	4·46	4·07	3·84	3·69	3·58	3·50	3·44	3·39	3·35	3·28	3·22	3·15	3·12	3·08	3·04	3·01	2·97	2·93
9	5·12	4·26	3·86	3·63	3·48	3·37	3·29	3·23	3·18	3·14	3·07	3·01	2·94	2·90	2·86	2·83	2·79	2·75	2·71
10	4·96	4·10	3·71	3·48	3·33	3·22	3·14	3·07	3·02	2·98	2·91	2·85	2·77	2·74	2·70	2·66	2·62	2·58	2·54
11	4·84	3·98	3·59	3·36	3·20	3·09	3·01	2·95	2·90	2·85	2·79	2·72	2·65	2·61	2·57	2·53	2·49	2·45	2·40
12	4·75	3·89	3·49	3·26	3·11	3·00	2·91	2·85	2·80	2·75	2·69	2·62	2·54	2·51	2·47	2·43	2·38	2·34	2·30
13	4·67	3·81	3·41	3·18	3·03	2·92	2·83	2·77	2·71	2·67	2·60	2·53	2·46	2·42	2·38	2·34	2·30	2·25	2·21
14	4·60	3·74	3·34	3·11	2·96	2·85	2·76	2·70	2·65	2·60	2·53	2·46	2·39	2·35	2·31	2·27	2·22	2·18	2·13
15	4·54	3·68	3·29	3·06	2·90	2·79	2·71	2·64	2·59	2·54	2·48	2·40	2·33	2·29	2·25	2·20	2·16	2·11	2·07
16	4·49	3·63	3·24	3·01	2·85	2·74	2·66	2·59	2·54	2·49	2·42	2·35	2·28	2·24	2·19	2·15	2·11	2·06	2·01
17	4·45	3·59	3·20	2·96	2·81	2·70	2·61	2·55	2·49	2·45	2·38	2·31	2·23	2·19	2·15	2·10	2·06	2·01	1·96
18	4·41	3·55	3·16	2·93	2·77	2·66	2·58	2·51	2·46	2·41	2·34	2·27	2·19	2·15	2·11	2·06	2·02	1·97	1·92
19	4·38	3·52	3·13	2·90	2·74	2·63	2·54	2·48	2·42	2·38	2·31	2·23	2·16	2·11	2·07	2·03	1·98	1·93	1·88
20	4·35	3·49	3·10	2·87	2·71	2·60	2·51	2·45	2·39	2·35	2·28	2·20	2·12	2·08	2·04	1·99	1·95	1·90	1·84
21	4·32	3·47	3·07	2·84	2·68	2·57	2·49	2·42	2·37	2·32	2·25	2·18	2·10	2·05	2·01	1·96	1·92	1·87	1·81
22	4·30	3·44	3·05	2·82	2·66	2·55	2·46	2·40	2·34	2·30	2·23	2·15	2·07	2·03	1·98	1·94	1·89	1·84	1·78
23	4·28	3·42	3·03	2·80	2·64	2·53	2·44	2·37	2·32	2·27	2·20	2·13	2·05	2·01	1·96	1·91	1·86	1·81	1·76
24	4·26	3·40	3·01	2·78	2·62	2·51	2·42	2·36	2·30	2·25	2·18	2·11	2·03	1·98	1·94	1·89	1·84	1·79	1·73
25	4·24	3·39	2·99	2·76	2·60	2·49	2·40	2·34	2·28	2·24	2·16	2·09	2·01	1·96	1·92	1·87	1·82	1·77	1·71
26	4·23	3·37	2·98	2·74	2·59	2·47	2·39	2·32	2·27	2·22	2·15	2·07	1·99	1·95	1·90	1·85	1·80	1·75	1·69
27	4·21	3·35	2·96	2·73	2·57	2·46	2·37	2·31	2·25	2·20	2·13	2·06	1·97	1·93	1·88	1·84	1·79	1·73	1·67
28	4·20	3·34	2·95	2·71	2·56	2·45	2·36	2·29	2·24	2·19	2·12	2·04	1·96	1·91	1·87	1·82	1·77	1·71	1·65
29	4·18	3·33	2·93	2·70	2·55	2·43	2·35	2·28	2·22	2·18	2·10	2·03	1·94	1·90	1·85	1·81	1·75	1·70	1·64
30	4·17	3·32	2·92	2·69	2·53	2·42	2·33	2·27	2·21	2·16	2·09	2·01	1·93	1·89	1·84	1·79	1·74	1·68	1·62
40	4·08	3·23	2·84	2·61	2·45	2·34	2·25	2·18	2·12	2·08	2·00	1·92	1·84	1·79	1·74	1·69	1·64	1·58	1·51
60	4·00	3·15	2·76	2·53	2·37	2·25	2·17	2·10	2·04	1·99	1·92	1·84	1·75	1·70	1·65	1·59	1·53	1·47	1·39
120	3·92	3·07	2·68	2·45	2·29	2·17	2·09	2·02	1·96	1·91	1·83	1·75	1·66	1·61	1·55	1·50	1·43	1·35	1·25
∞	3·84	3·00	2·60	2·37	2·21	2·10	2·01	1·94	1·88	1·83	1·75	1·67	1·57	1·52	1·46	1·39	1·32	1·22	1·00

$F = \dfrac{s_1^2}{s_2^2} = \dfrac{S_1/\nu_1}{S_2/\nu_2}$, where $s_1^2 = S_1/\nu_1$ and $s_2^2 = S_2/\nu_2$ are independent mean squares estimating a common variance σ^2 and based on ν_1 and ν_2 degrees of freedom, respectively.

Table A–5 (continued)

Upper 0·1% *points*

$\nu_2 \backslash \nu_1$	1	2	3	4	5	6	7	8	9	10	12	15	20	24	30	40	60	120	∞
1	4053*	5000*	5404*	5625*	5764*	5859*	5929*	5981*	6023*	6056*	6107*	6158*	6209*	6235*	6261*	6287*	6313*	6340*	6366*
2	998·5	999·0	999·2	999·2	999·3	999·3	999·4	999·4	999·4	999·4	999·4	999·4	999·4	999·5	999·5	999·5	999·5	999·5	999·5
3	167·0	148·5	141·1	137·1	134·6	132·8	131·6	130·6	129·9	129·2	128·3	127·4	126·4	125·9	125·4	125·0	124·5	124·0	123·5
4	74·14	61·25	56·18	53·44	51·71	50·53	49·66	49·00	48·47	48·05	47·41	46·76	46·10	45·77	45·43	45·09	44·75	44·40	44·05
5	47·18	37·12	33·20	31·09	29·75	28·84	28·16	27·64	27·24	26·92	26·42	25·91	25·39	25·14	24·87	24·60	24·33	24·06	23·79
6	35·51	27·00	23·70	21·92	20·81	20·03	19·46	19·03	18·69	18·41	17·99	17·56	17·12	16·89	16·67	16·44	16·21	15·99	15·75
7	29·25	21·69	18·77	17·19	16·21	15·52	15·02	14·63	14·33	14·08	13·71	13·32	12·93	12·73	12·53	12·33	12·12	11·91	11·70
8	25·42	18·49	15·83	14·39	13·49	12·86	12·40	12·04	11·77	11·54	11·19	10·84	10·48	10·30	10·11	9·92	9·73	9·53	9·33
9	22·86	16·39	13·90	12·56	11·71	11·13	10·70	10·37	10·11	9·89	9·57	9·24	8·90	8·72	8·55	8·37	8·19	8·00	7·81
10	21·04	14·91	12·55	11·28	10·48	9·92	9·52	9·20	8·96	8·75	8·45	8·13	7·80	7·64	7·47	7·30	7·12	6·94	6·76
11	19·69	13·81	11·56	10·35	9·58	9·05	8·66	8·35	8·12	7·92	7·63	7·32	7·01	6·85	6·68	6·52	6·35	6·17	6·00
12	18·64	12·97	10·80	9·63	8·89	8·38	8·00	7·71	7·48	7·29	7·00	6·71	6·40	6·25	6·09	5·93	5·76	5·59	5·42
13	17·81	12·31	10·21	9·07	8·35	7·86	7·49	7·21	6·98	6·80	6·52	6·23	5·93	5·78	5·63	5·47	5·30	5·14	4·97
14	17·14	11·78	9·73	8·62	7·92	7·43	7·08	6·80	6·58	6·40	6·13	5·85	5·56	5·41	5·25	5·10	4·94	4·77	4·60
15	16·59	11·34	9·34	8·25	7·57	7·09	6·74	6·47	6·26	6·08	5·81	5·54	5·25	5·10	4·95	4·80	4·64	4·47	4·31
16	16·12	10·97	9·00	7·94	7·27	6·81	6·46	6·19	5·98	5·81	5·55	5·27	4·99	4·85	4·70	4·54	4·39	4·23	4·06
17	15·72	10·66	8·73	7·68	7·02	6·56	6·22	5·96	5·75	5·58	5·32	5·05	4·78	4·63	4·48	4·33	4·18	4·02	3·85
18	15·38	10·39	8·49	7·46	6·81	6·35	6·02	5·76	5·56	5·39	5·13	4·87	4·59	4·45	4·30	4·15	4·00	3·84	3·67
19	15·08	10·16	8·28	7·26	6·62	6·18	5·85	5·59	5·39	5·22	4·97	4·70	4·43	4·29	4·14	3·99	3·84	3·68	3·51
20	14·82	9·95	8·10	7·10	6·46	6·02	5·69	5·44	5·24	5·08	4·82	4·56	4·29	4·15	4·00	3·86	3·70	3·54	3·38
21	14·59	9·77	7·94	6·95	6·32	5·88	5·56	5·31	5·11	4·95	4·70	4·44	4·17	4·03	3·88	3·74	3·58	3·42	3·26
22	14·38	9·61	7·80	6·81	6·19	5·76	5·44	5·19	4·99	4·83	4·58	4·33	4·06	3·92	3·78	3·63	3·48	3·32	3·15
23	14·19	9·47	7·67	6·69	6·08	5·65	5·33	5·09	4·89	4·73	4·48	4·23	3·96	3·82	3·68	3·53	3·38	3·22	3·05
24	14·03	9·34	7·55	6·59	5·98	5·55	5·23	4·99	4·80	4·64	4·39	4·14	3·87	3·74	3·59	3·45	3·29	3·14	2·97
25	13·88	9·22	7·45	6·49	5·88	5·46	5·15	4·91	4·71	4·56	4·31	4·06	3·79	3·66	3·52	3·37	3·22	3·06	2·89
26	13·74	9·12	7·36	6·41	5·80	5·38	5·07	4·83	4·64	4·48	4·24	3·99	3·72	3·59	3·44	3·30	3·15	2·99	2·82
27	13·61	9·02	7·27	6·33	5·73	5·31	5·00	4·76	4·57	4·41	4·17	3·92	3·66	3·52	3·38	3·23	3·08	2·92	2·75
28	13·50	8·93	7·19	6·25	5·66	5·24	4·93	4·69	4·50	4·35	4·11	3·86	3·60	3·46	3·32	3·18	3·02	2·86	2·69
29	13·39	8·85	7·12	6·19	5·59	5·18	4·87	4·64	4·45	4·29	4·05	3·80	3·54	3·41	3·27	3·12	2·97	2·81	2·64
30	13·29	8·77	7·05	6·12	5·53	5·12	4·82	4·58	4·39	4·24	4·00	3·75	3·49	3·36	3·22	3·07	2·92	2·76	2·59
40	12·61	8·25	6·60	5·70	5·13	4·73	4·44	4·21	4·02	3·87	3·64	3·40	3·15	3·01	2·87	2·73	2·57	2·41	2·23
60	11·97	7·76	6·17	5·31	4·76	4·37	4·09	3·87	3·69	3·54	3·31	3·08	2·83	2·69	2·55	2·41	2·25	2·08	1·89
120	11·38	7·32	5·79	4·95	4·42	4·04	3·77	3·55	3·38	3·24	3·02	2·78	2·53	2·40	2·26	2·11	1·95	1·76	1·54
∞	10·83	6·91	5·42	4·62	4·10	3·74	3·47	3·27	3·10	2·96	2·74	2·51	2·27	2·13	1·99	1·84	1·66	1·45	1·00

* Multiply these entries by 100.

This 0·1% table is based on the following sources: Colcord & Deming (1935); Fisher & Yates (1953, Table V) used with the permission of Messrs Oliver and Boyd; Norton (1952).

Index

About the Authors

Peter R. Stopher is professor of civil engineering at Northwestern University. He was educated at University College, London, where he received the B.Sc. in civil and municipal engineering in 1964 and the Ph.D. in traffic studies in 1967. From 1967 to 1968, he was a research officer with the Greater London Council. Subsequently, he has held faculty appointments at Cornell University, McMaster University (Ontario), and Northwestern University, specializing in urban transportation. On leave during 1977–78 he acted as a transportation adviser to the National Institute for Transport and Road Research of the Council for Scientific and Industrial Research in Pretoria, South Africa.

Dr. Stopher has been a consultant to a number of private firms and governmental agencies on various aspects of urban transportation planning, travel demand, and impacts of transportation facilities. He is the joint author with Dr. Arnim H. Meyburg of *Urban Transportation Modeling and Planning,* and *Transportation Systems Evaluation,* and joint editor with Dr. Arnim H. Meyburg of *Behavioral Travel-Demand Models.* He has also written a number of technical papers, principally in travel-demand modeling and in travel-time evaluation, as well as in urban goods movement, and in statistical and psychological methods. He is a member of several professional societies and committees in both the United States and the United Kingdom.

Arnim H. Meyburg is a professor of civil and environmental engineering at Cornell University. He was educated at the University of Hamburg, Germany, the Free University of Berlin, Germany, and Northwestern University. He received the M.S. in quantitative geography in 1968 and the Ph.D. in civil engineering (transportation) in 1971, both from Northwestern University. From 1968 to 1969, he was a research associate at the Transportation Center of Northwestern University. Dr. Meyburg has been a faculty member at Cornell University since 1969. He held a visiting appointment at the University of California at Irvine during the fall of 1975 and at the Technical University of Munich during the summer of 1976. During 1978–1979, he was a Humboldt Foundation Research Fellow at the Technical University of Munich.

Dr. Meyburg has also been a consultant to private industry and several governmental agencies. In addition to being joint author with Dr. Stopher of two earlier books and joint editor of one, he has written a number of technical papers in the subject areas of travel-demand modeling, urban goods movement, and transportation-systems analysis. He is a member of several professional societies and committees.